SNOW

A NATURAL HISTORY; AN UNCERTAIN FUTURE

Australian Alps National Parks Co-operative Management

Eight conservation reserves abut one another in the mainland high country area to form a continuous protected area of approximately 1.6 million hectares. The major reserves, Kosciuszko, Namadgi, Alpine and Snowy River National Parks, are well known to much of the community of south-eastern Australia. Brindabella National Park, Bimberi Nature Reserve, Scabby Range Nature Reserve and the Avon Wilderness are also becoming better known. Other conservation areas and resorts with alpine and sub-alpine environments on the mainland and in Tasmania also benefit from regular contact and information developed through the co-operative management approach to the Australian Alps national parks.

As well as containing Australia's highest peaks, the Australian Alps national parks also protect Australia's unique alpine flora and fauna, a rich aboriginal and non-aboriginal heritage and a highly valued recreation resource for many Australians, the majority of its mainland hydro-electricity generating resources, and primary water supplies for Canberra and mountain towns.

To assist conservation agencies to manage these precious resources in comparable and compatible ways, the Ministers responsible for the management of alpine areas in the Australian Capital Territory, New South Wales and Victoria in conjunction with the Commonwealth Environment Minister signed a Memorandum of Understanding in relation to the co-operative management of the Australian Alps (MOU) in July 1986. The MOU was revised and re-signed in 1989 when self-government in the ACT and the new Alpine National Park in Victoria led to amendments to the original document. It has been further amended and re-signed since then, the latest commitment being made in 1996.

ACT Parks and Conservation Service · Parks Victoria · National Parks & Wildlife Service NSW · Environment Australia

SNOW

A NATURAL HISTORY; AN UNCERTAIN FUTURE

Edited by

Ken Green

Australian Alps Liaison Committee
Canberra

AUSTRALIAN ALPS
NATIONAL PARKS

The National Library of Australia
Cataloguing-in-Publication entry:

Green, Kenneth, 1951– .
Snow, a natural history; an uncertain future.
Bibliography.
Includes index.
ISBN 0 949324 80 9.

1. Snow — Environmental aspects. I. Title.

577.586

Published January 1998

Published by
Australian Alps Liaison Committee
PO Box 636
Canberra 2601

Printed in Australia and distributed by
Surrey Beatty & Sons Pty Ltd
43 Rickard Road, Chipping Norton, NSW 2170

Preface

This book had its origins at an Australian Alps Liaison Committee field day at Providence Portal in the Snowy Mountains, where models of global warming and studies of the impacts of increased ultraviolet light due to the thinning of the ozone layer were presented. The potential impacts on the Australian Alps were examined in relation to the long-term implications for management. It became apparent from discussion that followed that there was concern about our ideas of managing the Australian Alps, and that in future we will be forced to look at management beyond our own parochial interests — perhaps beyond our own lifetimes. This then led to suggestions for a conference to bring together from Australia and overseas, experts in fields as diverse as meteorology, climate modelling, glaciology, pedology, zoology and botany in order to explore the possible consequences of global warming and increased levels of ultraviolet light on the natural environment of the alps. There have been several Australian studies of the possible impacts of global warming on the present snow cover; most relate to perceived threats to the ski industry in Australia, and few studies have set out to consider the implications for the natural environment. The conference scheduled for Jindabyne in February 1998 set out to do this and to provide a framework for future cooperative research and monitoring. First, however, there was a perceived need to document what we know about snow and the snow-country in Australia, and to compile an inventory of our present knowledge; the results of this stocktake are presented in this book.

Acknowledgements

This book has been written and published as a project of the Australian Alps Liaison Committee. As such I extend my appreciation to the Natural Heritage Working Group who saw fit to recommend funding for publication of this work and to the NSW National Parks and Wildlife Service for allowing me the time to bring it to a conclusion. I naturally thank the authors of the chapters who fitted the writing of their respective chapters into what for most was an already busy work schedule. I thank also the people who agreed to referee the chapters as they came off the production line: Ian Allison, Jamie Allnutt, Linda Broome, Bill Budd, Eric Colhoun, Graeme Enders, Mary Green, Kevin Hennessy, Ron Hoham, John Jenkins, Janette Lindsay, Hau Ling, Will Osborne, Catherine Pickering, Ramasamy Suppiah, Michelle Walter, Peter Whetton and Genevieve Wright. The index was compiled by Mary Green.

A number of figures had to be re-drawn for this publication and for this I thank Lorraine Oliver. I thank Nicole Chilcott for drawing the map which appears in the introduction. For other assistance with the book I thank Linda Byrne, Neville Byrne, Cate Gillies and Amy Horsell.

Ken Green
Heritage Unit
Snowy Mountains Region
National Parks and Wildlife Service

Contents

Authors

Linda Broome NSW National Parks and Wildlife Service, Southern Zone, PO Box 2115, Queanbeyan, New South Wales 2620. email linda.broome@npws.nsw.gov.au

Sara Broomhall Applied Ecology Research Group, University of Canberra, PO Box 1, Belconnen, Australian Capital Territory 2616.email broomhal@aerg.canberra.edu.au

Clem Davis Canberra Meteorology Office, Bureau of Meteorology, GPO Box 787, Australian Capital Territory 2601, email c.davis@bom.gov.au

Maxine Davis Applied Ecology Research Group, University of Canberra, PO Box 1, Belconnen, Australian Capital Territory 2616. email osborne@aerg.canberra.edu.au

Robert W. Galloway 68 Barada Crescent, Aranda, Australian Capital Territory 2614. Telephone: (06) 251 2501.

Roger Good NSW National Parks and Wildlife Service, Southern Zone, P.O. Box 2115, Queanbeyan, New South Wales 2620. email roger.good@npws.nsw.gov.au

Ken Green Heritage Unit, Snowy Mountains Region, NSW National Parks and Wildlife Service, PO Box 2228, Jindabyne, New South Wales 2627.
email ken.green@npws.nsw.gov.au

Kevin Kiernan Forest Practices Board, 30 Patrick St. Hobart, Tasmania 7000.
email kevink@fpb.tas.gov.au

Urs König CRES, Australian National University, Canberra, Australian Capital Territory. 0200/Department of Geography University of Zürich, Winterthurerstrasse 190, 8057 Zürich, Switzerland. email ukoenig@geo.unizh.ch

Harvey Marchant Australian Antarctic Division, Channel Highway, Kingston, Tasmania 7050. email harvey_mar@antdiv.gov.au

William Osborne Applied Ecology Research Group, University of Canberra, PO Box 1, Belconnen, Australian Capital Territory 2616. email osborne@aerg.canberra.edu.au

Jim Peterson Department of Geography, Monash University, Clayton, Victoria 3168. email Jim.Peterson@arts.monash.edu.au

Andrew Ruddell Antarctic CRC, University of Tasmania, GPO Box 252C, Hobart, Tasmania 7001. email andrew.ruddell@antcrc.utas.edu.au

Michelle Walter Heritage Unit, Snowy Mountains Region, NSW National Parks and Wildlife Service, PO Box 2228, Jindabyne, New South Wales 2627.
email michellewalter@hotmail.com

Ian Wardlaw CSIRO Division of Plant Industry, GPO Box 1600, Australian Capital Territory 2601. email Ian.Wardlaw@anu.edu.au

Peter Whetton CSIRO Division of Atmospheric Research, Private Bag 1, Aspendale Victoria 3195. email phw@dar.csiro.au

Introduction

K. GREEN

One of the unifying themes of the Australian Alps National Parks is that the core areas of these parks contain alpine or subalpine environments. These environments are extremely restricted within Australia, both in altitude and in space. The theoretical altitudinal range currently necessary for a permanent snow cover on mainland Australia is 2750-2800m (Slatyer *et al.* 1985), well above the highest point at 2228m. The alpine zone extends upwards for only 400 or so metres in New South Wales from the treeline at about 1850m with an even smaller altitudinal extent in Victoria. In terms of area, the winter snow cover is not extensive, with only 1200 km^2 above 1528 m averaging 60 days of snow cover per winter in NSW and the ACT combined (Slatyer *et al.* 1985) and a further 475 km^2 (above 1480 m) in Victoria (calculated from data in Slatyer *et al.* 1985). Combined, these areas cover only 0.02 % of mainland Australia. If areas with 90 days of snow cover per year are considered (above 1700m in NSW and 1655m in Victoria), Australia has only about one fortieth the snow covered landscape of Switzerland (Slatyer *et al.* 1985). The snow country of New South Wales and Victoria suffers also from a lack of continuity. The alpine areas are strung out like islands in an archipelago with the discontinuity further emphasised by the break between the alpine country of the mainland and Tasmania (Fig. 1). However, in terms of their exposure to snowfall events, Victorian and NSW alpine regions can really be considered as one overall region (Chapter 1).

Much of this mainland snow country consists of deeply-dissected uplifted mountain ranges and high plains that lie between about 1400 and 1800 m above sea level (Green and Osborne 1994). The highest summits are in the Snowy Mountains, occurring along an extensive, rolling alpine ridge usually referred to as the Main Range. To the north-west of this ridge, the land drops steeply to the Geehi River at 400 m above sea level. To the east, the land descends more gradually in a series of steps to the tableland of the Monaro at about 900 m above sea level. The Bogong Mountains and Bimberi Range, the latter in the ACT some 40 kilometres north of the Snowy Mountains, occur as northern outliers of subalpine snow-country. To the south and east of the Main Range are further isolated tracts of subalpine country including the Pilot, and across the upper Murray River, the Cobberas and Davies Plain.

In Victoria, the greatest concentration of snow-country occurs further to the south-west in three areas centred on Mt. Bogong, the Bogong High Plains and Mt. Hotham in a roughly north/south line separated by lower land at Bogong Creek Saddle and Cobungra Gap respectively, with a further break to the south before the Dargo High Plains. Fifty kilometres south-east of Mt. Hotham are rugged and deeply incised ranges centred on Mt. Howitt and stretching southwards for a further 45 kilometres to Mt. Wellington and Gable End.

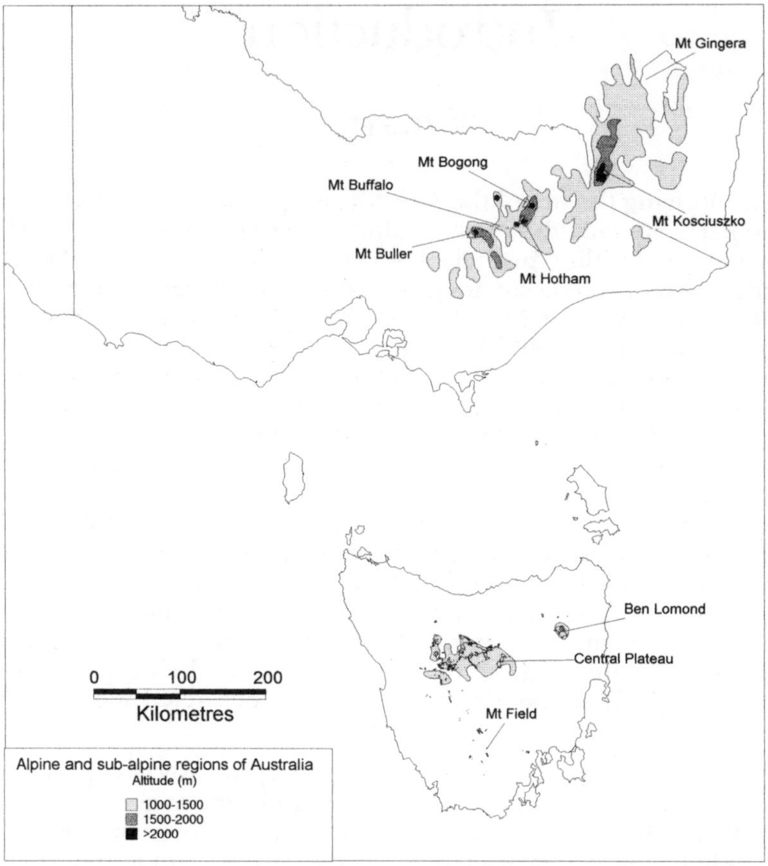

Fig. 1. Distribution of the high-country across Tasmania, Victoria, New South Wales and the Australian Capital Territory.

There are many subalpine outliers in Victoria including Mt. Skene, Mt. Buller and Mt. Stirling, Lake Mountain, Mt. Kent, and granitic massifs such as Mt. Buffalo and Mt. Baw Baw. In Tasmania most alpine peaks occur as isolates, except on the Central Plateau where peaks, including Mt. Ossa (1617 m), Barn Bluff (1559 m) and Cradle Mountain (1545 m) are linked by montane tableland at about 900 m altitude.

Natural communities within the alpine and subalpine areas have evolved special adaptations to cope with the environment and in many cases are dependent upon the presence of winter snow. Organisms inhabiting snow are living in an extreme environment, but one on which they are totally dependent (Chapter 5), whilst beneath the snow some plants may still be growing even at temperatures below 0°C (Chapter 7). One result of coping with snow has been the high degree of endemism among some life forms (Costin *et al.* 1979, Green and Osborne 1994). Despite interannual variation in snow cover (over the past 40 years there has been a 300% variation in snowpack - SMHEA 1993), there is on the mainland a degree of stability in snow cover. The extent of snow patches, for example, is highly variable from year to year (Chapter 4) but their occurrence is sufficiently regular to inhibit the normal vegetation cover, and snow patch sites can be readily identified even in the absence of snow.

These alpine and subalpine environments are, however, under threat from predicted global warming due to the enhanced greenhouse effect. Predictions by climatologists suggest other environmental changes such as increased storms, increased sea level, alteration to winds, rainfall patterns and even ocean currents (Brereton *et al.* 1995) but the main one concerning the mountains is increase in temperature (coupled with changes in precipitation) leading to the loss of snowcover. There is a natural variation in snow cover and it might be difficult to detect trends, but an examination of metre-days of snow, a measure which integrates the factors involved in snow cover shows that there has been a decrease since the 1950s (Chapter 3). In a worst case climatic scenario, predictions are that the snow country will decline to a small area around Mt. Kosciuszko by 2070 (Whetton 1996, and Chapter 11). What we are potentially facing in Australia is the loss, not just of one component of the alpine ecosystem, not just of snow, but the loss of the alpine ecosystem itself.

Using BIOCLIM (a bioclimate analysis and prediction system) to examine the fate of alpine areas with a rise in mean annual temperature of 3°C, Busby (1988) warned of the mass extinction of present alpine species. Brereton *et al.* (1995) examined individual species and warned that a 1°C rise would cause the loss of the Mountain Pygmy Possum (*Burramys parvus*). The crisis of global warming is likely to be compounded by a number of elements acting independently or synergistically. These will have significant impacts independent of the impact of global warming on the climate and will certainly act in concert to deepen any crisis that might occur. These additional impacts include increased UV-B radiation due to ozone depletion, human overpopulation, exotic plant and animal invasion; the whole problem being compounded by the isolation of these mountain tops. We must look further than just global warming in examining the fate of alpine species because of these other global threats. Brereton *et al.* (1995) examined the expected response of a number of species in south-eastern Australia to predicted global warming scenarios. They warned that their predictions, relying upon bioclimatic data, only covered the survival prospects of the species in the broadest sense because there were other factors involved (they specified mainly terrestrial ones). Even with an increase of 3°C, Brereton *et al.* (1995) predicted that a suitable bioclimatic envelope, albeit much reduced, would remain for the Alpine Tree Frog (*Litoria vereauxii alpina*). However, the present threat of increased levels of UV-B radiation might see the extirpation of this species long before such warming could come about (Broomhall 1997, Chapter 13).

An increasing human population has been mentioned, and the impacts of human interference must also be taken into account. Retreat of species to refugia with changes in climate has in the past been hampered by natural barriers such as rivers but now there are increased numbers of human barriers (Brereton *et al.* 1995). In the mountains these include cleared land and dammed waterways. One potential result that we must examine in predictions for the future is that the distribution of a species under a global warming scenario might become disjunct (Brereton *et al.* 1995). This may have positive as well as negative impact because the isolation of populations can lead to morphological and genetic divergence resulting in the evolution of two species from one (Wright 1969, Mayr 1970), as possibly happened with the two species of Corroboree Frogs separated by only 35 km (Osborne and Norman 1991). Unlike many mountain areas overseas, the Australian Alps have seen little settlement. Population pressures in the alps have to a certain extent been nullified by the acceptance that the value of the alps for water exceeds their value as grazing country. However, the greatest human pressure now must be tourism connected with the ski industry and how this reacts to the possibility of reduced snow cover is examined in Chapter 12.

Much of what is said about global threats may seem hypothetical, but we do already have local evidence that something untoward is happening. A particularly worrying example is provided by the decline in alpine and subalpine frogs. Frogs have both an aquatic and a terrestrial phase, they have skins permeable to water and are therefore very sensitive to environmental degradation. Throughout the world they are regarded as indicators of the health of the ecosystem. The Southern Corroboree Frog (*Pseudophryne corroboree*) once occurred in its hundreds in the Smiggin Holes region of the Snowy Mountains (Osborne 1989). However, monitoring over a ten-year period has shown a decline to a single individual south of the Snowy River in 1996 and none in 1997 (Osborne and Davis 1997, Hunter *et al.* 1997). Northwards, ponds that had frogs calling in their tens or hundreds fifteen years ago now have one or two or maybe even none. This is happening not only at ponds near developments or roads - this is happening in the heart of the Jagungal wilderness. Similarly, the Baw Baw Frog (*Philoria frosti*) was once common on the Baw Baw Plateau but is now only found in deep gullies off the side of the plateau in montane forest (Hollis 1995). The Alpine Tree Frog used to occur as high as Rawsons Pass in the Snowy Mountains in the mid to late 1970s. In the last two years, searches for them found none higher than at Sponars Inn at 1500 m (Hunter *et al.* 1997). This means that Alpine Tree Frogs have disappeared from the top 500m of their altitudinal range and now at their highest altitude are only found in turbid waters in artificial ponds.[1]

There have been a number of suggestions as to the cause of amphibian declines which are considered to be a worldwide problem. At high altitudes though, UV-B radiation is seen as one of the major culprits. The greatest extent of the hole in the ozone layer is in spring, and currently most plants and animals are protected beneath the snow for a part or all of that period (especially at the higher altitudes), therefore

[1]Two Alpine Tree Frogs were found calling at Charlottes Pass (1760 m) in November 1997 by David Hunter.

we cannot examine global warming in isolation but must also look at what flows on from predictions of less snow cover. Frogs are only the first indicators of the problem. We have no knowledge of how plants in the Australian Alps are being affected by increased UV-B radiation. UV-B radiation has been found to be biologically damaging at a depth of several metres in clear water (Schindler *et al.* 1996). This will have serious implications for the fauna and flora of glacial lakes and high altitude rivers, particularly if they are not covered by a layer of snow in the spring.

How predicted changes in climate will affect alpine ecosystems depends very much on whether change is more towards a best-case scenario (least temperature increase, greatest precipitation increase) or worst-case scenario (greatest temperature increase, greatest precipitation decrease) (Chapter 11). The effects of the first case (when the mainland alpine environment could become more similar to that in the Tasmania high country) can be seen in Chapter 9 (Figure 1). An altitudinal transect for the insectivorous marsupials *Antechinus swainsonii* and *A. stuartii* in south-eastern New South Wales shows an increase in trap success to about 1000 metres followed by a decline through to 1500m. The decline in numbers of *A. stuartii* is halted briefly at the 1501-1750m level while numbers of *A. swainsonii* increase. Green (1988) suggested that the cold wet areas of the montane forest below 1500m are not favoured by small mammals. Similar low trap success of 1.5 *A. swainsonii* per 100 trap nights was found on Mt. Field in Tasmania which is wetter than the Snowy Mountains in summer with wet and shallow snow (when it is present) in winter. The increase in altitude, and the lowering of temperature in the Snowy Mountains actually ameliorates conditions for mammals because once the snow cover is established they are protected in the subnivean space from the worst of the cold and wet (Chapter 8).

Alternatively there might be slight warming and decreased precipitation, but with temperatures in the past few years dropping to as low as -23°C at Charlottes Pass, even with global warming there will still be cold nights, cold weeks or cold months. We might see the paradoxical situation of global warming leading to the exposure of the animals and plants to colder conditions because there is less snow to insulate them (Chapter 6 and 8). This may cause mortality in animals such as the hibernating Mountain Pygmy-possum because, once the temperature in its hibernating site drops too low, it has to warm itself from a body temperature of just 2°C (Chapter 10). This has a high energy cost and if it occurs too many times through the winter then the Mountain Pygmy-possum has not got the energy store to survive through to spring (Chapter 10). In addition to direct effects on the alpine fauna, any upwards movement of the snowline will allow greater access by feral animals such as cats and rabbits, (Chapter 9) and increased spread of weeds (Chapter 6).

The impact of global change on the snow country is likely to have wider ramifications than just for the region. The Australian Alps are possibly the major water catchment in Australia. In New South Wales the Snowy Mountains feed the south-flowing Snowy River to the south-east, and through the Tooma and Swampy Plains Rivers contributes directly to the upper Murray River flowing to the west, while to the north the Tumut

River feeds into the Murrumbidgee River which also collects outflow from the ACT, on the western side via the Goodradigbee River and on the eastern side via the Cotter (Brown and Millner 1989). In Victoria the major runoff from the Alps is into the Murray via the upper Murray, Mitta Mitta, Kiewa, Ovens, Broken and Goulburn Rivers, with drainage to the south reaching the sea through the Snowy, Tambo, Mitchell, Thomson, Latrobe and Yarra Rivers (Brown and Millner 1989). Whilst there have been studies of hydrology to which snow is seen as important it has generally only been seen as a contributor to the accumulation of water — the form in which it falls has not been seen as important. There has been little interest shown in the importance of snowpack to the hydrology of the streams on the part of electricity-generating authorities because it is really the total amount (and quality) of runoff (including both surface and groundwater flows) that is important to power generation rather than how the water falls. In NSW, for example, there is considered to be adequate storage in Lake Eucumbene to balance any change in timing of peak flows (Brown and Millner 1989) and therefore there is no need for storage of the water in the form of snow. With the smaller storage for the Kiewa Valley Scheme in Victoria, however, the storage of water as snow may be more important.

It is known that reservoirs raise winter water temperatures and lower summer temperatures (Brown and Millner 1989) but the impact of global warming will possibly have a greater effect on stream ecology upstream. "There are few, if any, alpine regions in the world that have such an extensive array of hydrologic and meteorologic data available as the Snowy Mountains" (Brown and Millner 1989). However, "there have not . . . been many studies that have utilised it" (Brown and Millner 1989). Additionally, very few investigations have been undertaken to examine the physical properties of "Australian snow" (Chapter 2). One of the aims of this book is to stimulate alpine research and the interaction between snow and the hydrology of the streams would be a good starting point.

This book is, ultimately, about snow; what it is, how it is formed, how it lies, where it lies and for how long, what is the variability and how it affects the landscape, how the snow determines what animals and plants can exist in the alps and how they survive. The collection of papers here provides a benchmark against which to measure future change.

REFERENCES

Brereton, R., Bennett, S. and Mansergh, I., 1995. Enhanced greenhouse climate change and its potential effect on selected fauna of south-eastern Australia: a trend analysis. *Biol. Cons.* **72:** 339–354.

Broomhall, S. D., 1997. "Comparative effects of ultraviolet-B (UV-B) radiation on two sympatric species of Australian anurans, *Litoria verreauxii alpina* and *Crinia signifera*". Unpublished Honours thesis, University of Canberra.

Brown, J. A. H. and Millner, F. C., 1989. Aspects of the meteorology and hydrology of the Australian Alps. Pp. 297–329 *in* "The Scientific Significance of the Australian Alps" ed by R. B. Good. Australian Alps Liaison Committee.

Busby, J. R., 1988. Potential impacts of climate change on Australia's flora and fauna. Pp. 387–398 *in* "Greenhouse: Planning for Climate Change" ed by G. I. Pearman. CSIRO, Melbourne.

Costin, A. B., 1954. "A Study of the Ecosystems of the Monaro Region of New South Wales." Government Printer, Sydney.

Costin, A. B., Gray, M., Totterdell, C. J. and Wimbush, D. J., 1979. "Kosciusko Alpine Flora." CSIRO Collins Melbourne/Sydney.

Green, K., 1988. "A study of *Antechinus swainsonii* and *Antechinus stuartii* and their prey in the Snowy Mountains". Unpublished Ph.D. thesis. Australian National University, Canberra.

Green, K. and Osborne, W. S., 1994. "Wildlife of the Australian Snow-Country". Reed, Sydney.

Hollis, A. J., 1995. Reassessment of the distribution, abundance and habitat of the Baw Baw Frog *Philoria frosti* Spencer: preliminary findings. *Vic. Nat.* **112:** 190–201.

Hunter, D., Osborne, W. S. and Smith, M. J., 1997. Distribution and abundance of the Alpine Tree Frog (*Litoria verreauxii alpina*) in the Australian Alps National Parks: Report on the first season's survey (1996-97). Report to the Australian Alps National Parks Liaison Committee, Canberra.

Mayr, E., 1970. "Populations, Species and Evolution." Harvard University Press, Cambridge, Massachusetts.

Osborne, W. S., 1989. Distribution, relative abundance and conservation status of Corroboree Frogs, *Pseudophryne corroboree* Moore (Anura: Myobatrachidae). *Aust. Wildl. Res.* **16:** 537–547.

Osborne, W. S. and Davis, M., 1997. Long-term variability in temperature, precipitation, and snow cover in the Snowy Mountains: is there a link with the decline of the Southern Corroboree Frog (*Pseudophryne corroboree*)? Report to the NSW National Parks and Wildlife Service, Jindabyne.

Osborne, W. S. and Norman, J. A., 1991. Conservation genetics of Corroboree Frogs, *Pseudophryne corroboree* Moore (Anura: Myobatrachidae): population subdivision and genetic divergence. *Aust. J. Zool.* **39:** 285–297.

Schindler, D. W., Curtis, P. J., Parker, B. R. and Stainton, M. P., 1996. Consequences of climate warming and lake acidification for UV-B penetration in North American boreal lakes. *Nature* **379:** 705–708.

Slatyer, R. O., Cochrane, P. M. and Galloway, R. W., 1984. Duration and extent of snow cover in the Snowy Mountains and a comparison with Switzerland. *Search* **15:** 327–331.

SMHEA 1993. "Snowy Precipitation Enhancement Project, Draft Environmental Impact Statement." SMHEA, Cooma.

Whetton, P. H., Haylock, M. R. and Galloway, R., 1996. Climate change and snow-cover duration in the Australian Alps. *Climatic Change* **32:** 447–479.

Wright, S., 1969. "Evolution and the Genetics of Populations. Volume 2. The theory of Gene Frequencies." University of Chicago Press, Chicago.

PART 1

Physical aspects of snow

CHAPTER 1

Meteorological aspects of snow

CLEMENT J. DAVIS

SNOW FORMATION AND DEVELOPMENT

SNOW is a naturally occurring atmospheric element that can have large effects on flora, fauna and human activities over long periods of the year. It forms through the accumulation of ice crystals in the atmosphere until they are large enough to fall against gravity, reaching the ground before melting to liquid water. Therefore, an understanding of snow needs to begin with why and how it forms in the atmosphere and a recognition of the synoptic situations that bring snow. As snow develops through atmospheric motion, an idea of cloud formation and ice crystal development within these clouds is also required.

Cloud formation

Clouds form as a result of the rising of moist air. As air rises, it cools through expansion due to the lowering of air pressure with increasing height above ground, leading to an increase in relative humidity as the air temperature and dew-point temperature converge, with saturation occurring when they equalize. Further cooling of the air then produces supersaturated conditions with the excess water vapour condensing into tiny suspended water droplets that we see as cloud. Further development into precipitation is unlikely unless foreign substances are present that can attract these particles and result in the formation of droplets large enough to fall against gravity.

Cooling of the cloud below 0°C leads to the development of super-cooled water. This is water that exists in liquid form even though the surrounding temperature of the cloud may be well below 0°C. This type of water can exist in a pure environment to a temperature as low as –39°C. However, it does not often exist in large quantities in the atmosphere much below a temperature of –25°C due to the presence of foreign particles such as salt, sulphuric acid and dust, which act as nuclei for ice crystal formation. These particles are known as ice nuclei.

TABLE 1

Definition of types of ice crystals in the atmosphere (from Mason, 1971)

Crystal Type	Remarks
Plates	Also combinations of plates with or without very short connecting columns
Stellar Crystals	Also parallel stars with very short connecting columns
Columns	And combinations of columns
Needles	And combinations of needles
Spatial Dendrites	Spatial combinations of feathery branchlike crystals
Capped Columns	Columns with plates on either (or one) side
Irregular particles	Irregular aggregates of microscopic crystals
Graupel (soft hail)	Isometric shape, central crystal cannot be recognized
Ice Pellets	Ice shell, inside mostly wet
Hail	

Formation of Ice crystals

Clouds containing super-cooled water occur commonly in the atmosphere, particularly in areas of strong up-motion. The introduction of ice nuclei into an environment containing super-cooled water causes the super-cooled droplets to freeze on contact with these nuclei to produce ice crystals. Not only do these particles cause the formation of ice crystals, but they also influence crystal shape due to their hygroscopic properties as well as their surface tension characteristics. Other major factors affecting the growth of ice crystals include the ambient environmental temperature and humidity and the process of diffusion and thermal conduction to and from the nuclei; all of which indicate that the physics of ice development is highly complex. A classification scale has been designed which lists 10 different types of ice crystals and is summarized in Table 1.

Formation of snowflakes

Snowflakes develop through the growth of ice crystal aggregates. This process is controlled by the collision of the crystals, the aggregation efficiencies of these crystals, and by their relative motions as a result of their aerodynamic properties (e.g. different crystals move at different speeds). Experiments into ice crystal growth rate in the atmosphere indicate that while pressure affects ice crystal growth inversely (i.e. greater growth at lower pressures), the main influence is temperature, with maximum growth occurring around –15°C over a wide range of pressures. A breakdown of ice crystal formation and temperature as listed in the Australian Bureau of Meteorology training notes is indicated in Table 2.

When ice crystals combine to form snowflakes, there is no set pattern, resulting in the fact that no two snowflakes have identical structures. Ice crystals most prominent

TABLE 2.

Ice crystal formation as a function of atmospheric temperature (from Bureau of Meteorology Training notes).

Temperature	Basic Crystal structure	Types of crystal at slight water super-saturation
0°C to –4°C	Plate-like	Thin hexagonal plates
–4°C to –10°C	Prism-like	Needles (–4° to –6°C)
		Hollow columns (–5° to –10°C)
–10°C to –22°C	Plate-like	Sector plates (–10° to –12°C)
		Dendrites (–12° to –16°C)
		Sector plates (–16° to–22°C)
–22°C to –50°C	Prism-like	Hollow columns

in the formation of snowflakes are the dendrite or branch-like crystals, however, needle and plate formations may also form into snowflakes. The adhesion of ice crystals to form snowflakes depends on the structure of the crystal surfaces and the ambient temperature, with the aggregation of crystals into snowflakes more likely to occur when ice crystals move into environmental temperatures higher than –10°C.

From this information, it can be seen that snow formation is most likely to occur at cloud temperatures above –16°C as this is the lowest temperature at which dendrites, the most common ice crystals for snow, seem to form. Knowledge of cloud type and temperature and water vapour content and how they relate to ice crystal formation can therefore be used in forecasting snow development.

SNOW CLIMATOLOGY IN AUSTRALIA

Australian Snow Regions

There are three regions in Australia that receive regular snow during the winter months — Tasmania, the Victorian alpine region located in northeast Victoria and the NSW alpine region located in southern NSW and the ACT. In fact, the Victorian and NSW alpine regions can really be considered as one overall region as they are very close together and are part of the same geographic feature, the Great Dividing Range. The snow season in Australia is neither long nor regular and tends to be confined to winter and early spring. The major influences on the Australian snow season are as follows:

Geographic position of Australia

Australia is located between the sub-tropical ridge and a broad-scale belt of westerly winds to the south, which contain low-pressure systems. As a result, the colder conditions from the westerly winds, which produce the snow situations, tend to occur during the colder months of the year, due to the northward movements of these systems. Annual variations in the movements of these broad belts, and the local pressure within them, therefore can have a great bearing on the type of snow season from one year to the next.

Height of the Australian alpine regions

The Australian alpine regions are in fact quite low with the highest peak (Mt. Kosciuszko) only being 2228 m above sea level. This fact, when combined with overall geographic location, results in very sensitive atmospheric temperature conditions over the region, where only small changes in the weather patterns from one season to the next can have a big bearing on the type of snow season being experienced.

Mean Winter Broad-scale Circulation

According to Budin (1985), the following average meteorological pattern is found over Australia during the winter period:

a. A sub-tropical ridge lies across central Australia between 28–30°S.

b. A long wave trough is located near southwest Western Australia (the long wave trough is indicative of the position where cold fronts are likely to peak in development).

c. The mean wind speed at 850 hPa (approx. 1500 m) over highland areas is westerly at 6 m/s.

d. The dew-point temperature (T_d) at 850 hPa is lower over southeast Australia than over southwest Australia at the same latitude.

e. Free air temperature at 850 hPa is around 2°C over the NSW alpine region with the 0°C isotherm lying along latitude 45°S in eastern Australian longitudes.

The major mainland areas of winter snow cover have an elevation between 1370m and 2228m, this means that the normal temperature of these areas is around 0°C for most of the winter and hence any precipitation is more likely to be in the form of snow during this period.

These investigations indicate that the variations in the latitude of the sub-tropical ridge, and the longitude of the long wave trough, from the mean positions can have a marked influence on the type of season and snow depth. If the sub-tropical ridge is located further north and/or the long wave trough further east, then the snow season tends to be good, because this situation means that the cold fronts will still be developing and moving northeast as they pass over southeast Australia and the alpine regions. However, if the sub-tropical ridge is further south and/or the long wave trough further west, the season tends to be poor, with the cold fronts reaching their peak further to the west and weakening and moving southeast as they pass through the region.

Although Budin (1985) found a positive correlation between El Niño events and maximum snow depth, it was a weaker signal than for precipitation, and an El Niño situation may not necessarily result in a poor snow season. An explanation for this occurrence is that during an El Niño episode there is usually less rain (which has more of a damaging effect on snow cover) and with less rain, skies tend to be less cloudy. This results in sunnier daytime conditions which has less effect on melting snow due to the snow's high albedo (i.e. the ability to reflect solar radiation) and cold, dry night time conditions which allows for any softened snow to re-freeze again overnight. Thus if a general snow cover develops in these periods, even if it is not as thick as normal, then it is likely to remain for a considerable time.

Analyses of snow data

While alpine weather records in Australia are very patchy and incomplete an overall trend of snow events for the mainland regions has been obtained from analyzing all available data from both the mainland alpine regions and from Tasmania.

The individual records for the mainland stations used in the analyses are listed in Table 3. For the purpose of these analyses the definition of a snow day was where any one or more stations in the particular region reported snow on any given day. The information on Tasmania was supplied by the Bureau of Meteorology, Tasmanian Regional Office (M. Jones pers. comm.) and is constructed from snow reports extracted from the Bureau of Meteorology, Tasmanian Monthly Weather Reports over the period 1957–1991. Unfortunately, these records have missing data for the years 1959–1960, 1964 and 1969, and therefore snow events overall could be understated. This is particularly the case with regard to the Tasmanian Highlands where there has been no long-term recording station and in this area some of the reports are anecdotal.

TABLE 3

Snow data for Australian mainland alpine regions used in analyses
(Snowy Mountains Region 1970–1996, Victorian Alps 1978–1991)

Site (altitude of station)	Record History
Snowy Mountains–Charlotte Pass (1755m)	July 1970–October 1975. October 1993–December 1996
Snowy Mountains–Perisher Valley (1735m)	July 1976–September 1996. Observations outside June/September period not as comprehensive
Snowy Mountains–Thredbo Village (1380m)	May 1970–June 1987 continuous. October 1987–May 1996, data only for periods from October/mid June
Snowy Mountains–Crackenback Ski Station (1957m)	January 1970–October 1983 continuous. June 1984–October 1995, data only for periods from mid June to early October
Snowy Mountains–Cabramurra (1420m)	January 1970–December 1996 continuous
Vic. Alps–Mt. Hotham (1750m)	December 1977–July 1990*
Vic. Alps–Mt. Buffalo (1372m)	January 1978–October 1979*
Vic. Alps–Mt. Buller (1707m)	April 1985–December 1991*
Vic. Alps–Falls Creek (1510m)	July 1986–September 1990*

*Note data available for only one station from November 1979–March 1985

Analyses of the mainland data sets indicate that there is a very close correlation between the monthly percentage occurrences of snow in the Snowy Mountains region and the Victorian Alps. This correlation is even more pronounced when the data for the Snowy Mountain region are compared with the Victorian Alps over the same period (1978–1991). Therefore, the Victorian and the Snowy Mountains alpine areas can be treated as a single region in terms of the number of events and occurrences, with nearly 90% of all snow events occurring between the months of May and October. The most regular snow occurrences are between June and September with only a low number of events in the period January to March, as would be expected, with February being the month least likely to have snow.

By comparison, the data for Tasmania indicate that only around 74% of snow occurrences happen during the May and October period. Again, only few snow events are likely to occur during January and March with February the least likely month for snow. As most of the Tasmanian highlands are below 1500 m and with the general maritime climate of the island, there is usually rapid melting of snow in most seasons, and significant snow remains on the highlands only when there is a prolonged cold dry spell or frequent snow events. The results of these analyses are set out in Figures 1(a) and 1(b).

MAIN SNOW PRODUCING SITUATIONS

Snow occurs when the ice crystals making up the snowflakes do not melt before reaching the ground. Therefore, atmospheric temperature and moisture content, height of the Freezing Level (FZL) and conditions near the ground are all important in determining snowfall. Colquhoun (1978) summarized the four main snow-producing mechanisms in the Australian Alps. These mechanisms are:

Upper "Cold Pools"

These features tend to produce the heaviest snowfalls, but may only persist for a day or two at a time and occur irregularly during the season. An upper cold pool in late May or in June can provide a good base of snow that may last for the rest of the season, provided there are continual regular falls for the remainder of the season. Usually in a good year there may only be two or three of these systems occurring during the season.

Indications of possible cold pools can be obtained from the Mt. Gambier temperature trace (Mt. Gambier is located in South Australia, approx. 800 km WSW of the Snowy Mountains region). If the Mt. Gambier 700 hPa (approx. 3000 m) temperature (T_{700}) ≤ -5°C, then snow is likely whilst if the $T_{700} \leq -10$°C then heavy snowfalls are likely in the alpine region provided the cold pool continues to move over southern Australia. These cold pool situations may not initially have a surface pattern that can be recognized as a snow situation. Subsequent development of a slow moving surface low pressure system either over the region or in the Tasman Sea can then occur resulting in strong winds and the maintenance of cold conditions for several more days. Figure 2 shows an example of a cold pool situation.

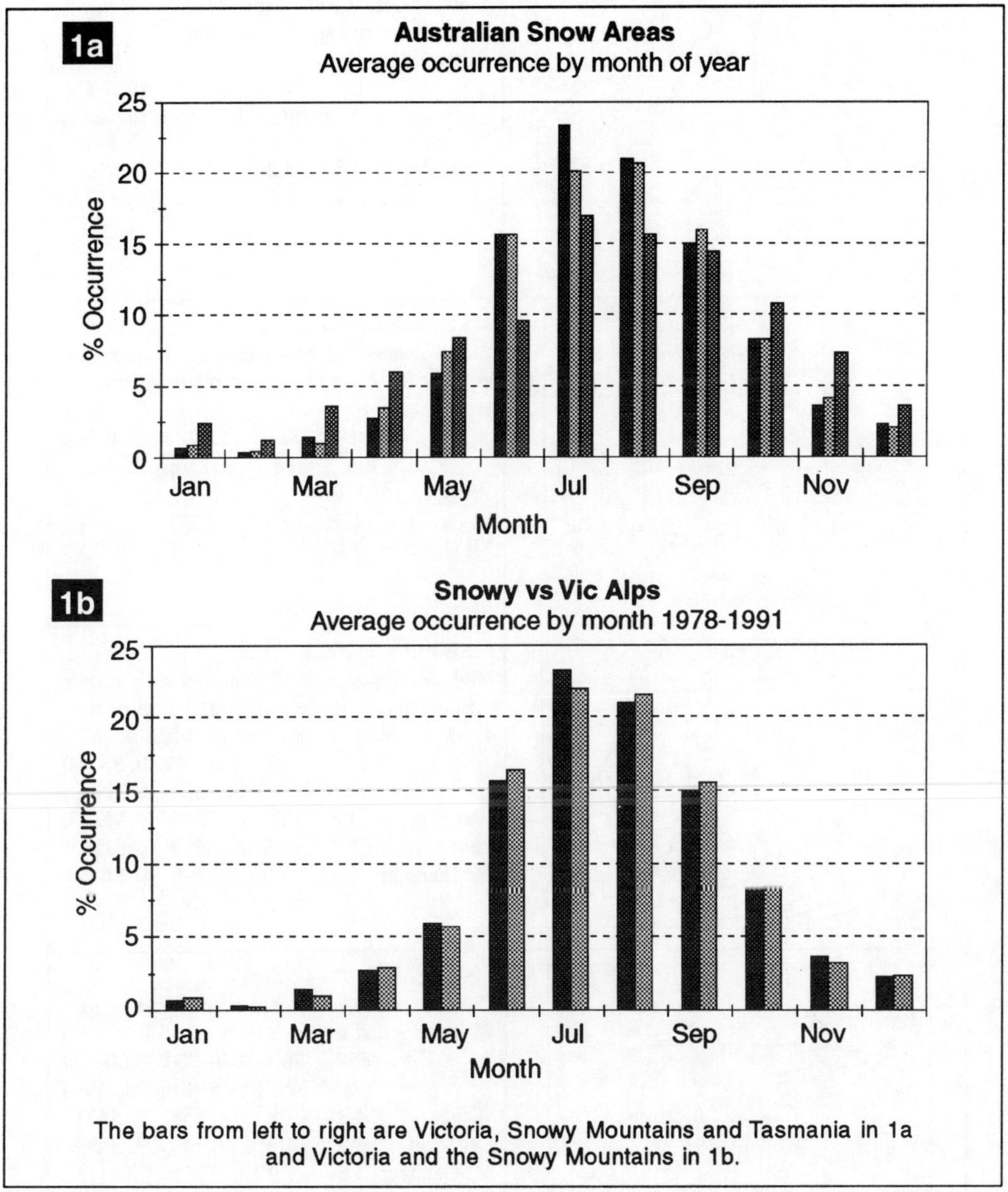

Figure 1. (a) Average percentage occurrences of snow days by month of total annual snow days for Australian snow regions. (b) Comparison of average monthly occurrences between the Snowy Mountain region and the Victorian Alps 1978–1991.

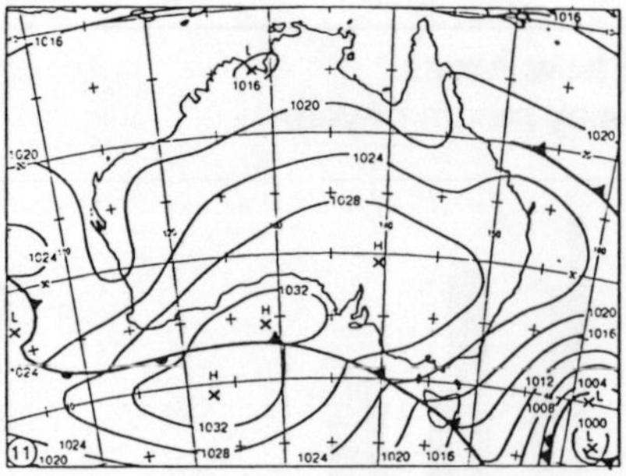

MSL Analysis 00z 11Sept
Cold front moving up from south of Tasmania.

	850 hPa T	700 hPa T	850-700 thickness	1000-500 thickness
Wagga	0.9	-1.9	1571	5479
Laverton	0.8	-3.3	1556	5450
Mt Gambier	0.6	-4.0	1542	5437

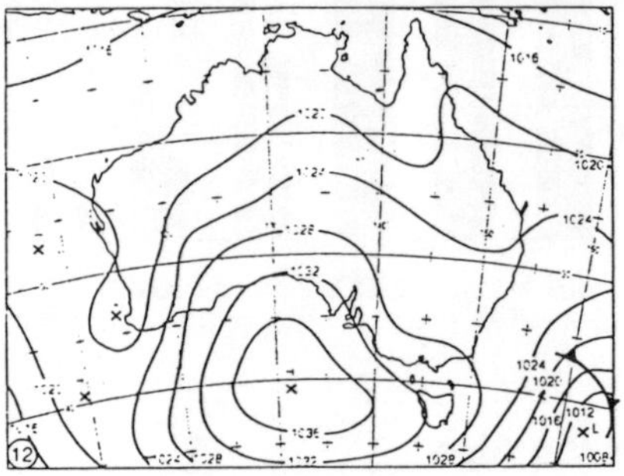

MSL Analysis 00z 12 Sept
Ridge developing over Tasmania. Upper cold air over SE Australia with snow in Alpine areas.

	850 hPa T	700 hPa T	850-700 thickness	1000-500 thickness
Wagga	1.1	-5.3	1539	5417
Laverton	-1.3	-9.4	1533	5344
Mt Gambier	-2.8	-11.9	1513	5323

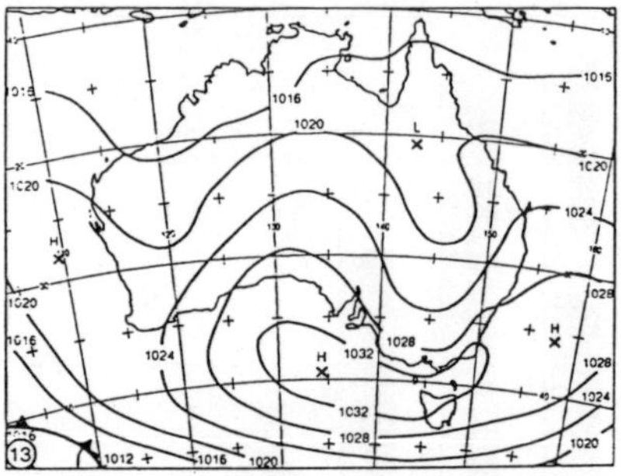

MSL Analysis 00z 13 Sept
High develops south of continent with trough deepening over eastern Australia in response to upper cold pool. Further snowfalls occur.

	850 hPa T	700 hPa T	850-700 thickness	1000-500 thickness
Wagga	4.8	-7.1	1547	5427
Laverton	2.7	-8.9	1537	5390
Mt Gambier	2.0	-7.2	1538	5404

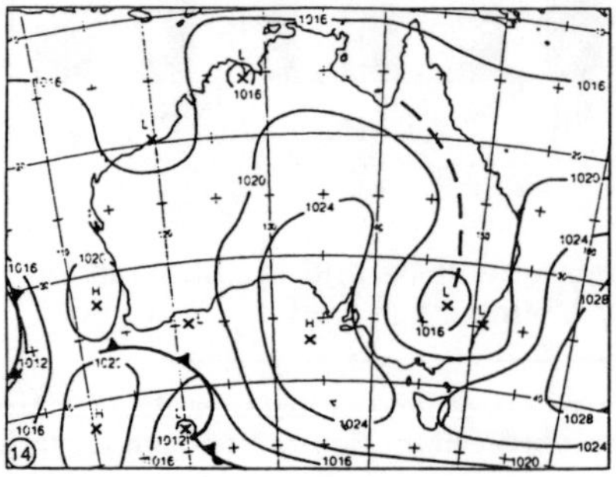

MSL Analysis 00z 14 Sept
Cut off surface low forms over Alpine region with continued snow.

	850 hPa T	700 hPa T	850-700 thickness	1000-500 thickness
Wagga	4.3	-3.8	1557	5477
Laverton	3.2	-4.7	1549	5439
Mt Gambier	2.5	-5.9	1545	5416

Figure 2. Example of upper "cold pool" snow situation. September 11–17 1993.

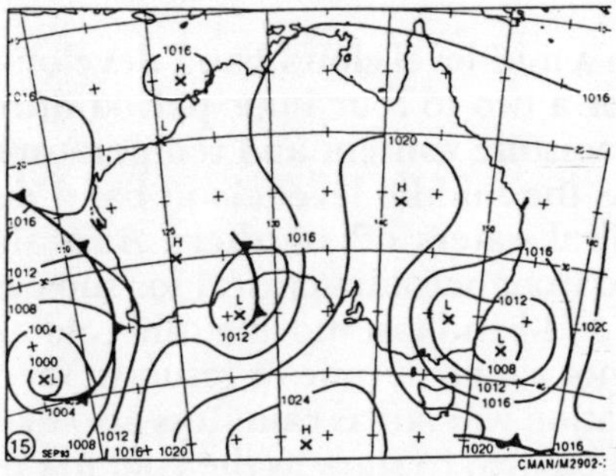

MSL Analysis 00z 15 Sept
Cut off low pressure system only very slowly moving east with advection of more cold air from south causing further snow in Alpine region.

	850 hPa T	700 hPa T	850-700 thickness	1000-500 thickness
Wagga	2.0	-7.3	1541	5415
Laverton	2.8	-5.2	1548	5445
Mt Gambier	1.3	-5.8	1547	5419

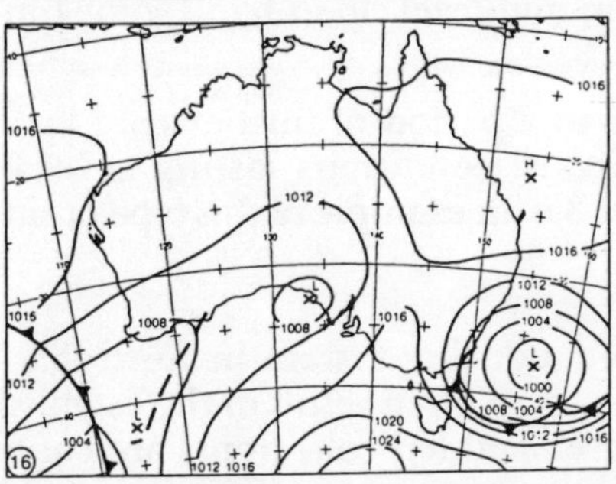

MSL Analysis 00z 16 Sept
Low moves into Tasman Sea, but cold pool still cold enough to produce more snow in Alpine areas.

	850 hPa T	700 hPa T	850-700 thickness	1000-500 thickness
Wagga	4.3	-4.8	1555	5451
Laverton	1.0	-4.7	1549	5435
Mt Gambier	3.8	-5.3	1551	5422

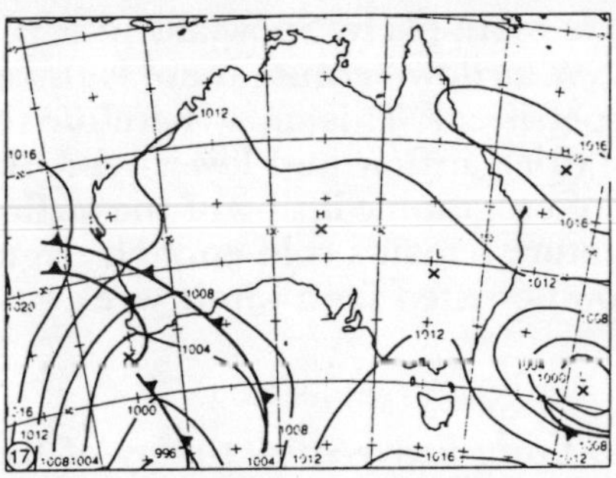

MSL Analysis 00z 17 Sept
Low moves further east but upper air still cold enough to cause snow.

	850 hPa T	700 hPa T	850-700 thickness	1000-500 thickness
Wagga	2.5	-5.8	1549	5433
Laverton	2.2	-6.2	1548	5425
Mt Gambier	2.3	-5.6	1547	5433

Cold pre-frontal low level NW airflow

These snow situations tend to occur when a mid-level cloud-band develops ahead of a cold front. Good snowfalls can occur over a two to four hour period during the passage of the cloud-band. Because of high moisture content and temperatures close to 0°C, this snow tends to fall as wet snow. As the middle level cloud-band develops from water vapour whose source is the tropical waters off northern Australia, it is relatively warm and thus the timing of the cold front becomes critical for snow in these situations. If the front is slow moving or the cloud-band fast moving, the cold air-mass undercutting the middle level moisture may not arrive in time to result in favourable conditions for snow in which case the precipitation will fall as rain. It is also likely that these situations can start by producing rain, then turn to snow as the cold low level air pushes in from the west and undercuts the warm mid-level cloud-band some time after precipitation has commenced.

This situation can often be the precursor to a period of further cold fronts and westerly winds bringing more snow and blizzard conditions lasting several days, depending on the broad-scale circulation. Figure 3 is an example of this type of situation.

Westerly airflow

This situation is the most common snow-producing system in Australia. It can cause persistent snow, spanning several days, as it can often result in the establishment of a zonal pressure pattern where fast-moving embedded cold fronts are continually passing over southeast Australia. With a particularly strong westerly regime, blizzard conditions which could last for several days are more likely. Snowfalls in a W to SW airflow tend not to be as heavy as in a W to NW airflow because there is usually less moisture available, but the snow can be more persistent. This snow therefore tends to be a lot drier and more powdery, due to the drier airflow and lower temperatures. Snowfall from this type of situation is a result of orographic lifting of the airflow over the mountains, provided the air-mass temperature remains cold enough. Figure 4 is an example of this type of situation where snow persisted for a whole week.

Southerly Cold Outbreaks

These systems are the ones most likely to bring heavy snow to any of the snow regions of Australia and can result in snow occurring well into northern NSW, depending on the extent and severity of the outbreak. Several of these events have been studied in some detail by Stern and Taylor (1985), Elliot (1989) and Jones and McCulloch (1992) whilst Shanahan (1967) summarized three situations where snow occurred on the Central and Northern Tablelands of NSW.

Generally, if the airflow is S to SW, the moisture levels are lower, and the snow tends to be dry and powdery. Snowfalls are mostly limited to within 36 hours of the event. S to SW airflow means that Tasmania and Victoria are more affected whilst in a SE airflow, the eastern Snowy Mountain region is affected. Figure 5 is an example of this type of situation.

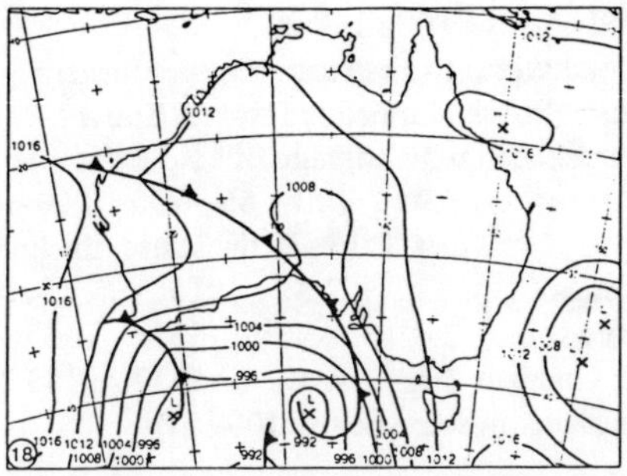

MSL Analysis 00z 18 Sept
Cold front extending across central Australia moving through Bight.

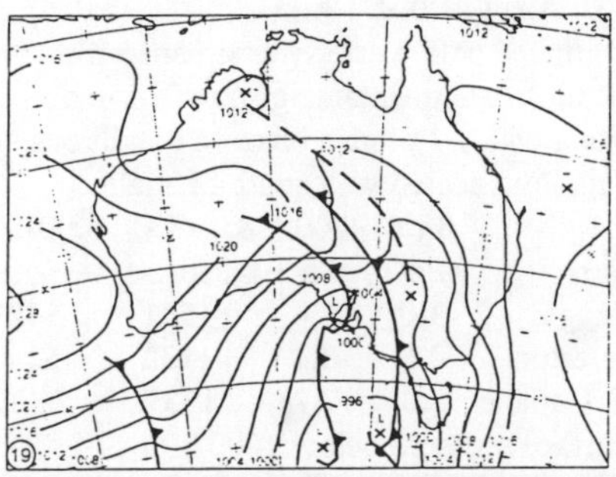

MSL Analysis 00z 19 Sept
Because of extensive northward component of front, a warm northerly airstream has developed bringing an in feed of moist air from tropics. Would expect a middle level cloudband and rain to develop ahead of front, turning to snow as the cold air behind the front undercuts the warmer air aloft.

	850 hPa T	700 hPa T	850-700 thickness	1000-500 thickness
Wagga	10.8	1.6	1594	5584

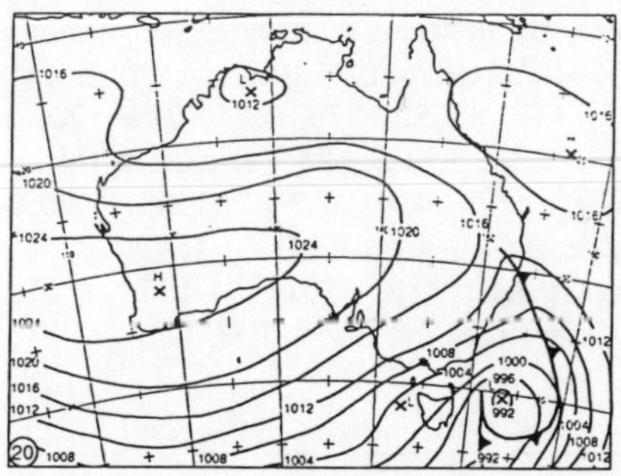

MSL Analysis 00z 20 Sept
Colder SW airstream established behind front with further snow.

	850 hPa T	700 hPa T	850-700 thickness	1000-500 thickness
Wagga	0.9	-7.8	1536	5444

Figure 3. Example of pre-frontal snow situation. September 18–20, 1993.

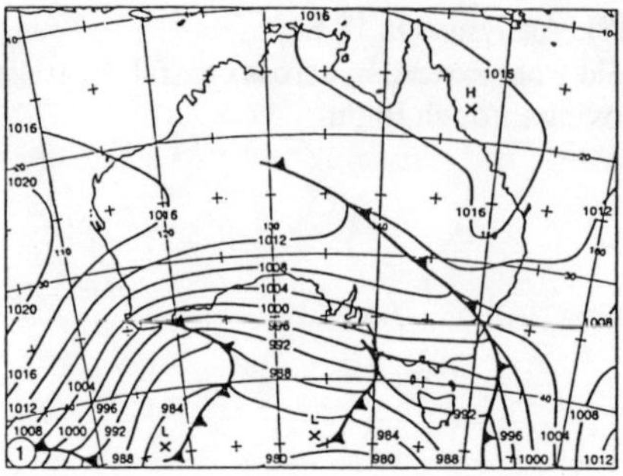

MSL Analysis 00z 1 Aug

Broad westerly airstream across southern Australia with embedded fronts. Snow commenced with passage of cold front.

	850 hPa T	700 hPa T	850-700 thickness	1000-500 thickness
Wagga	3.3	-2.8	1561	5478
Laverton	2.0	-8.4	1536	5364*
Mt Gambier	2.3	-8.1	1537	5372*

*surface pressure below 1000 hPa

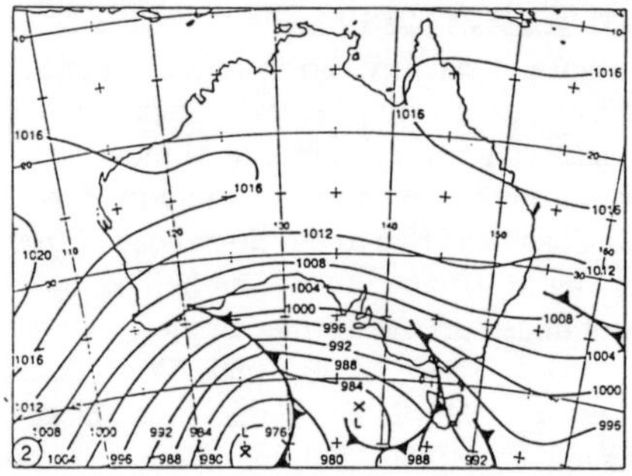

MSL Analysis 00z 2 Aug

Continued cold westerly airstream with further fronts maintaining snowfalls and strong winds. Possible blizzard conditions. Main low located well south of Bight.

	850 hPa T	700 hPa T	850-700 thickness	1000-500 thickness
Wagga	3.0	-6.4	1541	5423
Laverton	0.6	-8.1	1533	5387*
Mt Gambier	2.0	-7.7	1537	5337*

*surface pressure below 1000 hPa

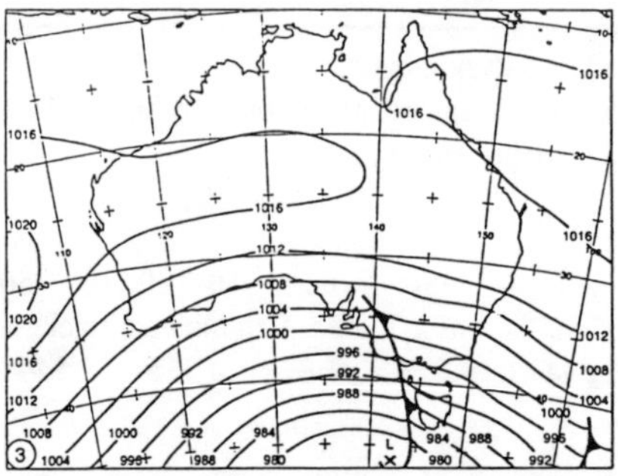

MSL Analysis 00z 3 Aug

Westerly airstream maintained with continued snowfalls reported.

	850 hPa T	700 hPa T	850-700 thickness	1000-500 thickness
Wagga	4.4	-7.3	1543	5535
Laverton	1.6	-7.4	1540	5404*
Mt Gambier	0.9	-9.6	1531	5371*

*surface pressure below 1000 hPa

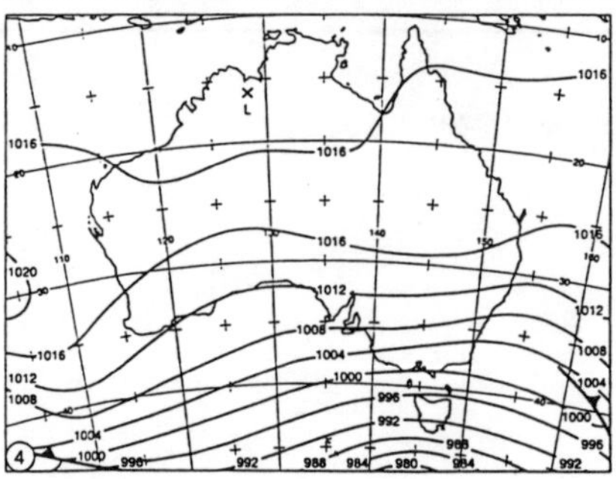

MSL Analysis 00z 4 Aug

Cold westerly regime maintained with further snow.

	850 hPa T	700 hPa T	850-700 thickness	1000-500 thickness
Wagga	1.8	-6.7	1541	5401
Laverton	-0.6	-10.1	1542	5366
Mt Gambier	-1.0	-10.4	1522	5343

Figure 4. Example of persistent westerly air-stream snow situation. August 1–8, 1996.

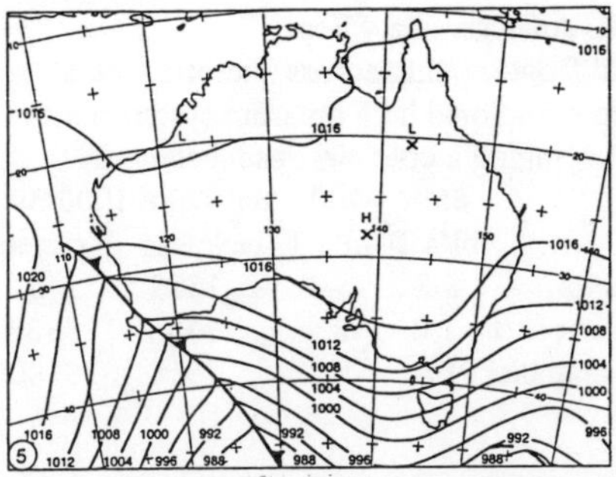

MSL Analysis 00z 5 Aug
Cold front moving into WA with high pressure ridge over central Australia. Westerly airstream in Alpine area still bringing snowfalls.

	850 hPa T	700 hPa T	850-700 thickness	1000-500 thickness
Wagga	-0.4	-7.8	1529	5371
Laverton	-1.4	-7.8	1526	5362
Mt Gambier	n/a			

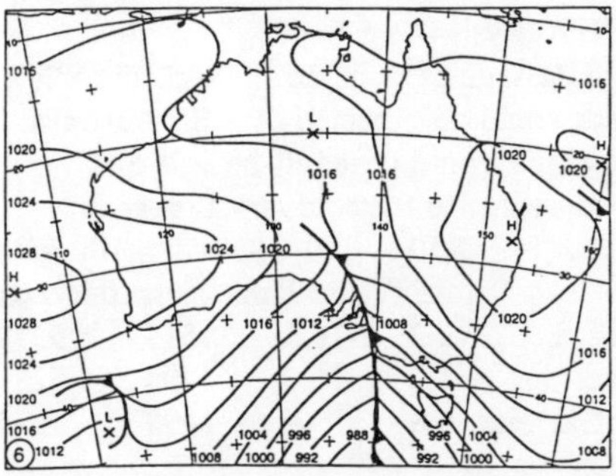

MSL Analysis 00z 6 Aug
Cold front approaching SE Australia with extra moisture being advected from north ahead of front causing further snow

	850 hPa T	700 hPa T	850-700 thickness	1000-500 thickness
Wagga	2.0	-7.6	1538	5440
Laverton	2.7	-7.9	1538	5401
Mt Gambier	3.0	-7.1	1543	5428

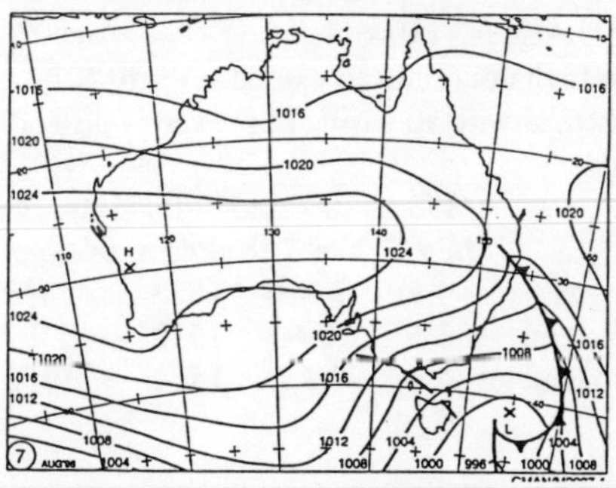

MSL Analysis 00z 7 Aug
Very cold airstream behind front maintaining snow situation.

	850 hPa T	700 hPa T	850-700 thickness	1000-500 thickness
Wagga	-2.0	-11.4	1519	5341
Laverton	-1.6	-12.6	1504	5297
Mt Gambier	-3.0	-13.9	1505	5293

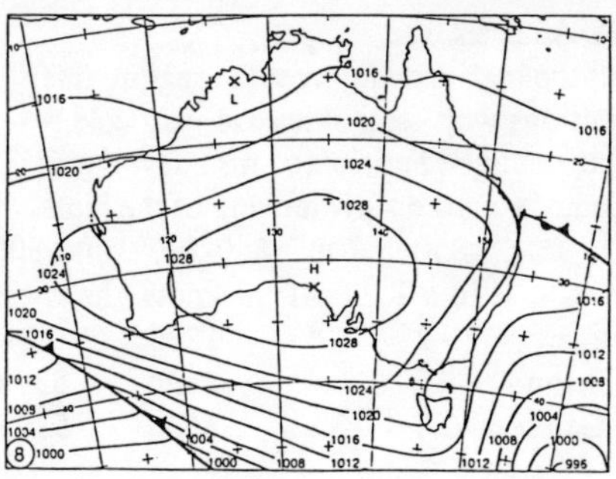

MSL Analysis 00z 8 Aug
The last of the fronts has moved through with a high pressure ridge starting to develop further south. Snowfalls easing.

	850 hPa T	700 hPa T	850-700 thickness	1000-500 thickness
Wagga	-3.8	-13.9	1504	5275
Laverton	-3.8	-13.4	1506	5378
Mt Gambier	-3.3	-6.6	1530	5400

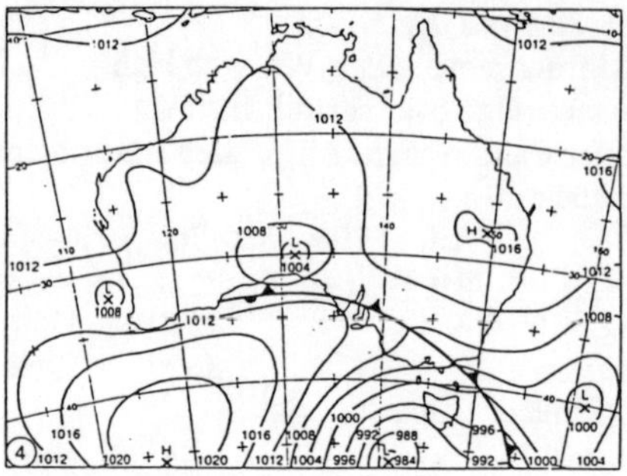

MSL Analysis 00z 4 Sept

Cold front moving across Victoria with a deep meridional high pressure system south of WA bringing a cold airstream behind front.

	850 hPa T	700 hPa T	850-700 thickness	1000-500 thickness
Wagga	5.7	-5.3	1555	5467
Laverton	-1.0	-8.4	1545	5388
Mt Gambier	-1.9	-8.3	1518	5380

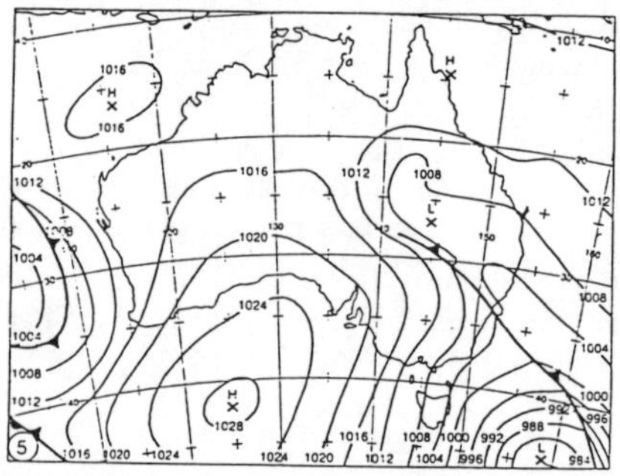

MSL Analysis 00z 5 Sept

With passage of front, strong and cold S'ly winds could be expected over SE Australia. Snowfalls would generally be light and powdery due to the cold, dry airstream.

	850 hPa T	700 hPa T	850-700 thickness	1000-500 thickness
Wagga	-2.2	-10.3	1513	5337
Laverton	-5.2	-15.9	1495	5231
Mt Gambier	-4.8	-15.4	1495	5276

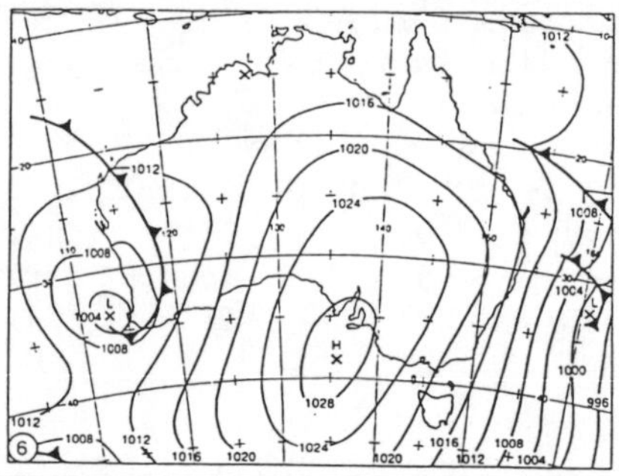

MSL Analysis 00z 6 Sept

The high continues to extend towards SE Australia with an easing in the snowfalls and winds.

	850 hPa T	700 hPa T	850-700 thickness	1000-500 thickness
Wagga	-4.8	-9.6	1523	5440
Laverton	-3.5	-10.6	1512	5327
Mt Gambier	-4.8	-8.1	1514	5350

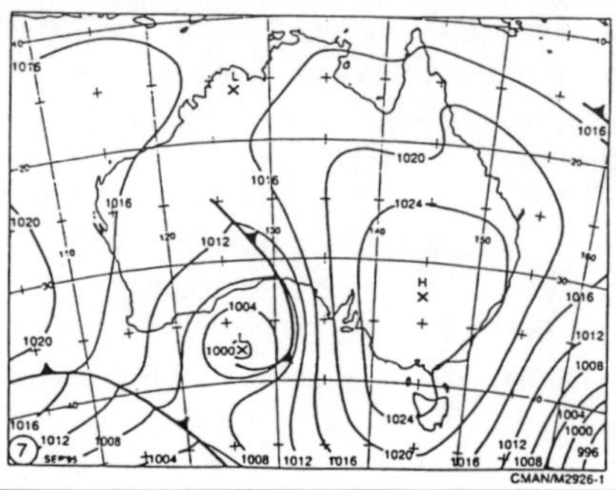

MSL Analysis 00z 7 Sept

With the high moving over the region, fine conditions with clear skies and very cold nights could be anticipated for a few days depending on the slow moving of the high.

	850 hPa T	700 hPa T	850-700 thickness	1000-500 thickness
Wagga	0.2	-8.4	1526	5386
Laverton	-0.3	-8.1	1526	5378
Mt Gambier	-2.7	-5.8	1536	5395

Figure 5. Example of a cold outbreak snow situation. September 4–7, 1995.

While the above situations are a summary of the major snow producing mechanisms in the Australian area, they are not necessarily separate events and may either occur together or follow one another in succession. Figure 6 is an example of a multiple synoptic category snow producing sequence.

As well as the different regions having different snow situations, those events in Tasmania are even more complicated due to the island's topography with different areas within the State receiving snow in different circumstances. Thus if one type of pattern predominates through the season, one area can receive a lot of snow while another area may miss out altogether.

FORECASTING SNOW

Indicators for forecasting snow

The major problems with forecasting snow are in determining the difference in precipitation between rain and snow and in assessing the likely amount of snow. Because snow, due to its crystalline structure, is less dense than its liquid form (water), relating measurement of snowfall to the equivalent amount of water can be difficult particularly as each snow event may result in different ratios of ice to water. For example, wet snow tends to contain more water than does powder snow and therefore will result in a thinner snow layer for the same total moisture content. Overall, an estimate of 10:1 is used as a conversion factor (i.e. 1 cm of snow = 1 mm of water). The major indicators that forecasters are looking for in forecasting snow are the following:

a. The height difference between atmospheric pressure levels (thickness value).

In computer models, pressure levels are often used as a mathematically convenient way of expressing the vertical extent of the atmosphere. The differences in height between these levels can alter depending on the atmospheric conditions. Thus forecast values of these height differences can provide a clue to the likely conditions of the atmosphere. The standard pressure level for the surface is 1000 hPa, for 1500 m (approx. 5000 ft) it is 850 hPa, for 3000 m (approx. 10,000 ft) it is 700 hPa and for 5000 m (approx. 18,500 ft) it is 500 hPa. The height difference between 1000 hPa and 500 hPa is then described as the 1000–500 hPa thickness value with all these values being expressed in metres. The value often used in the prediction of snow is a 1000–500 hPa thickness of 5400m.

b. Water vapour concentration in the atmosphere, particularly around the 700 hPa level.

This parameter provides an indication of the available water vapour for possible precipitation as without sufficient water vapour, no precipitation will occur.

c. Temperature and dew point temperature conditions near the ground.

These conditions are the most apparent indicators of the possibility of snow. Provided there is precipitation, if there is a sufficiently thick, cold and dry layer near

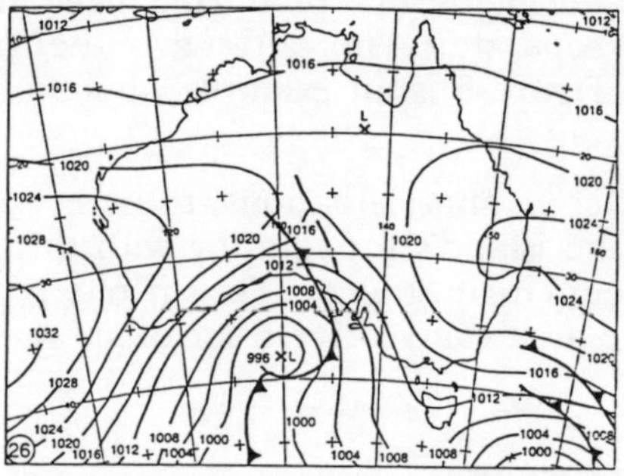

MSL Analysis 00z 26 Jun
Low pressure system developing in Bight with weakening ridge over eastern Australia.

	850 hPa T	700 hPa T	850-700 thickness	1000-500 thickness
Wagga	4.2	-4.2	1553	5489
Laverton	3.0	-5.2	1546	5454
Mt Gambier	3.8	-3.3	1551	5464

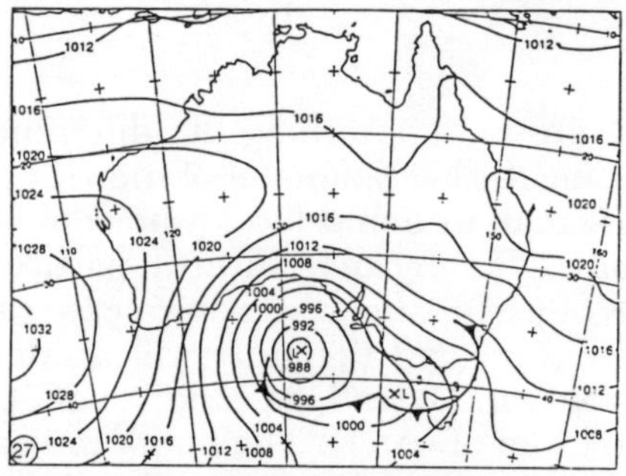

MSL Analysis 00z 27 Jun
Cold pre-frontal NW airstream

	850 hPa T	700 hPa T	850-700 thickness	1000-500 thickness
Wagga	3.8	-4.8	1534	5461
Laverton	2.0	-7.3	1538	5395
Mt Gambier	1.0	-8.1	1535	5399

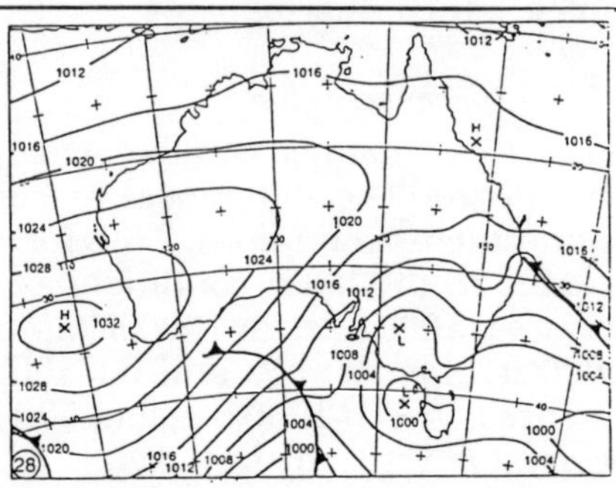

MSL Analysis 00z 28 Jun
Upper cold pool extending over region.

	850 hPa T	700 hPa T	850-700 thickness	1000-500 thickness
Wagga	1.9	-9.9	1530	5371
Laverton	-0.4	-10.4	1525	5365
Mt. Gambier	0.6	-7.3	1536	5389

Figure 6. Example of a combination of snow-producing events. 26 June-01 July 1995.

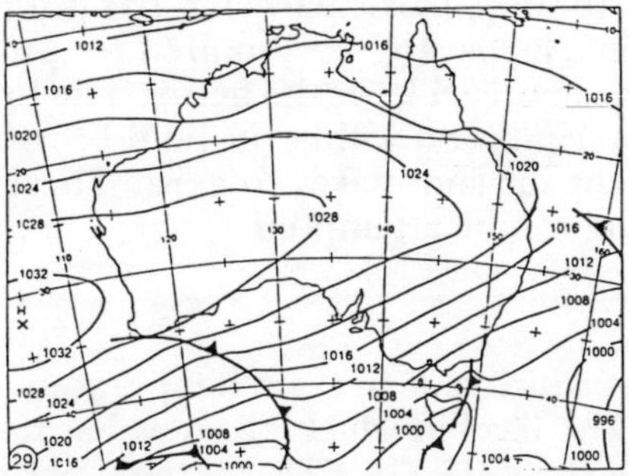

MSL Analysis 00z 29 Jun
W/SW airstream develops. Strong winds and snow in Alpine region.

	850 hPa T	700 hPa T	850-700 thickness	1000-500 thickness
Wagga	-0.6	-8.8	1529	5401
Laverton	-2.3	-13.1	1511	5319
Mt Gambier	-2.0	-4.7	1529	5402

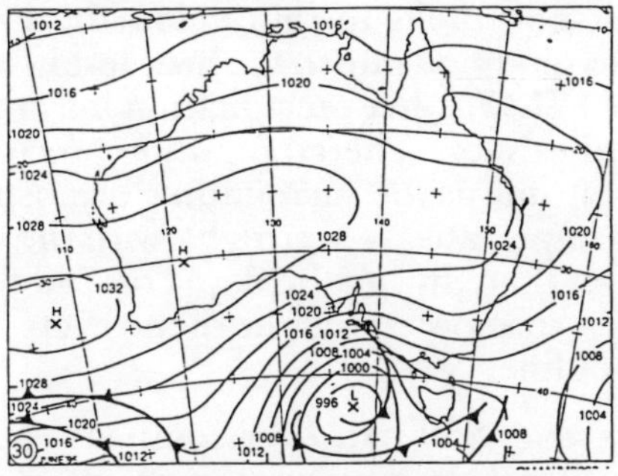

MSL Analysis 00z 30 Jun
Embedded front with further cold pool approaching from west.

	850 hPa T	700 hPa T	850-700 thickness	1000-500 thickness
Wagga	-1.1	-5.9	1539	5411
Laverton	0.6	-9.9	1531	5373
Mt Gambier	1.3	-9.4	1528	5377

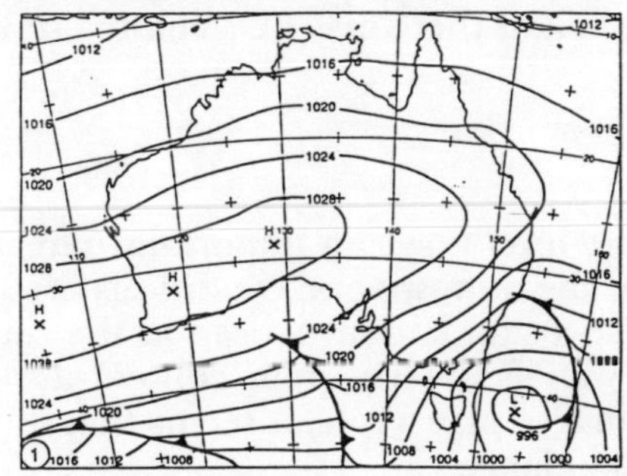

MSL Analysis 00z 1 Jul
Passage of front and with strong SW winds and continued snow.

	850 hPa T	700 hPa T	850-700 thickness	1000-500 thickness
Wagga	-2.3	-10.3	1522	5388
Laverton	0.3	-9.8	1528	5363
Mt Gambier	1.0	-5.8	1539	5406

the ground, the precipitation can revert back to snow, particularly if it has just melted. Generally surface temperatures and dew points below 1°C are required for pure snow - the colder and the drier the conditions of the low level layer the better. The lower the temperature, obviously the less likely the precipitation will be in liquid form whilst drier conditions provide a better environment for snowflakes to exist. These values also determine whether artificial snow-making can be attempted.

d. Freezing Level. (FZL)

This is the height of the atmosphere at which the free air temperature is 0°C. Below this level the snow will start to melt. The level at which all snow has melted is considered to be around 300 m (1000 ft) below the FZL. The FZL can therefore provide an indication of the probable snow line. As snow melts to rain below the FZL, the temperature of the surrounding layer becomes modified due to the heat lost in melting the snow, and thus results in a lowering of the FZL. Further precipitation may then not melt as quickly, and actually reach the ground as snow. Therefore, while precipitation may begin as rain, it could turn into snow over time as the surrounding atmosphere is modified. Factors such as time of day are also important, as overnight, with the drop in temperatures, the FZL can quite easily be lower than during the day. This can result in situations where rain during the day can turn to snow overnight then revert back to rain during the day, particularly near the snowline.

As the FZL is located around, or just above, 850 hPa over southeast Australia during winter, the exact determination of this level can play an important role in the decision to forecast snow or rain and at what height there may be a change from snow to rain.

e. Temperature of the atmosphere

The temperature of the 700 hPa pressure level plays an important part in snow forecasting, as it can provide information on the atmospheric conditions around the level of ice crystal formation. This level is also a standard level used in the computer models, and therefore temperature data as well as water vapour content are readily available. Although 500 hPa temperatures can also play a role, it seems that this lower level is more critical and can provide better guidance.

f. Wind

Vertical wind profiles are important in determining the ascent or descent of the broad-scale flow and hence the instability of the atmosphere and thus the likelihood of precipitation. However, surface wind strength and direction can have more of an effect in determining where snow falls rather than the amount that falls. As snowflakes are extremely light, the wind can easily blow them around. Hence, snow will build up on leeward areas and will be thinner on windward slopes. This also causes problems in determining just how much snow has fallen as the cover will not be uniform.

g. Local effects

Local effects have considerably greater impact on the weather in the alpine region than in less mountainous terrain. This topography can determine wind strength and direction (and hence ascent of air), as well as the amount of sunshine at any point, which in turn can affect both temperature and cloud development. As these conditions can also change very quickly during the day, there is a need for regular observations both temporally and spatially to be able to determine these effects. Knowledge of these local effects can therefore only be gained by experience built up over many years of observation.

Rain versus Snow

There have been many attempts to try to determine whether precipitation will fall as rain or snow. Predictors fall into two categories — long-term predictors and short-term predictors.

Long-term predictors

Long-term predictors tend to include atmospheric thickness values, middle level temperature fields and the height of the FZL (i.e. broad-scale considerations). Boyden (1964) analyzed the different thickness values between standard pressure levels then available as snow predictors - the major thickness value used at the time being the 1000–500 hPa thickness value.

Generally the type of precipitation is determined by conditions in the lowest levels of the atmosphere (e.g. the condition of the layer just above the surface). Thus the 1000–500 hPa thickness values are rather crude as up to 9/10ths of this layer has no bearing on the problem, taking in much of the atmosphere not required for this process. The justification for its use is that this value is a standard value derived from computer models. Its success is related to the fact that during precipitation the temperature lapse rates in the lower troposphere remain consistent and thus is moderately well indicated by the mean temperature of a much deeper layer. The 1000–700 hPa layer would then be considered to be a more accurate parameter as it excludes most of this irrelevant layer, but it may not be available on a regular basis.

However, in the Australian context, because the snowfields are mostly above 850 hPa, any predictor including values for the 1000 hPa level is partially irrelevant. The use of the thickness value of the 850–700 hPa layer as an indicator for snow forecasting has been investigated by Henry (1971) and Heppner (1982). These investigations indicate that the value for this thickness should not be greater than 1550m for forecasting snow. This value is also readily extracted from the computer models now in use. An empirical study of this predictor in Canberra for the 1994 season indicates that this could be an excellent first estimate as there were no cases studied where snow occurred when this thickness value was more than 1560 m as measured by the Wagga Wagga temperature trace. No data are available on either the false alarm ratio or of the times rain occurred rather than snow for occasions this thickness value was below 1550 m.

Short-term predictors

Short-term predictors need to be more sensitive and hence have a greater emphasis on current trends and local effects. Booth (1970) investigated the use of dew point as a snow predictor in England. He was looking for a predictor that was readily available and easily measured from hour to hour and according to Boyden (1964), the surface temperature was unreliable as a predictor for snow because:

1. Surface temperature depends on many factors including cloud cover, whether there is a snow cover or whether the ground is frozen.

2. Surface temperature changes rapidly as rain turns to snow and extrapolated surface temperatures give little indication of when this is likely to occur.

Thus while surface temperature is not a good indicator, Booth (1970) considered the dew point temperature (T_d) fits the criteria better as it is an indicator of air-mass change and it can easily be tracked on the analysis. In his study, he found the following indicators to hold:

- Snow never fell when $T_d \geq 3°C$
- Rain never fell when $T_d \leq -3°C$
- The type of precipitation changed markedly with $T_d = 0°C$ for rain and $T_d = -1°C$ for showers
- Snow was unlikely when $T_d \geq 1°C$
- There was >50% chance of snow when Dew Point Depression (DPD) $\leq \{-2xT_d\} + 1$ in rain conditions.
- These data give no reference to either the intensity or duration of the snowfall.

IS THE ALPINE CLIMATE CHANGING?

There have been a number of studies carried out on the climate of the Australian snow country, the duration of snow cover and possible effects of atmospheric warming. These studies include those by Slatyer *et al.* (1985), Galloway (1988), Ruddell *et al.* (1990), Duus (1992), Haylock *et al.* (1994), Whetton *et al.* (1996) and see Chapter 11, all of these tend to consider the duration of the snow cover rather than the actual snow events (i.e. the occurrence of snow days). The analysis undertaken here considers the actual identified snow events and compares this information with the upper air data from the Bureau station at Wagga Wagga and looks at the changes that have occurred over time. Although radio-sonde instrumentation changed in 1988, the change of temperature probes used in the radio-sondes does not appear to have resulted in any noticeable difference in the temperature readings taken at the 850 hPa and 700 hPa levels.

The data for the mainland Alpine regions used in the analysis consist of observations from the Victorian Alps (1978–1991) and the Snowy Mountains region (1970–1996 including the subset 1978–1991 for comparison with the Victorian Alps). Figure 7 (a) shows time series analyses for the Snowy Mountain region for the 6 month period May/October and for the 4 month period June/September from 1970 to 1996 while Figure 7 (b) shows similar analyses for the Victorian Alps from 1978 to 1991. Although these time series are relatively short, there does appear to be a decreasing trend in the Snowy Mountains region data over the 27 years analysed while there is no apparent trend evident in the Victorian analysis. However, as Figure 8 (a) indicates, this trend is not apparent in the Cabramurra snow data which actually indicate a slight increase in snow days compared to the decrease for the Snowy Mountains region as a whole. This increasing trend in snow seems to be occurring despite an increasing trend in both the Cabramurra maximum and minimum temperatures for the May/Oct period as shown in Figure 8 (b).

There are several possible explanations for these differences:

1. The incomplete data set for the higher altitude stations in the Snowy Mountains region may be resulting in an under-recording of snow events and therefore biasing the results. The data set for the Victorian Alps covers only a very short period and therefore is no real indicator of whether any trend is occurring. Further work is necessary on the Snowy Mountains region data set to determine how representative this set actually is over the period of record. However, the fact that both the Snowy Mountains region and the Victorian Alps give almost identical percentages of occurrence by month, even when compared over all years of records, would suggest that most of the snow events during the period concerned have been identified.

2. If there are fewer snow events occurring due to atmospheric warming, then in marginal cases where snow was recorded at higher altitudes but only rain at Cabramurra, it may now be raining at these higher altitudes instead of snowing.

3. Atmospheric warming may be leading to a shortening of the snow season at higher altitudes and hence fewer snow events. This shortening may be occurring at both ends of the season.

4. The occurrences of snow events are not only related to temperature but to wind speed and direction, the amount of atmospheric moisture and the type of situation producing snow, which may not be related to temperature. Synoptic situations over this period may therefore be favouring snow events on western ranges rather than southern ranges of the Snowy Mountains.

5. This is a comparison of a group of stations against a single station and a single station is more likely to be affected by changes in observing practices and other inhomogeneities, which may account for any trends.

An analysis of the closest Bureau upper air station at Wagga Wagga was undertaken to determine whether any such anticipated changes were reflected in upper air temperatures. This analysis indicates that there is some evidence for an increasing trend in the upper air temperatures over this time period. As can be seen from Figure 9,

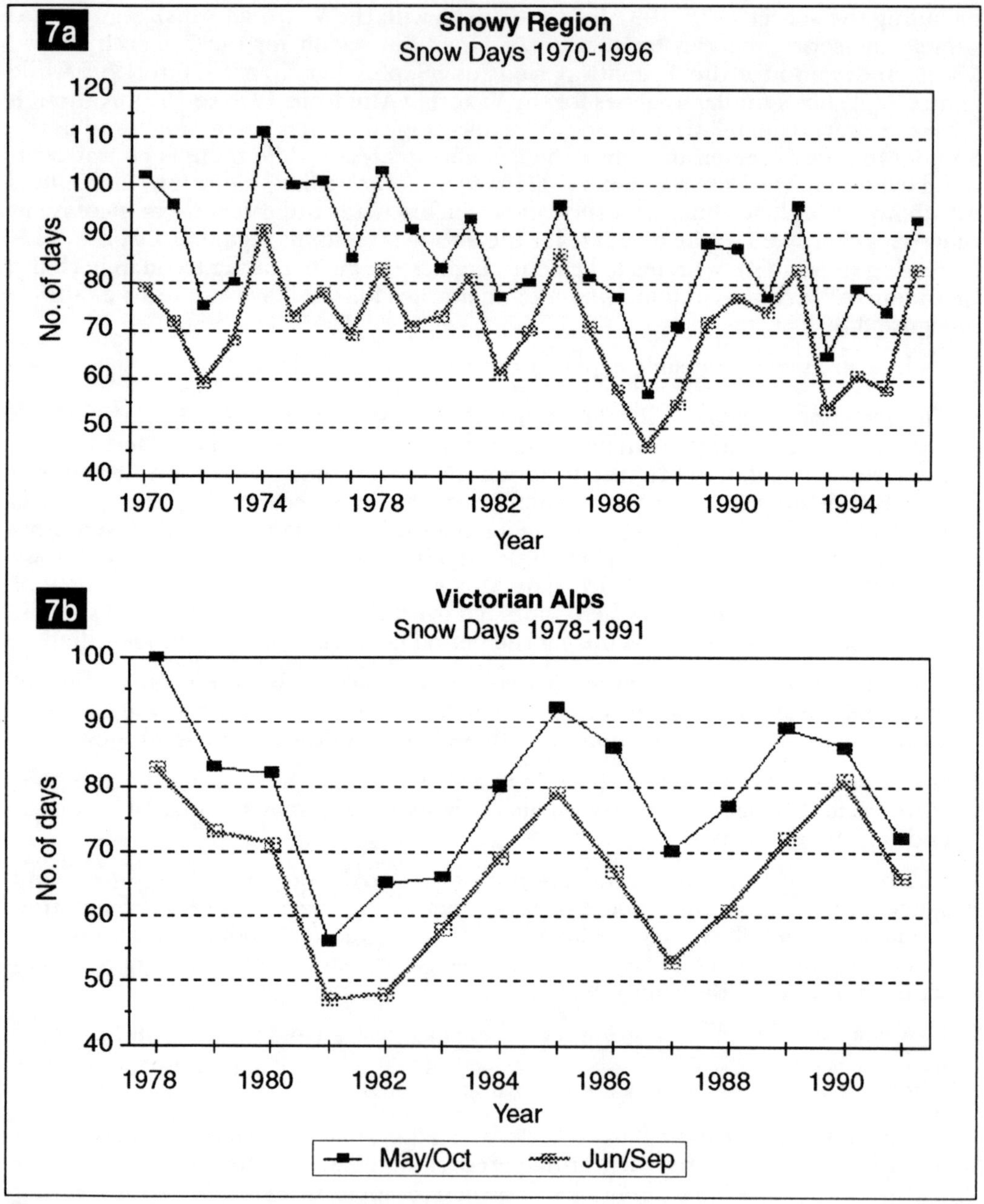

Figure 7. (a) Time series analyses. May-Oct and Jun-Sep snow days, Snowy Mountain region 1970–1996. (b) Time series analyses. May-Oct and Jun-Sep snow days, Victorian Alps 1978–1991.

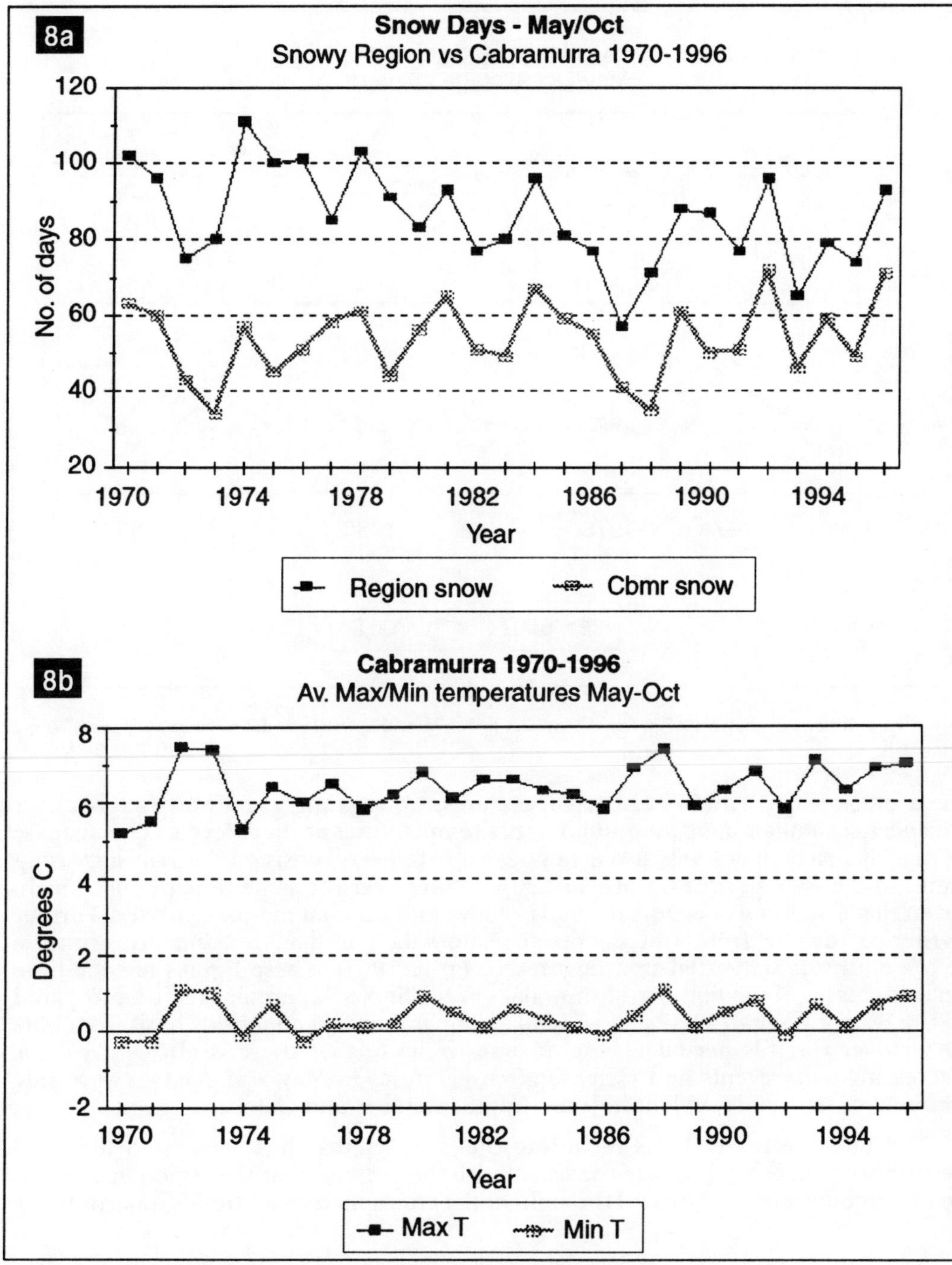

Figure 8. (a) Time series analyses. May-Oct snow days for Cabramurra compared to Snowy Mountain region 1970–1996. (b) Average May-Oct maximum and minimum temperatures for Cabramurra 1970–1996.

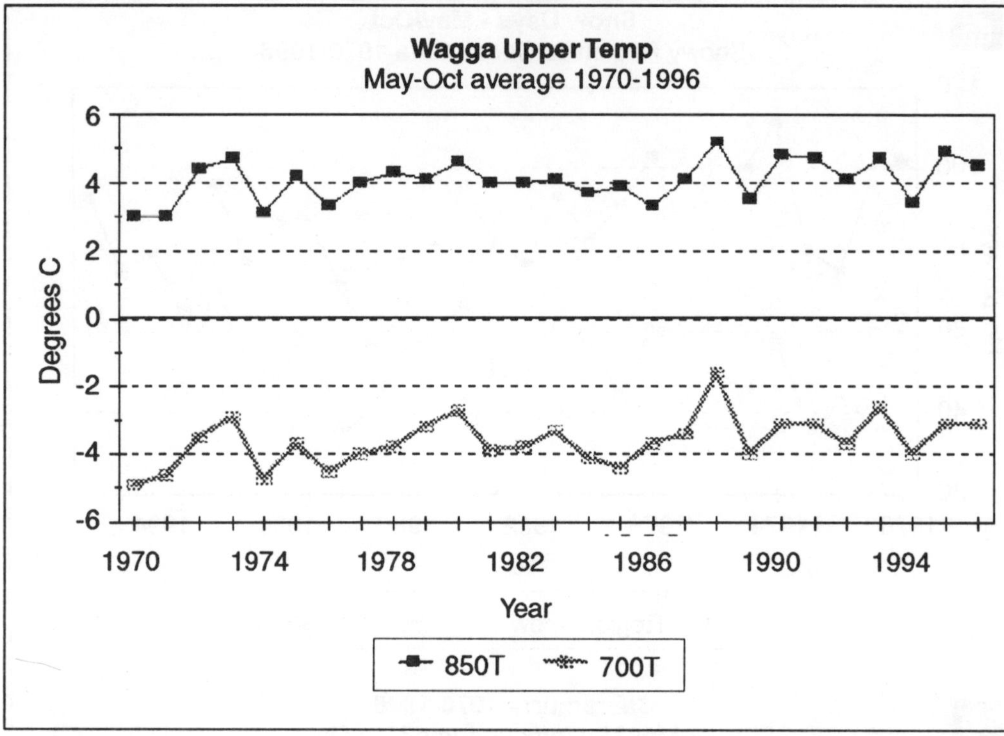

Figure 9. Time series analyses of average May-Oct 850 hPa and 700 hPa temperatures at Wagga Wagga 1970–1996.

the average 850 hPa and 700 hPa temperatures at Wagga Wagga for May-October over this period show an increasing trend in the temperatures at these levels. The increase in the 850 hPa temperatures at Wagga Wagga is in fact very consistent with the increasing trend in the Cabramurra surface air temperatures over the same time period, which, given that Cabramurra is close to 1500 m above MSL, is what may be expected. Further evidence for the trends of decreasing snow days and increasing atmospheric temperatures is suggested from Figures 10 (a) to 10 (f). These figures are the time series analyses of the number of snow days in the Snowy Mountains region compared to the Wagga Wagga 700 hPa and 850 hPa temperatures on a monthly basis from May to October for this period. These analyses indicate that there is also a trend for decreasing snow events and increasing temperatures in May and August, with only minimal or no trends evident in June, July, September and October.

Despite the reservations regarding the data set that has been used for snow occurrences, these analyses are consistent with the premise that the region appears to be undergoing a warming trend throughout the atmosphere, with trends towards fewer snow events per year. A shortening of the deep season may be also occurring when the data for August are analyzed in detail. Figures 11 (a) and 11 (b) are the time series analyses of the snow events in both the first and second halves of August plotted against the Wagga Wagga 850 hPa and 700 hPa average August temperatures. While the first half of August shows trends similar to July, the second half plot indicates a tendency to have a greater variation in the number of snow days per year, particularly since 1982, and increasing trend in both 850 hPa and 700 hPa temperatures at Wagga Wagga.

These analyses indicate strong correlation between decreasing snow days and increasing upper air temperatures at Wagga Wagga in May, with less strong evidence for similar trends both at in the latter half of August and in October. Very few changes in trends are evident for the months of June, July and September.

Calculations of the regression values for the lines of best fit for these time series are listed in Table 4 along with values from the Student t test which provide an indication as to the statistical significance of these trends. There are two assumptions made in relation to these calculations. These are:

1. That there is independence between yearly values
2. That a normal distribution exists

For this range (27 years), the t value for the 95% confidence level is 2.056. Thus any t value more than this value indicates a very high level of significance. The values from these lines of best fit indicate that while there is an apparent trend of rising atmospheric temperatures throughout May to October over this time period with generally decreasing snow events over the region, statistically the most significant values are those involving May. The tests for the other months are more inconclusive.

The warming trend over the period is also consistent with the increasing trend of the average annual temperatures over Australia during this period as shown in Figure 12.

In Tasmania, trends tend to indicate that there has been a steady decrease in snow falling below an elevation of 450 m over the 30 years since 1962. However, there does not appear to be any significant changes in the 850 hPa and 500 hPa temperatures, the height of the Freezing Level nor the 1000–500 hPa thickness values (M. Jones pers. comm.).

Whether any inferred climate change indicated by these analyses could be linked to an enhanced Greenhouse Effect is very much open to question. This is because climate change which could be produced by other natural processes, such as ENSO (internal to the climate system) or volcanic activity or planetary movements (external to the climate system), may be superimposed on or swamp any change due to this effect. Also the period under study is only over a relatively short time scale and a more extensive study over a longer period is required to determine if this apparent change is relevant overall.

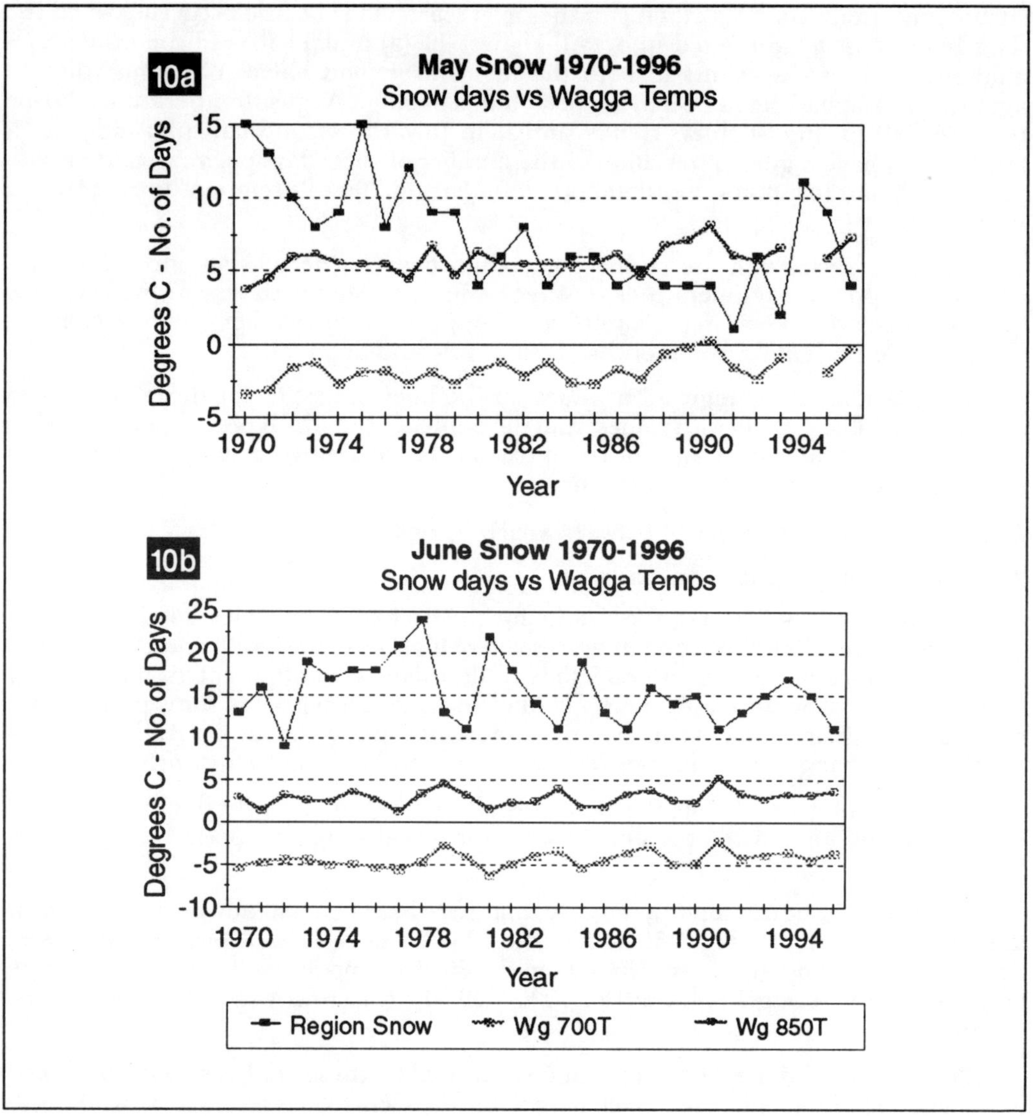

Figure 10. (a,b) Snow days in Snowy Mountain region compared to May-Jun 850 hPa and 700 hPa temperatures at Wagga Wagga 1970–1996 for the months May to October.

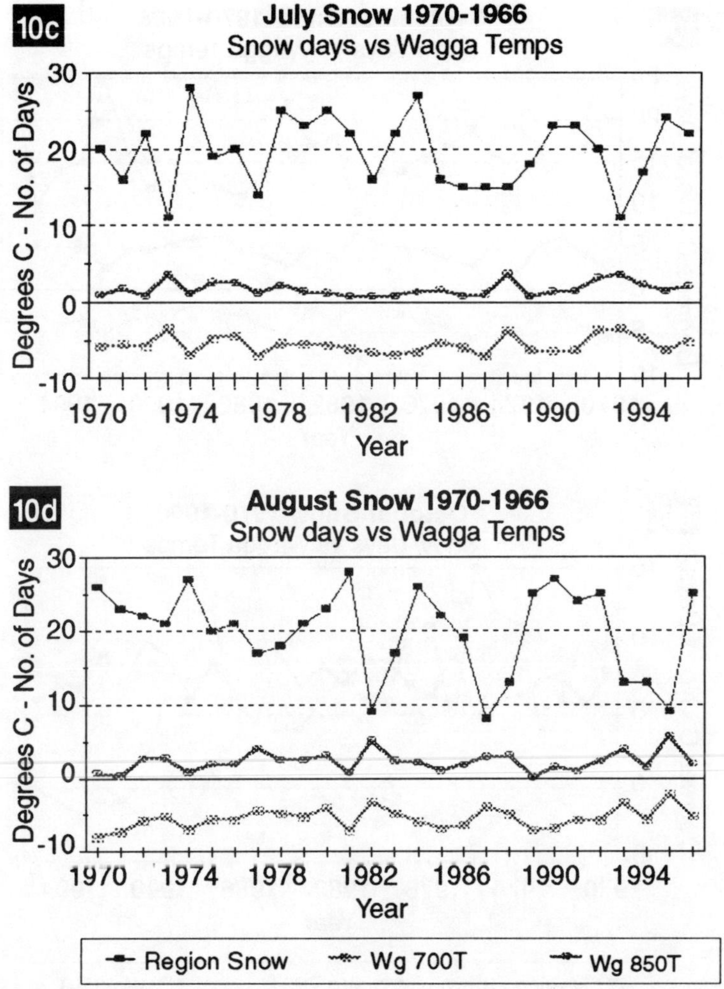

Figure 10. (c,d) Snow days in Snowy Mountain region compared to Jul-Aug 850 hPa and 700 hPa temperatures at Wagga Wagga 1970–1996 for the months May to October.

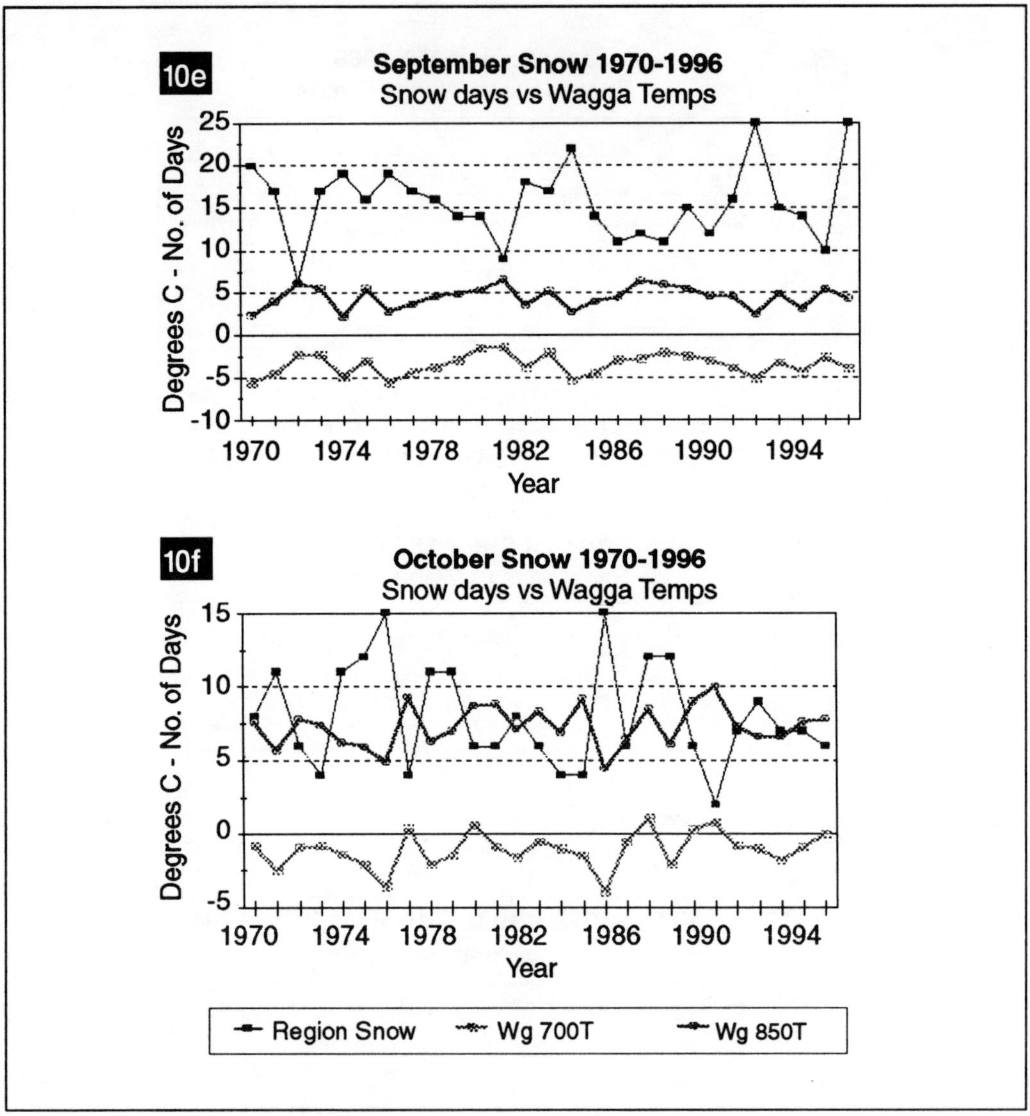

Figure 10. (e, f) Snow days in Snowy Mountain region compared to Sep-Oct 850 hPa and 700 hPa temperatures at Wagga Wagga 1970–1996 for the months May to October.

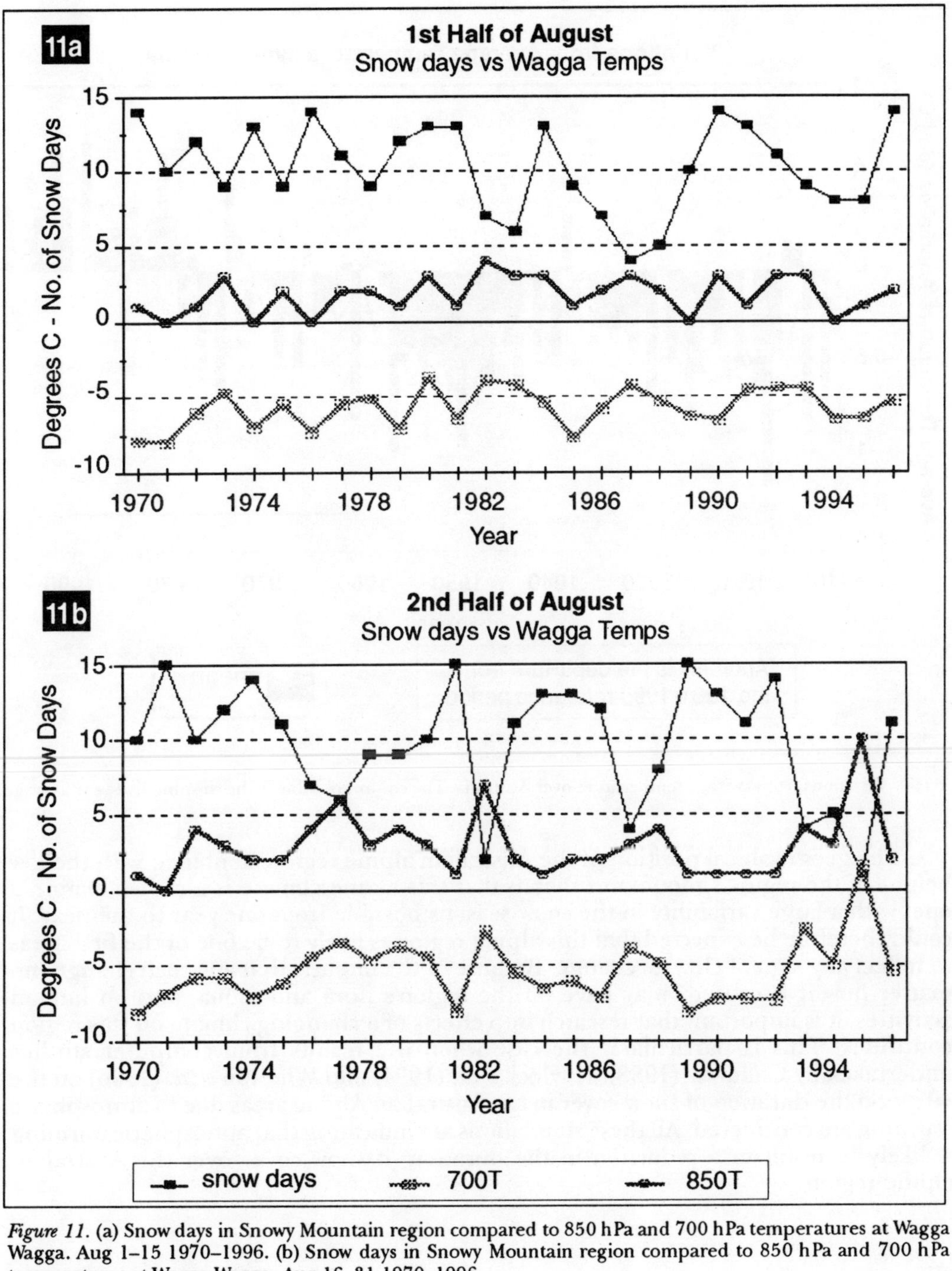

Figure 11. (a) Snow days in Snowy Mountain region compared to 850 hPa and 700 hPa temperatures at Wagga Wagga. Aug 1–15 1970–1996. (b) Snow days in Snowy Mountain region compared to 850 hPa and 700 hPa temperatures at Wagga Wagga. Aug 16–31 1970–1996.

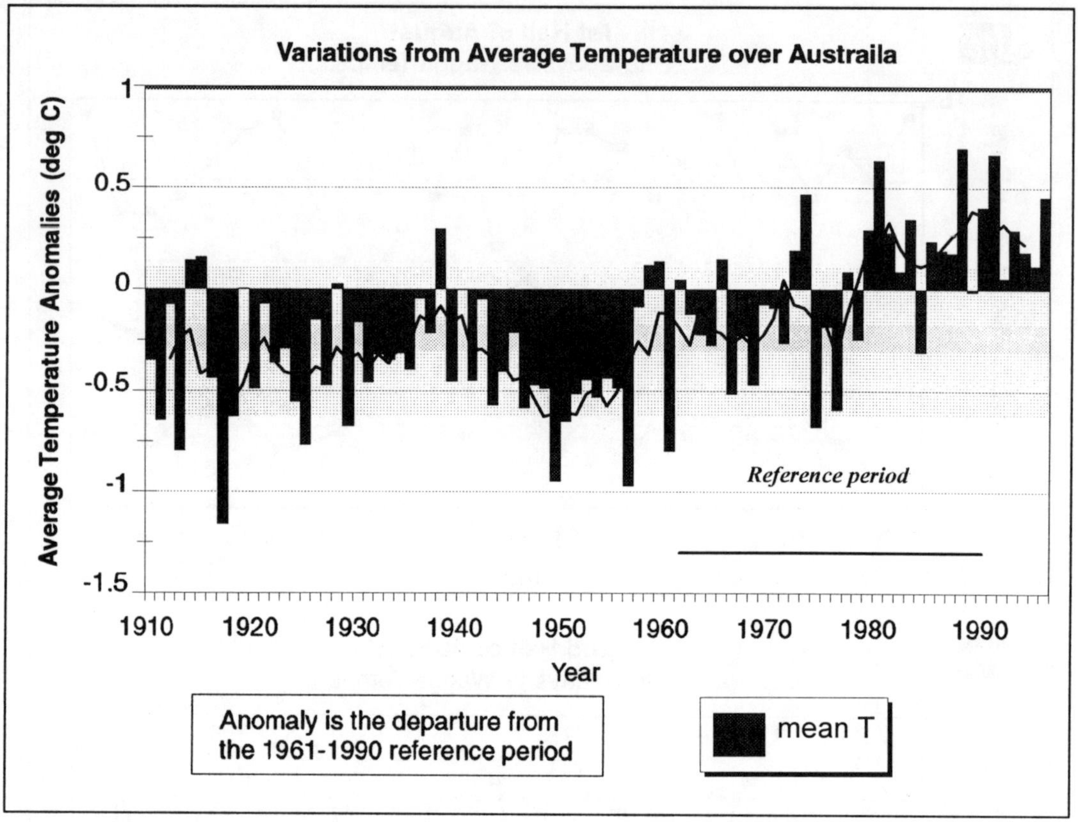

Figure 12. Variations from average temperature over Australia. The continuous line is the running five-year average.

The geographical position of the Australian alpine region combined with the low height of the region's mountains means that this alpine climate is a rather marginal one, with a large variability in the snow seasons possible from one year to the next. It could therefore be expected that this alpine region is likely to be one of the first areas to reflect any sign of climate change. Because of the impacts that any such change (no matter how it is caused) may have on the region's flora and fauna, and on human activities, it is important that research into effects of a changing climate on this region continues. This is particularly the case when the results from computer studies undertaken by Galloway (1988), Haylock *et al.* (1994) and Whetton *et al.* (1996) on the effect on the duration of snow cover in the Australian Alpine areas due to atmospheric warming are considered. All these simulations are indicating that atmospheric warming is likely to result in a reduction in the duration of snow cover over the Australian alpine region.

TABLE 4.

List of regression values for the time series analyses. (*May 1994 data not available)

	Y	Trend (°C/year)	1970 from formula	1996 from formula	Difference (°C)	Error of trend	t value
WG 850 Temp	May/Oct	0.032	3.6	4.5	0.9	0.014	2.26
	June/Sep	0.024	2.5	3.1	0.6	0.016	1.46
	May*	0.070	4.9	6.8	1.9	0.022	3.23
	June	0.034	2.5	3.4	0.9	0.023	1.5
	July	0.021	1.4	2	0.6	0.024	0.91
	August	0.035	1.8	2.7	0.9	0.034	1.05
	1–15 Aug	0.028	1.3	2	0.7	0.030	0.95
	16–31 Aug	0.035	2.4	3.3	0.9	0.054	0.66
	Sept	0.015	4.2	4.6	0.4	0.032	0.47
	October	0.032	6.9	7.7	0.8	0.034	0.94
WG 700 Temp	May/Oct	0.042	–4.2	-3.1	1.1	0.016	2.55
	June/Sep	0.036	–5.2	–4.3	0.9	0.017	2.11
	May*	0.063	–2.6	-0.9	1.7	0.021	3.02
	June	0.052	–5.1	-3.7	1.4	0.022	2.38
	July	0.016	–5.8	–5.4	0.4	0.028	0.57
	August	0.069	–6.3	–4.5	1.8	0.033	1.85
	1–15 Aug	0.050	–6.4	–4.8	1.6	0.050	1.66
	16–31 Aug	0.069	–6.3	–4.5	1.8	0.052	1.32
	Sept	0.018	–3.8	–3.3	0.5	0.032	0.58
	October	0.039	–1.6	-0.6	1.0	0.030	1.32
Region Snow Days	May/Oct	-0.736	95.4	76.3	–19.1	0.284	–2.59
	Jun/Sep	-0.346	75.1	66.1	–9.0	0.270	–1.28
	May	-0.304	11.2	3.3	–7.9	0.072	–4.23
	June	-0.122	16.9	13.7	–3.2	0.091	–1.34
	July	-0.025	20.0	19.3	-0.7	0.116	-0.22
	August	-0.208	22.7	17.3	–5.4	0.143	–1.45
	1–15 Aug	-0.081	11.3	9.2	–2.1	0.072	–1.11
	16–31 Aug	-0.126	11.4	8.1	–3.3	0.098	–1.29
	Sept	0.009	15.5	15.7	0.2	0.113	0.08
	October	-0.086	9.1	6.8	–2.3	0.086	–1.00
Cabramurra	May/Oct	0.196	51.3	56.4	5.1	0.248	0.79

REFERENCES

Auer, A. H., 1974. The rain versus snow threshold temperatures. *Weatherwise* **27:** 67.

Boyden, C. J., 1964. A comparison of snow predictors. *Met. Mag. Lond.* **93:** 353- 365.

Booth, B. J., 1970. Dew point temperature as a snow predictor. *Met. Mag. Lond.* **99:** 363–367.

Budin, G. R., 1985. Interannual variability of Australian snowfall. *Aust. Met. Mag.* **33:** 145–159.

Bureau of Meteorology, Australia. Clouds and Precipitation Training Notes.

Bureau of Meteorology, Australia. Annual Climate Summary 1996.

Colquhoun, J. R., 1978. Snowfall on the New South Wales Snowy Mountains. Tech. Rep. 25 Bureau of Meteorology, Australia.

Duus, A. L., 1992. Estimation and analysis of snow cover in the Snowy Mountains between 1910 and 1991. *Aust. Met. Mag.* **40:** 195–204.

Elliott, P. R., 1989. The Tasmania Snowstorm 24 and 25 July 1986. Met. Note. 187 Bureau of Meteorology, Australia.

Erickson, M. C., Dallavalle, J. P. and Jensenius, J. S. Jnr., 1993. Comments on "Snow versus Rain: Looking Beyond the 'Magic' Numbers". *Weather & Forecasting* **8:** 542–544.

Galloway, R. W., 1988. The Potential impact of climate changes on Australian ski fields. Pp. 428–437 *in* "Greenhouse: Planning for Climate Change" ed G. I. Pearman. CSIRO/Brill, Melbourne.

Haylock, M. R., Whetton, P. H. and Desborough, C., 1994. Climate Change and Snow Cover Duration in the Victorian Alps. EPAV Publication No 403 EPAV, Melbourne.

Henry, C. D., 1971. The Role of 850/700 mb Thickness in Short Range Prediction of Heavy Snow on the Prairies. Technical Memoranda, Environment Canada, Atmospheric Environment Service.

Heppner, P. O. G., 1992. Snow versus rain: looking beyond the "magic" numbers. *Weather & Forecasting* **7:** 683–691.

Mason, B. J., 1971. "The Physics of Clouds." Clarendon Press.

Fletcher, N. H., 1962. "The Physics of Rainclouds." Cambridge University Press, Cambridge.

Jones, M. C. and McCulloch, S. R., 1992. Tasmanian Snowstorm 27 June 1990. Met. Note 200. Bureau of Meteorology, Australia.

Ruddell, A. R., Budd, W. F., Smith, I. N., Keage, P. L. and Jones, R., 1990. "The South East Australian Alpine Climate Study." Report by University of Melbourne for the Alpine Resorts Commission.

Shanahan, B. W., 1967. Three Cold Snaps on the Northern and Central Tablelands of New South Wales during 1966. Met. Note No. 10 Bureau of Meteorology, Australia.

Slatyer, R. O., Cochrane, P. M. and Galloway, R. W., 1985. Duration and extent of snow cover in the Snowy Mountains. *Search* **15:** 11–12.

Taylor, C. M. and Stern, H., 1982. A Cold Outbreak Over Victoria. Met. Note No. 130 Bureau of Meteorology, Australia.

Whetton, P. H., Haylock, M. R. and Galloway, R., 1996. Climate Change and Snow Cover Duration in the Australian Alps. *Climate Change* **32:** 447–479.

CHAPTER 2

The physical properties of snow

ANDREW RUDDELL

INTRODUCTION

SNOW is a porous aggregation of mainly solid atmospheric precipitation consisting of ice crystals, air, impurities and sometimes water. If the deposition on the ground surface persists in spite of processes of evaporation and melt (ablation) then a *snow cover* results. A temporary snow cover exists for several days, while weeks or months of consecutive snowfalls will result in a seasonal snow cover which then gradually disappears during the snowmelt season as part of an annual cycle. From the time of its formation at sub-zero conditions in the atmosphere, through the metamorphism and finally ablation of the snow-cover, the texture and other properties of snow are continuously changing. This process of 'ageing' is a result of variations in temperature distribution in the snow cover which are a product of energy exchange with the atmosphere. In turn, many of the properties of snow are extremely dependent on its temperature, density, and therefore, the metamorphosis of its crystalline structure.

The majority of studies that have contributed to our understanding of this subject have been undertaken in the colder and drier snow regimes of the northern hemisphere (e.g. Kuz'min 1961, Mellor 1962, Bader 1962, Male 1980). In the Australian alpine region, which is generally above 1400m (Ruddell *et al.* 1990), the high frequency of above-freezing conditions has led to a snow cover which is often infiltrated by melt and rain water. The presence of liquid water has a significant effect on the processes of snow metamorphism and melt. Very few investigations have been undertaken in this region which examine the physical properties of 'Australian snow', and it is hoped that the rather brief overview presented here may stimulate the development of a local awareness.

THE FORMATION OF SNOW CRYSTALS IN THE ATMOSPHERE

Snow originates in clouds at below freezing temperatures (see Chapter 1). The physical properties of the snow crystal that arrives at the ground are the product of its life history in the atmosphere, which may be just simple ice crystal formation or it may include numerous complex modifications to shape, mass, density, crystal structure, etc.

As moist air cools, due to uplift (either orographic or convective) or radiation cooling, water vapour condenses on minute nuclei (aerosols) to form cloud droplets of approximately 10 μm radius. Condensation nuclei are about two orders of magnitude smaller than cloud droplets, are quite abundant with typical concentrations of about 1000 cm^{-3}, and originate mainly from blowing dust, oceanic salts, air pollution, volcanic dust and organic particles. The resulting cloud droplets, do not necessarily freeze at 0°C but may be supercooled to as low as –40°C. Ice nucleation occurs by the freezing of droplets on contact with aerosols; the direct deposition of vapour on aerosols; or the freezing of droplets on an already existing ice nucleus. A rapid increase in the number of crystals may occur through small ice splinters which have become detached by air currents and collisions with other crystals.

The processes of nucleation are particularly important for the incorporation of impurities into the snow cover. Some characteristics of predominant crystals found in various types of clouds are given in Table 1. The major structural types of snow are depicted in Figure 1 (LaChapelle 1969).

Following nucleation, snow crystal growth occurs by the process of sublimation, aggregation of crystals or by riming (Fig. 2). Sublimation is taken here as being the change in phase of vapour to a solid, without passing through the liquid phase. Riming is the freezing of cloud droplets onto a solid object.

Sublimation: In sublimation, vapour is directly deposited on ice nuclei. Once a droplet has frozen it grows rapidly at the expense of the remaining water droplets because the vapour pressure at the ice surface is less than that above the water surface.

TABLE 1

Crystal types found in various types of clouds. (after Weickmann in Mellor, 1964)

Layer	Cloud type	Temperature range	Predominant crystal types	Approx. crystal dimensions
Lower troposphere	Nimbostratus, stratocumulus, stratus	0 to –15°C	Thin hexagonal plates	50μ to 0.5 mm diameter, 10 to 20 μ thick
			Stellar dendritic crystals	0.5 to 5 mm diameter
Middle troposphere	Altostratus altocumulus	–15 to –30°	Thick hexagonal plates	200μ diameter
			Prismatic columns, single prisms and twins	200μ long
Upper troposphere	Isolated cirrus	<–30°C	Clusters of primatic columns (with cavities)	1 mm diameter
	Cirrostratus		Single complete prisms, thick plate crystals	100μ long

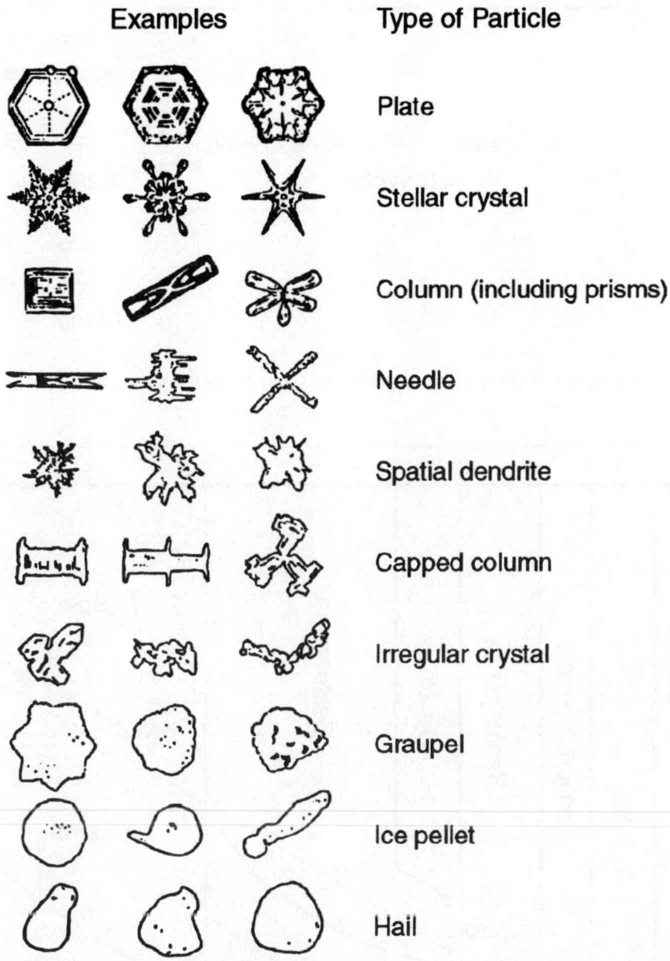

Fig. 1. A summary of the major structural categories of snow crystals. Modified from LaChapelle (1969).

The basic shape of an ice crystal depends on the temperature at formation, but subsequent growth patterns are determined by the degree of supersaturation (Fig. 3). Predominant growth occurs on either the crystal *c*-axis or *a*-axis, resulting in crystals which have the form of plates, needles, hollow prisms, sector plates, or dendrites.

Aggregation: Maximum aggregation occurs at about 0°C when adhesive force is high. This results in snowflakes, which may also be produced by sintering and interlocking of crystals. Sintering is the transfer of material by sublimation or diffusion to the points of contact between particles, to form bonds.

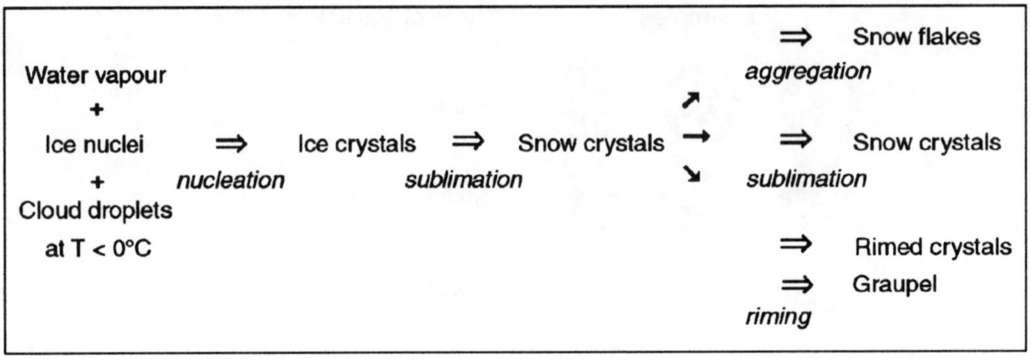

Fig. 2. Schematic diagram of the development of different types of snow.

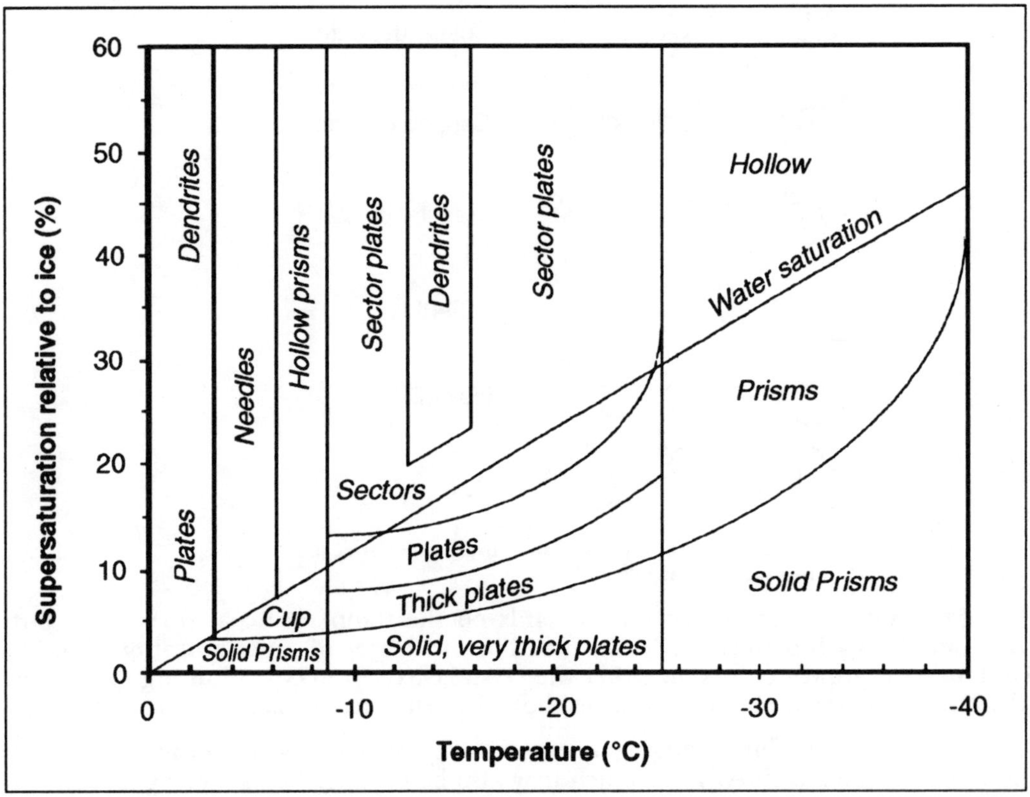

Fig. 3. The variation of crystal form with temperature and supersaturation relative to flat ice surface at time of formation ("Mason's scheme"). From Mellor (1964).

Riming: Riming is growth by accretion, collision or adhesion of cloud particles to the surface of the ice nucleus. This results in rimed crystals or graupel.

The amount of snowfall over a given region will not only be determined by the frequency of vigorous weather systems and their interaction with topography, but also the effect that temperature will have on the possible amount of water vapour that can be held in the air (or absolute humidity). In Figure 4 it is shown that air at –20°C, and 100% saturation, can only hold 18% of the water vapour possible at freezing point. Some sudden cold outbreaks, while bringing a light covering of snow to low elevations, will rarely supply a significant amount to the alpine regions. This may be attributed to the low absolute humidity and air temperature.

As the snow crystal descends to the ground surface it is possible that it will undergo considerable transformation as it passes through different temperature and humidity regimes. This may include evaporation and melting of the crystal. If the snowfall is accompanied by high winds, crystals are broken into fragments favourable for close packing in the surface deposit. The deposited snow may be further modified by wind and by the impact of bounding particles. This produces a high initial density, often greater than 0.3 Mg m^{-3}, and often leads to the formation of sastrugi and snow dunes. Under calm, dry conditions stellar crystals, which may be aggregated into snowflakes, are likely to settle as a loose mass with density of about 0.1 Mg m^{-3} or less.

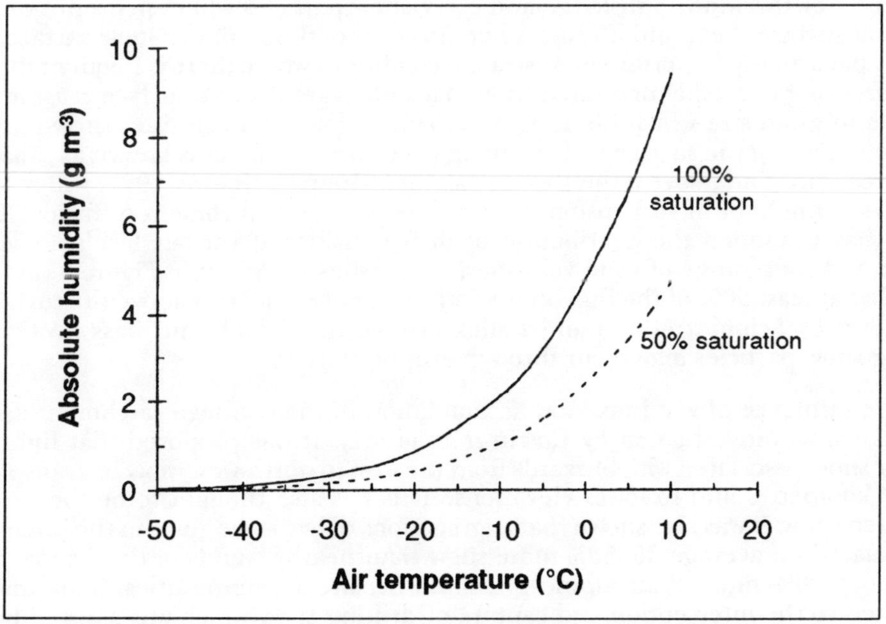

Fig. 4. The variation of absolute humidity with air temperature for 100% and 50% saturation.

SNOW COVER

The distribution of snow cover depth is largely influenced by elevation, slope, exposure, and vegetative cover. As the snow cover develops, its depth will generally increase with elevation. This is due to the increase in snow as a proportion of the total precipitation with lower temperatures, as well as the decrease in melt rate. Deviations from this pattern will occur mainly due to topographic effects. These include the removal of snow by wind on high and exposed summits and its re-distribution to lower elevations on the lee side. Avalanching may also re-distribute snow to lower and warmer elevations, however, this phenomenon is not common in the Australian Alps due to the relatively small areas of steep slopes and a relatively small precipitation of snow. In this region, the greater intensity of radiation received by slopes with a northerly aspect often results in the depletion of snow being significantly greater than for south facing and even level surfaces. The exposure to warm winds may also have a significant effect on snowmelt.

Much of Australia's alpine region consists of gently undulating topography (e.g. Bogong High Plains, Snowy Mountains, etc.) which is favourable for the transport of fresh snow by strong winds. This results in the deflation of snow from rises and the development of large drifts in sheltered depressions. Occasionally, large cornices may develop on the leeward side of windswept ridges such as Mt. Feathertop in the Victorian Alps.

The drifting of snow has been reviewed by Radok (1977) and Male (1980). The propensity of the wind to dislodge and carry snow particles will depend not only on sufficient surface shear and lift forces but also the condition of the snow surface. The latter is particularly important in Australian conditions where there is frequent melting and refreezing of freshly deposited snow. This often results in a surface crust and an increase in grain size which hinders snow drift. In low wind velocities snow particles may bound along the surface with a curved trajectory in a process known as saltation. An increase in wind above a threshold of about 8–16 m s^{-1} (Radok 1977) will result in particles being held in suspension through the process of turbulent diffusion. Budd (1966) has examined the distribution of drift densities at various levels above the surface and for a range of wind velocities. The results are given in Figure 5 and they show that at least 90% of the flux occurs within a few centimetres above the surface. It was shown by Schmidt (1972) and Radok (1977), that significant mass decrease of drifting snow particles may occur through evaporation.

The influence of wind on snow accumulation in the Guthega catchment (Snowy Mountains) was investigated by Costin *et al.* (1961). It was proposed that light, dry powder snow, associated with blizzards from the west to north-west, may be transported several kilometres and to lower elevations. This was also thought to be the case for deposited snow. It was also shown that aggregations of live snow gums in the catchment accumulated on average 30–50% more snow than nearby stands of dead snow gums and about 20% more than alpine grassland/herbfield communities. This may be attributed to the interception and capture of drifting snow which was assisted by the reduction of wind velocity and the production of locally turbulent conditions above

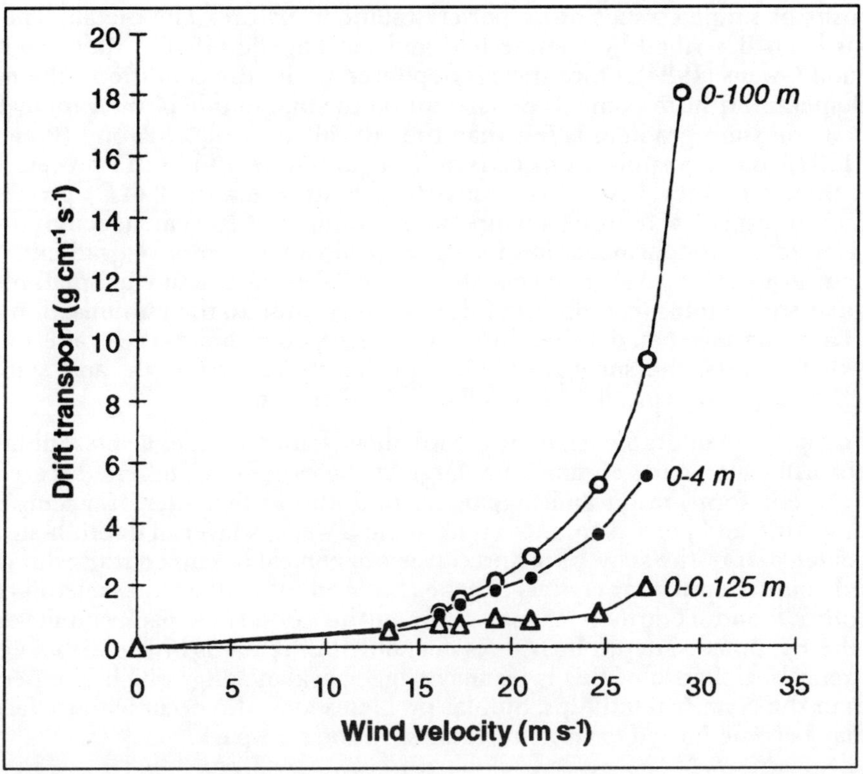

Fig. 5. The integrated drift transport as function of wind velocity for a range of heights above the snow surface. From Budd (1966).

breaks in the tree canopy. Furthermore, the tree cover retarded melt due to protection from solar radiation and warm winds. By retaining the snowdrift at higher and cooler elevations, it was proposed that snowmelt was sufficiently lagged to be of significant benefit to hydro-electric generation.

The drifting of snow is a major problem for roads and buildings in the alpine environment and much applied research has been undertaken which deals with these problems. The use of fences or forests to create wind eddies and capture drifting snow has many applications including the supply of snow for recreational skiing.

SNOW METAMORPHISM

Following the deposition of snow on the ground and the formation of a snow pack, changes in snow particles shape and size continue according to their environment. The snow particles are now considered as a matrix of ice particles called grains, which may consist of single crystals or be polycrystalline in nature. The metamorphism of snow has been described by Sommerfeld and LaChapelle (1970), Male (1980) and Prowse and Owens (1984). Once snow is deposited under dry conditions, the manner in which metamorphism occurs is dependent on the magnitude of the vapour flux. If the vapour pressure gradient is less than the critical value of 5 hPa m^{-1} (Prowse and Owens 1984), metamorphism proceeds as an equi-temperature (ET) process; if it is greater, then metamorphism occurs as a temperature gradient (TG) process. For temperatures just below freezing a temperature gradient of 10°C m^{-1} through the snow profile is used as an approximation for the critical vapour pressure gradient. With a low vapour gradient, star-shaped crystals or snowflakes re-structure as small rounded or irregular snow grains (equilibrium form), as a response to the minimisation of free energy. Larger grains then develop through sintering which bonds the grains together. During this process, the snow pack will gain density and strength, and stability is increased. Grains are typically 0.1 to 0.5 mm in diameter.

With high vapour gradients, moisture will move towards colder layers. Sublimation of vapour will cause some crystals to enlarge at the expense of others. Faceted snow crystals (kinetic form) result and may be up to 5 mm in diameter. If a temperature gradient > 20°C m^{-1} persists for more than about a week, a layer of depth hoar results in the colder part of the snow pack and consists of conical or cup crystals with distinct faces and voids. Depth hoar crystals may be classified into either the skeleton or solid type. Both ET and TG processes may occur in the same snow pack simultaneously. Unlike the ET process, depth hoar does not settle but remains brittle with a very low shear strength. Collapse of this layer may trigger avalanching, which is especially a problem in the New Zealand Alps. Similar problems may also occur with surface hoar which may become buried to form a weakness in the snowpack.

If the snow pack is at 0°C, melt-freeze (MF) metamorphism occurs. Grain growth proceeds at a faster rate than for ET metamorphism (Perla 1978). Because equilibrium temperature increases with grain size, a heat flux from larger to smaller particles occurs due to the resultant temperature gradient. The mean grain size increases with time as larger grains grow at the expense of smaller grains. With an increase in water saturation, the heat flux and grain growth rate are increased. An increase in grain size will result in an increase in snow permeability (Shimizu 1970). At first MF metamorphism will tend to strengthen the pack, but later when meltwater is sufficient to flow between layers, slab avalanching can occur.

MECHANICAL CHARACTERISTICS

Following the deposition of new snow there is an initial rapid densification which then decreases to a slower rate. The compactive viscosity of snow and its relationship with stress, density and temperature, has been investigated by Mellor (1975). The mechanical strength of dry snow and its resistance to deformation is strongly related to the development of intergranular bonding by sintering. The rate at which this hardening occurs increases with temperature. It is also dependent on crystal size and shape.

Cohesive snow may be regarded as a compressible visco-elastic material (Mellor 1964). Snow will respond elastically to moderate loads which have been applied for a very short period of time (a few seconds). However, if the load is sustained, the initial elastic deformation is followed by a continuous strain. The viscosity of snow is non-linear with respect to stress, and at a limiting level, an acceleration in the increase of strain rate sets in (tertiary creep). It is thought that this point initiates failure and is important in the release of avalanches. Stress analysis and failure mechanics of snow slabs has been examined by Perla (1980).

When snow crystals are transported or re-distributed by wind, or come in contact with air at an ambient temperature above freezing, their shape and other physical properties are affected, and they are deposited with higher densities than the parent materials. Studies in the Caucasus by Kuvaeva *et al.* (1975) have demonstrated the density variation due to wind. During calm conditions the snow surface had a density of 0.12 Mg m^{-3}, but after several hours of strong winds with drifting snow (15–17 m sec^{-1}, gusting to 25 m sec^{-1}), a surface crust had developed with a density of 0.30 Mg m^{-3}. The temperature was about –9°C. Sufficient frictional heat is generated by the impact of the mobile snow particles to produce melt, which subsequently refreezes. Furthermore, temperature and pressure gradients close to the surface enable the movement of water vapour in porous snow to cause sintering and further densification.

Following deposition and the formation of a snow cover, the density will generally increase with depth due to compression from above. However, considerable variation is caused by windcrusts, the percolation and refreezing of melt and rain water and other metamorphic processes. The densities of important types of snow deposits are given in Table 2. The rapid metamorphism of fresh snow in the Australian alpine region, due to the high frequency of melt-refreeze events, results in snow densities which are very rarely less than 0.1 Mg m^{-3}. During the melt season, which in Australia often occurs during spring, the snow cover attains a homogenous density distribution with values in the range of 0.5–0.6 Mg m^{-3}.

The variation of density with time for the State Electricity Commission of Victoria (SECV) snow pole transect on the Bogong High Plains is given in Figure 6. Immediately following a snow fall the density of snow may be as low as 0.16 Mg m^{-3}, however, it often increases up to 0.3 Mg m^{-3} within a few days. In late Feb. 1964, snow density was measured at 10 cm intervals in a 2 m snow pit dug into a large snow drift on Mt. Twynam (J. Bennett, pers. comm.). The average density was 0.59 Mg m^{-3}, snow particles were well rounded with grain sizes varying from 0.2 to 1 mm.

TABLE 2
Densities of various snow types

Description	Density (g cm^{-3})
Wild snow†	0.01–0.03
New snow (calm conditions)†	0.05–0.07
Damp new snow	0.1–0.2
Settled snow	0.2–0.3
Wind packed snow	0.3–0.4
Firn	0.4–0.83
Ice	0.83–0.92
Other:	
Wet snow and firn	0.5–0.83
Depth hoar and cup crystals	0.1–0.35

†Little is known of these types of snow in Australian conditions

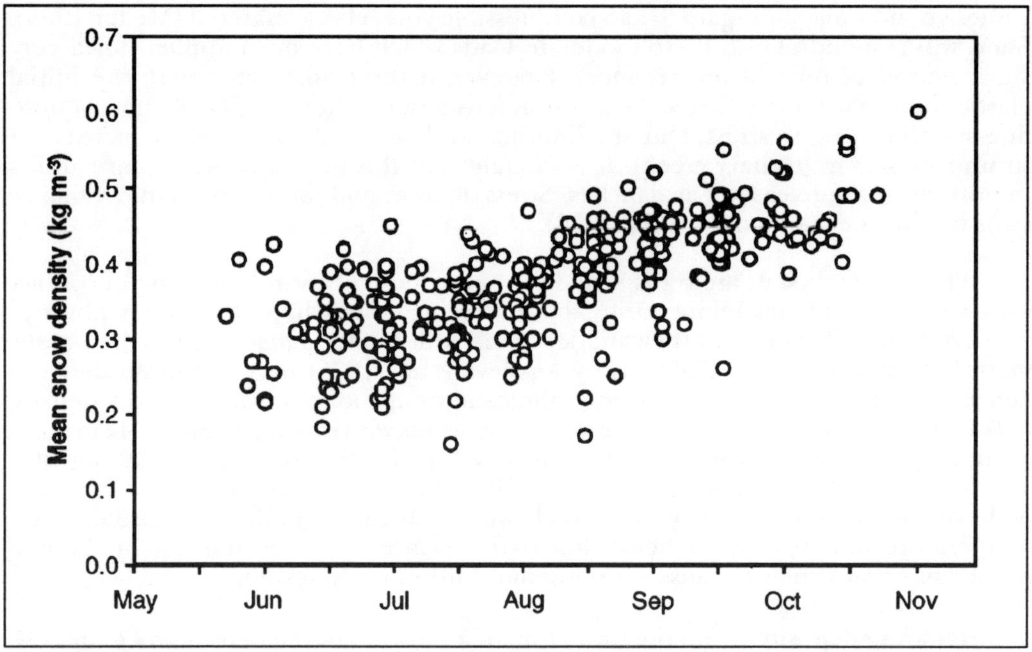

Fig. 6. Average density of snow cover as measured at the SECV Rocky Valley snow course, Bogong High Plains 1935–89. The low densities recorded late in the season are probably due to cold southerly outbreaks. From Ruddell *et al.* (1990).

Due to the variation in the density of snow, it is necessary for comparative purposes to refer to its water equivalent (w.e.). This is obtained by multiplying the depth of the snow sample or snow cover by its mean density.

The void ratio or porosity (n) of snow is defined as the percent volume of voids to total volume of dry snow, ie.

$$n = \left(1 - \frac{\rho}{\rho_i}\right) \times 100 \qquad (1)$$

where ρ is the density of the snow sample and ρ_i is the density of pure bubble-free ice (0.917 Mg m^{-3}). The porosity of dry new snow can be up to 95% while for bubble-free ice it is zero. If water is present it will tend to fill the voids.

THERMAL PROPERTIES

Snow

The specific heat of snow is the same as that of ice and is fairly constant at a value of 2009 J kg^{-1} K^{-1} (0.5 cal g^{-1} K^{-1}), which is half that of water and about twice that of air. The effective thermal conductivity of snow (k_s) is strongly related to its density and micro-structure. Sturm *et al.* (1997) have compiled a number of determinations of k_s

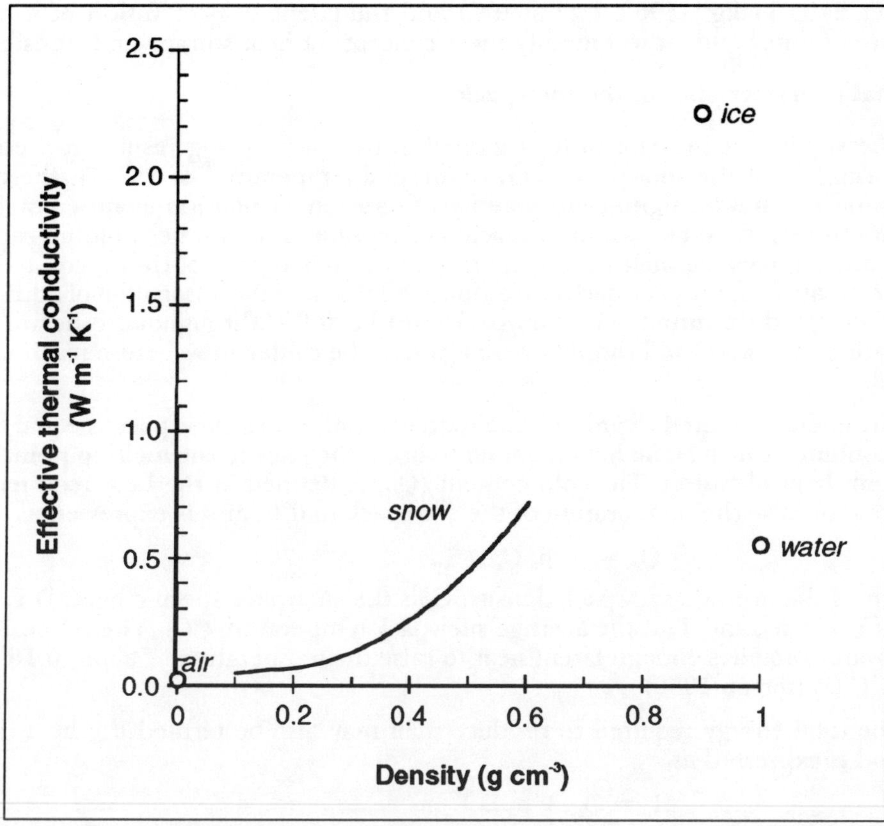

Fig. 7. Effective thermal conductivity of dry snow (solid line) from Sturm et al. (1997). The values for air, ice and water are also given.

and the average value of these are given in Figure 7. Sulakvelidze (1954, cited in Mellor 1964) found that the conductivity of wet snow is about two or three times greater than for dry snow. The thermal diffusivity (k_d) of snow is defined as

$$k_d = k_s / \rho_s C_s \tag{2}$$

where ρ_s is snow density and C_s is the specific heat.

The latent heat of snow is taken as that of the constituent ice particles. For many situations where a phase change occurs at 0°C the latent heat of fusion is taken as 335 kJ kg^{-1} (80 cal g^{-1}). This means that for pure ice a heat gain of about 3347 kJ m^{-2} (80 langleys) melts an equivalent of 1 cm of water at 0°C. The latent heat of fusion for ice at –10°C is 312 kJ kg^{-1} (74.5 cal g^{-1}) (Mellor 1964). The latent heat of sublimation of ice at 0°C is 2830 kJ kg^{-1} (676 cal g^{-1}) and this includes the latent heat of vaporization of water 2495 kJ kg^{-1} (596 cal g^{-1} at 0°C) and the latent heat of fusion of ice. The presence of water and air will modify these concepts of heat storage and transfer.

Thermal characteristics of the snowpack

The significant increase in heat gained in the early spring results in the rapid transformation of the snowpack to an isothermal temperature of 0°C. Furthermore, since snow can hold a significant quantity of water in suspension against gravity by capillary forces, the snowpack must reach its maximum liquid-water holding capacity before any appreciable melt runoff can take place. When both of these requirements have been satisfied, the snowpack is in a ripe condition and the absorption of additional energy will produce runoff. The snowpack must be at 0°C throughout, otherwise any percolating melt water will simply freeze again in the colder snow before reaching the ground.

The energy required to produce meltwater runoff from a snow pack is equal to the "cold content", which is the heat required to bring the pack to the melting point, plus the latent heat of fusion. The cold content (Q_{cc}) is defined as the heat required per unit area, to raise the temperature of the snowpack to 0°C, and is expressed as

$$Q_{cc} = -\rho_p C_s D T_p \tag{3}$$

where ρ_p is the average snowpack density, C_s is the snowpack specific heat, D is snow depth (+ve, cm), and T_p is the average snowpack temperature (°C). The refreezing of 1 g of water provides enough latent heat to raise the temperature of about 0.16 kg of snow 1°C (Patterson 1980).

The total energy required to produce melt may also be termed the heat deficit (H_d) and is expressed as

$$H_d = Q_{cc} + \rho_p D L_{ms} \tag{4}$$

where L_{ms} is the latent heat of melting of the snow. When the snowpack has some free water (~ 3–5%), L_{ms} will be slightly less than the latent heat of melting for ice (L_m). The ratio of these two quantities is referred to as the thermal quality of the snow (θ). This is expressed as

$$\theta = \frac{H_d}{\rho_p D L_m} = \frac{(L_{ms} - C_s T_p)}{L_m} \tag{5}$$

for dry snowpacks ($T_p < 0$) liquid water will not be present. Thus $L_{ms} = L_m$, and $\theta > 1$. For ripe snowpacks with some water content, $L_{ms} < L_m$ and by definition $T_p = 0$; and therefore $\theta < 1$. In the latter case, the latent heat of melting of snow must be proportional to its percentage of free water content (W) and therefore

$$\frac{L_{ms}}{L_m} = 1 - \left(\frac{W}{100}\right) \tag{6}$$

or

$$W = 100\,(1 - \theta) \tag{7}$$

When the snowpack is ripe it is holding water to its capacity and is isothermal with $T_p = 0$. In this situation, liquid water may exist in a hygroscopic form (which consists of a thin film of water on the snow crystal surface) and as capillary water (held by the menisci of the pores due to surface tension). With further melt of the ice matrix, the snowpack holding capacity is exceeded and water, under the action of gravity, will percolate downwards to the soil. If the rate of soil infiltration is exceeded, a slush layer is formed between the soil and snow and runoff begins. Once the melting temperature has been established throughout the snowpack it will tend to persist even with rather low air temperatures and the latter will merely lead to the freezing of a thin top crust.

In the Australian Alps, the daily maximum air temperature of even the highest peaks, may often reach above 0°C during the winter. As a result, the snowpack is often at or slightly less than 0°C. Temperatures measured at 50 cm intervals in a 2m snow pit dug on the summit of Mt. Buller (1809 m ASL) on 20 August 1964, were between –1 and –1.5 °C. About 10 days later (1 September) another pit of similar depth was dug and temperatures were at melting point throughout (G. Weller *et al.* pers. comm.). The free water content of snow at Spencers Creek has been measured by Radok *et al.* (1961) using electrical and freezing calorimeters. Local variability of the thermal quality of snow was large and the observed values of free water ranged from about 3–28%. The variation may be attributed to such factors as grain size, depth in the snowpack, and sampling time during the day. The measurement of free water content by calorimetric methods is difficult in the field, however, a quick and more convenient technique which uses time domain reflectometry has recently been outlined by Stein *et al.* (1997).

SNOW MELT

Energy balance

The rate at which a snow cover will melt is dependent on the energy balance at its surfaces.

$$Q_m = Q^* + Q_H + Q_e + Q_p + Q_G + Q_{cc} \qquad (8)$$

where: Q_m is the heat flux utilised in the snowmelt process, Q^* is the net radiation flux, Q_H is the sensible heat flux, Q_e is the latent heat flux, Q_p is the precipitation heat flux, Q_G is the conductive heat flux between the soil surface and subsurface, and Q_{cc} is the cold content. All components of Eq. 8 are expressed in units of W m^{-2}.

Radiative properties and radiation balance

The amount of solar radiation supplied to the snowpack depends not only on the radiation from the sun but also depends on the nature of the snowpack. The latter will influence the reflection and the depth of penetration of solar rays. The net radiation flux at the snow surface is expressed as

$$Q^* = S^* + L^* = S\downarrow (1 - A) + (L\downarrow - L\uparrow) \quad (9)$$

where S^* is the net shortwave radiation, L^* is the net longwave radiation, $S\downarrow$ is the incoming solar radiation (direct and diffuse), A is the snow surface albedo to shortwave radiation, $L\downarrow$ is the incoming atmospheric longwave radiation, $L\uparrow$ is the outgoing (terrestrial) longwave radiation.

$S\downarrow$ is present only during the daytime and nearly all of its energy is in the wavelength range of 0.3 to 8 μm. The amount of incoming radiation that is absorbed by the snow surface will depend on its albedo (A), which is defined as the ratio of the reflected radiation ($S\uparrow$) to the incident flux ($S\downarrow$). ie.

$$A = S\uparrow / S\downarrow \quad (10)$$

For dry snow, about 10% of $S\downarrow$ penetrates to a depth of 0.25 m in the snowpack and 1% reaches 1 m (Weller and Schwerdtfeger 1970).

Reflectivity and albedo

The albedo is a major factor in the amount of solar radiation absorbed by the snow and is therefore an important factor in determining the rate of snowmelt. It is also a property of snow that is particularly useful for remote sensing applications. Due to its large variation, fresh snow is easily distinguished from old snow and snow free surfaces. Furthermore, the age of snow and its increase in density can be empirically determined (Hall and Martinec 1985). This allows the water content to be estimated and is especially useful for the calculation of runoff for irrigation and hydro-electric generation.

The albedo (A) may be considered as the integrated reflectance over the short-wavelength spectrum. Depending on the type of measuring instrument, the limits of integration are often in the range of 0.3–3 μm. Albedo is inversely proportional to the square root of the grain size and R will decrease slightly with increasing wavelength (Bohren and Barkstrom 1974). The albedo of snow varies widely from about 100% for fresh new snow to about 20–30% for dirty wet snow. An albedo of 50% may be assumed for melting snow (U.S. Army Corps of Engineers 1956). Albedo depends on:

i) Properties of the reflecting surface (ie. texture, moisture content, particulate concentration),

ii) Structure and thickness of the snow cover. As the snow cover thins, the underlying surface has an increasing influence.

iii) Angle of incidence of solar radiation (height of sun, exposition and angle of inclination of the snow surface),

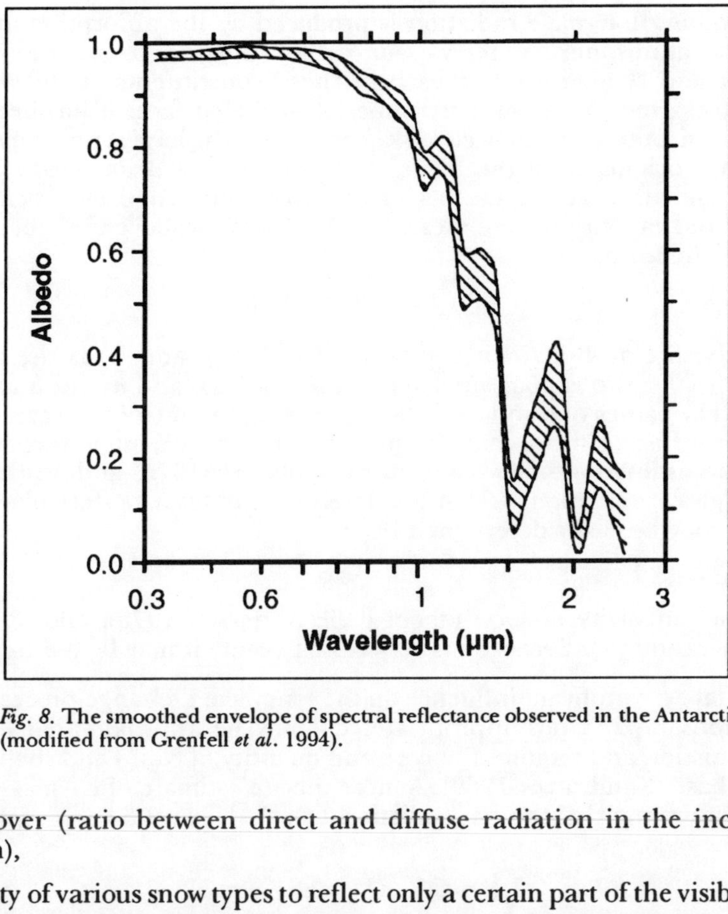

Fig. 8. The smoothed envelope of spectral reflectance observed in the Antarctic (modified from Grenfell *et al.* 1994).

iv) Cloud cover (ratio between direct and diffuse radiation in the incoming total radiation),

v) The ability of various snow types to reflect only a certain part of the visible spectrum.

The spectral reflectivity of snow (R) is dependent on factors such as: grain size, shape and orientation; impurities; near-surface liquid content; depth and surface roughness as well as solar elevation (Fig. 8). The reflectance of freshly fallen dry snow is very high in the visible wavelengths, but this will decrease with age. The decrease is greater in the longer wavelengths (near-infrared).

Long-Wave Radiation

The net longwave radiation at the snow surface is the difference between downward atmospheric radiation (L↓) and the upward flux from the snow surface (L↑). The net radiation is usually positive, and some heat is lost from the snow surface. Although the intensity of this component is small when compared to peaks of incoming shortwave radiation during the day, the total over a 24 hour period is significant.

The incoming longwave radiation is produced by the absorption and emission by clouds and atmospheric water vapour, carbon dioxide, ozone and other gases. The amount and temperature of each of these constituents at different heights throughout the atmosphere will determine L↓ at the surface. It has been shown by Geiger (1965) that under average clear sky conditions, the largest portion reaching the Earth's surface originates in the lowest 100m. Following Brunt (1952), this allows atmospheric radiation at the Earth's surface to be modelled empirically from air temperature and vapour pressure measured at 1.5 or 2 m above the ground, using an expression of the form:

$$L\downarrow = \sigma T_a^4[a+b\sqrt{e}] \qquad (12)$$

where σ is the Stefan-Boltzman constant (W m^{-2} K^{-4}), and T_a is the absolute air temperature (K), e the vapour pressure of the air (Pa), and *a* and *b* are empirical parameters. The values of *a* and *b* will vary according to time of the year and position on the Earth's surface. Further consideration of atmospheric temperature and humidity profiles in the estimation of L↓ is given by Brutsaert (1975) and Marks (1979). In terms of longwave radiation (L↑) snow closely resembles a perfect black body and therefore L↑ may be easily determined from

$$L\uparrow = \varepsilon \sigma T_s^4 \qquad (13)$$

where ε is the emissivity of snow (about 0.99; Anderson 1976), and T_s is the snow surface temperature (K). Because ε is very close to unity it may be omitted.

Clouds have a significant influence on the longwave exchange process and under such conditions Eq.2 is not appropriate. L↓ may be approximated by assuming blackbody emission and requires the type and quantity of cloud and the temperature of the cloud base (Kondratyev 1969). A more precise estimate of L↓ may be obtained by direct measurement of the net radiation Q*, A, L↑ and S↓ (e.g. Ishikawa *et al.* 1992).

$$L\downarrow = Q^* - (1 - a)\, S\downarrow + \sigma T_s^4 \qquad (14)$$

Topographical variations such as valley sides can obscure a significant portion of the sky and therefore reduce the incoming atmospheric radiation. In an alpine location, this may be compensated by the terrestrial radiation from the valley walls. To estimate the radiation flux from the unobscured portion of the sky, a view factor may be used (Lee 1962, Marks 1979).

Sensible and latent heat exchange due to turbulent energy transfer

In addition to net radiation, the turbulent transfer of sensible (Q_h) and latent heat (Q_e) are major influences on snow melt. The turbulent energy transfer of Q_h and Q_e between the snow surface and the atmosphere are generated by air temperature and moisture gradients. In mountainous regions, wind turbulence is enhanced by various topographic and radiative effects (Barry 1992) and complicates the quantification of these energy fluxes. The fluxes of sensible and latent heat have been reviewed by Kuz'min (1961) and Male (1980) and are briefly examined here. The fluxes can be approximated by

$$Q_h = \rho_a C_p k_h (T_a - T_s) \tag{15}$$

$$Q_e = \rho_a L_v k_e (0.622 / P_a)(e_a - e_s) \tag{16}$$

where ρ_a is air density (Kg m^{-3}), C_p is the specific heat of air at constant pressure, L_v is the latent heat of vaporisation of water, k_h is the exchange coefficient for sensible heat, k_e is the exchange coefficient for latent heat, T_a is the specified near surface air temperature and e_a is the near-surface vapour pressure. During snow melt, the snow surface temperature (T_s) can be assumed to be at 0°C, and the snow surface vapour pressure (e_s) can often be obtained by assuming that the air in contact with the snow is saturated. If $e_a > e_s$ a vapour transfer occurs from air to snow resulting in condensation or frost formation, if $e_a < e_s$ then evaporation occurs. For a water-vapour system L_v has a value of 2495 kJ kg^{-1} and for a snow-vapour system the value is 2830 kJ kg^{-1}. The ratio Q_h/Q_e is known as the Bowen's ratio.

Under neutral atmospheric conditions the exchange coefficients for sensible and latent heat may be assumed to be equivalent and are expressed as:

$$k_h = k_e = k^2 u / [\ln (z / z_o)]^2$$

where k is the von Karman constant (0.4), u is the wind velocity at height z (m) and z_0 is roughness height (m). If the atmospheric conditions are not neutral then the exchange coefficients should be adjusted using stability functions such as that described by Male (1980) and Thom (1975). On the Ivory Glacier in the New Zealand Alps, Hay and Fitzharris (1988) were able to make a good estimate of the exchange coefficient using observed melt and the energy balance (Eq. 8).

With a dwindling snow cover, the local advection from adjoining patches of bare ground should be considered. In such cases the use of Eqs. 15 and 16 may be inappropriate as they only consider the vertical flux. The energy balance models reviewed by Male (1980), which incorporate horizontal fluxes, may be more appropriate.

Conductive heat transfer

Rainfall

Often the temperature of rain is near freezing point and the heat transfer to the snow pack is quite small. On rare occasions, when temperature is high, rainfall may constitute a significant heat source. It has a significant effect on the ageing process of the snow. The total flux may be expressed as:

$$Q_p = \rho_w C_w P_i (T_p - T_s) \times 10^{-3} \tag{17}$$

where ρ_w is the density of water (kg m^{-3}), C_w is the specific heat of water (4.169 kJ kg^{-1} K^{-1}), P_i is the rainfall rate (mm s^{-1}), T_p is the rainfall temperature (K), and T_s is the snow surface temperature (K). Anderson (1976) suggests that the wet-bulb air temperature will give a close approximation of rainfall temperature.

Ground heat flow

Ground conduction melt may be attributed to the upward transfer of thermal energy, stored in the ground during the preceding autumn and summer. This may cause melt to occur during the winter period when melt at the surface is negligible. Heat transfer by ground conduction (Q_g) may be expressed as

$$Q_g = k_s \, dT/dz \tag{18}$$

where k_s is the thermal conductivity of soil (W m^{-1} K^{-1}) and dT/dz is the temperature gradient (K m^{-1}) perpendicular to the soil surface. This heat flux is often very low, but may become significant when integrated over long periods. During the spring snowmelt soil surface temperatures may be below or at freezing point and the ground may act as a heat sink.

Empirical methods

A calculation of the energy balance requires detailed meteorological data and in many cases these are not readily available. Air temperature is a good indicator of the energy available for snow and ice melt (Ishikawa *et al.* 1992, Braithwaite 1981) and can often be used to provide an empirical relationship (e.g. Ruddell *et al.* 1990). Best results are often obtained by averaging temperatures over a period of a month or more. Sometimes the dependence on temperature can be improved by using the positive-degree-day method (e.g. Hoinkes and Steinacker 1975). This uses an integration of the positive mean daily temperatures over a given period of melt and ignores those temperatures which are negative. Further elaboration of the empirical approach may consider the relationship of melt with other parameters such as rain, wind and humidity.

MEASUREMENT OF VARIOUS PHYSICAL CHARACTERISTICS OF SNOW BY REMOTE SENSING METHODS

The collection of data such as snow depth and density over large alpine regions can be time-consuming and difficult. Satellites provide a convenient means by which such data may be frequently collected over large areas. This does not negate the need for ground observations, as they are necessary for the calibration and verification of satellite-derived estimates. The monitoring of snow cover by remote sensing has been reviewed by Hall and Martinec (1985), Foster and Chang (1993), and Rango (1993). The latter study provides an assessment of the advantages of various electromagnetic spectral bands which are applicable to the remote sensing of snow cover (see Table 3). While the mapping of passive micro-wave emissions has considerable advantages for

TABLE 3
The responses of some relative sensor bands to various snowpack properties (from Rango, 1993)

Snow property	Sensor band: Gamma rays	Visible/near infra-red	Thermal infra-red	Microwaves
Snow covered area	Low	High	Medium	High
Depth	Medium	If very shallow	Low	Medium
Water equivalent	High	If very shallow	Low	High
Stratigraphy	No	No	No	High
Albedo	No	High	No	No
Liquid water content	No	Low	Low	High
Temperature	No	No	Medium	Low
Snowmelt	No	Low	Low	Medium
Snow-soil interface	Low	No	No	High
Additional factors				
All weather capability	No	No	No	Yes
Current best spatial resolution from space platform	Not possible	10 m	100 m	25 km passive 10 m active

the estimation of water storage by the snowpack, especially for alpine regions which are often covered by cloud, the poor resolution is a problem. However, this is likely to improve with further developments (Rango 1993). Microwave emissions are strongly influenced by various electrical properties which determine the dielectric constant of a material (Ambach and Denoth 1975, Foster *et al.* 1984)

REFERENCES

Ambach, W. and Denoth, A., 1975. On the dielectric constant of wet snow. *International Association of Hydrological Sciences,* publication no. **114:** 136–142.

Anderson, E. A., 1976. A point energy and mass balance model of a snow cover. *NOAA Technical Report NWS 19.*

Bader, H., 1962 Snow as a material. *in* "Cold Regions Science and Engineering" ed by F. J. Sanger. Part II, Sect. B. U.S. Army Cold Regions Research and Engineering Laboratory, Hanover, New Hampshire.

Barry, R. G., 1981. "Mountain weather and climate." Methuen, London.

Bohren, C. F. and Barkstrom, B. R., 1974 Theory of the optical properties of snow. *J. Geophys. Res.* **79:** 4527–4535.

Braithwaite, R. J., 1981 On glacier energy balance, ablation, and air temperature. *J. Glaciol.,* **27:** 381–391.

Brunt, D., 1952. "Physical and Dynamical Meteorology." Cambridge Univ. Press, London.

Brutsaert, W., 1975 On a derivable formula for long-wave radiation from clear skies. *Water Resource Res.,* **11:** 742–744.

Budd, W. F., 1966 Glaciological studies in the region of Wilkes, Eastern Antarctica, 1961. *ANARE Sci. Rep.* Series A (IV), Glaciology. Publication No. 88. 152pp.

Costin, A. B., Gay, L. W., Wimbush, D. J. and Kerr, D., 1961. Studies in catchment hydrology in the Australian Alps. III. Preliminary snow investigations. C.S.I.R.O., Division of Plant Industry Technical Paper No. 15. 31pp.

Foster, J. L. and Chang, A. T. C., 1993. Snow cover. Pp. 361–370. *in* "Atlas of satellite observations related to global change." ed by R. J. Gurney, J. L. Foster and C. L. Parkinson. Cambridge University Press.

Foster, J. L., Hall, D. K., Chang, A. T. C. and Rango, A., 1984. An overview of passive microwave snow research and results. *Reviews in Geophysics and Space Physics,* **22:** 195–208.

Geiger, R. 1965. "The Climate Near the Ground." Harvard University Press.

Grenfell, T. C., Warren, S. G. and Mullen, P. C., 1994 Reflection of solar radiation by the Antarctic snow surface at ultraviolet, visible, and near-infrared wavelengths. *J. Geophys. Res.,* **99(D9):** 18,669–18,684.

Hall, D. K. and Martinec, J., 1985. "Remote Sensing of Ice and Snow." Chapman and Hall.

Hawkins, F. F., 1985. Equilibrium-line altitudes and paleaoenvironment in the Merchant Bay area, Baffin Island, N. W. T., Canada. *J. Glaciology* **31:** 205–212.

Hay, J. E. and Fitzharris, B. B., 1988. A comparison of the energy-balance and bulk-aerodynamic approaches for estimating glacier melt. *Glaciology,* **34:**145–153.

Hoinkes, H. and Steinacker, H., 1975. Hydrometeorological implications of the mass balance of the Hintereisferner, 1952/53 to 1968/69. *International Association of Hydrological Sciences,* publication no. 104, pp 144–149.

Ishikawa, N., Owens, I. F. and Sturman, A. P., 1992. Heat balance characteristics during fine periods on the lower parts of the Franz Josef Glacier, South Westland, New Zealand. *Int. J. Climat.* **12:** 397–410.

Kondratyev, K. Ya., 1969. "Radiation in the Atmosphere." Academic Press.

Kuvaeva, G. M., Sulakvelidze, G. K., Chitadze, V. S, Chotorlishvili, L. I. and El'Mesov, A. M. 1975. *Fizicheskie Svoistva Snezhnogo Pokrova Bol'shogo Kavkaza. (Physical Properties of Snow Cover of the Greater Caucasus.)* Translated from Russian. Glaciology Section IX of the Program of International Geophysical Year, No.17. Published for the U.S. Department of Agriculture, Forest Service and the National Science Foundation, Washington, D.C. by the Indian National Scientific Documentation Centre, New Delhi. 309pp.

Kuz'min, P. P., 1961. "Melting of Snow Cover." Translated from Russian. Leningrad. Israel Program for Scientific Translations, Jerusalem 1972.

LaChapelle, E. R., 1969. "Field Guide to Snow Crystals." University of Washington Press.

Lamb, H. H., 1955. Two-way relationship between the snow or ice limit and 1,000 - 500 mb thicknesses in the overlying atmosphere. *Quart. J. Roy. Met. Soc.* **81(348):** 172–189.

Lee, R., 1962. The theory of the equivalent slope. *Month. Weather Rev.,* **4(90):** 165–166.

Male, D. H., 1980. Seasonal snow cover Pp. 305–395. *in* "Dynamics of Snow and Ice Masses." ed by S. C. Colbeck. Academic Press.

Marks, D., 1979 An atmospheric radiation model for general alpine application. Pp. 167–178 *in* "Modelling Snow Cover Runoff" ed by S. C. Colbeck and M. Ray. U.S. Army Cold Regions Research and Engineering Laboratory, Hanover, New Hampshire.

Mellor, M., 1964. Properties of snow. *in* "Cold Regions Science and Engineering" ed by F. J. Sanger. Part III, Sect. A3C. U.S. Army Cold Regions Research and Engineering Laboratory, Hanover, New Hampshire.

Patterson, S., 1980. "The Physics of Glaciers" Pergamon.

Perla, R. I., 1980. Avalanche release, motion and impact. Pp. 397–456. *in* "Dynamics of Snow and Ice Masses." ed by S. C. Colbeck. Academic Press.

Prowse, T. D. and Owens, I. F., 1984. Characteristics of snowfalls, snow metamorphism, and snowpack structure with implications for avalanching, Craigieburn Range, New Zealand. *Arct. Alp. Res.* **16:**107–118.

Radok, U., 1977. Snow drift. *J. Glaciol.* **19:** 123–139.

Radok, U., Stephens, S. K. and Sutherland, K. L., 1961. On the calorimetric determination of snow quality. *Int. Assoc. Hydrol. Sci.* **54:** 132–134.

Rango, A., 1993. Snow hydrology processes and remote sensing. *Hydrol. Pro.* **7:** 121–138.

Ruddell, A. R., Budd, W. F., Smith, I. N., Keage, P. and Jones, R., 1990. "The South East Australian Alpine Climate Study." A report by the Meteorology Dept., University of Melbourne for the Alpine Resorts Commission.

Schmidt, R. A., 1972. Sublimation of wind-transported snow – a model. U.S.Dept. of Agriculture. Forest Service. Research Paper RM–90.

Stein, J., Laberge, G. and Lévesque, D., 1997. Monitoring the dry density and the liquid water content of snow using time domain reflectometry (TDR). *Cold Reg. Sci. Tech.* **25:** 123–126.

Sturm, M, Holmgren, J., König, M, and Morris, K., 1997. The thermal conductivity of seasonal snow. *J. Glac.* **43:** 26–41.

Thom, A. S., 1975. Momentum, mass and heat exchange of plant communities. Pp 57–109 *in* "Vegetation and the Atmosphere 4, Principles" ed by J. L. Monteith. Academic Press, London.

U.S. Army Corps of Engineers 1956. "Snow Hydrology." North Pacific Division, Portland, Oregon.

Weller, G. and Schwerdtfeger, P., 1970. Thermal properties and heat transfer processes of the snow of the central Antarctic Plateau. *Int. Assoc. Hydrol. Sci.* **86:** 284–298.

CHAPTER 3

Temporal and spatial variation in snow cover

WILLIAM S. OSBORNE, MAXINE S. DAVIS and KEN GREEN

INTRODUCTION

THE Australian Alps are subjected to variable snow cover throughout the winter, and sometimes during the summer months. While it is not uncommon for a light temporary cover of snow to be present in the summer months following a brief but heavy snowstorm, it is from late autumn to early spring, when temperatures are consistently low, that precipitation at the higher elevations occurs predominantly as snow. The duration of this snow cover ranges between a few weeks below the winter snowline to as long as five or six months at higher elevations and, in some years, may persist as restricted snowpatches for over 12 months in sheltered glacial cirques. In this chapter we examine the temporal and spatial aspects of snow cover in the Victorian Alps and the Snowy Mountains and briefly review related studies undertaken in this region.

Extent and duration of snow cover

Snow cover is not distributed evenly across the alps, either temporally or spatially (Table 1). Variability occurs in the duration, depth and timing of snow cover at different localities and from year to year due to regional and localised climatic and topographical variations (Brown and Millner 1989). Chapter 1 in this volume provides a detailed discussion of the factors which contribute to the formation of snow.

Two of the most important factors contributing to the maintenance of snow cover are temperature and precipitation. However, these are not constant across the region, varying with altitude, latitude and distance from the coast. Generally in the alps, precipitation increases and temperature falls with increased elevation. Large variations in precipitation and snow cover can occur over relatively short distances due to sharp changes in the topography. For example, in the Snowy Mountains Slatyer *et al.* (1984)

TABLE 1:
The mean areal extent and duration of snow cover in the Snowy Mountains and Victorian High Country (from Slatyer *et al.* 1984)

Duration of cover	Snowy Mountains		Victorian High Country	
	Elevation	Area	Elevation	Area
60 days	1528 m	1201 km²	not estimated	
90 days	1701 m	405 km²	1655 m	160 km²
120 days	1875 m	105 km²	1828 m	11 km²

found that there was a consistent linear relationship between altitude and snow cover on an altitudinal transect from Thredbo Village (1373m) to the Crackenback Range (1950m). Snow duration varied from 26 days at the bottom to 128 days at the top, with large standard deviations for each of their five sample points.

Factors related to topography, such as elevation, slope and wind exposure are also important determinants of the distribution of precipitation (Basist *et al.* 1994). The pattern of precipitation for the Victorian alpine region has been described by Ruddell *et al.* (1990), who noted that topography is a major control on the amount of precipitation received. Greatest falls are received at high elevations with a north-westerly orientation. The orientation and shape of mountain ranges affects airflow so that the relative position of individual sites greatly influences the amount of precipitation that is received. In winter and spring, a moist, westerly air stream brings most precipitation to the region. The alignment of the ranges causes differences in the amount of precipitation, wind and sun exposure between the western and eastern slopes.

Brown and Millner (1989) provide a detailed discussion of the meteorology and hydrology of the Australian Alps. The maximum precipitation occurs in the winter and spring months, and the minimum in the late summer months. However, the ratio of summer to winter precipitation varies markedly from west to east across the alps (Brown and Millner 1989). The ratio is lower on the western side which receives heavy falls of rain and snow which are increased by orographic lifting. There is also a north-south variation in precipitation, with southern meteorological stations receiving relatively higher precipitation in early autumn and late spring than northern stations. The winter snowline is at a lower elevation on the southern slopes of the Alps in Victoria compared to the Bimberi Range just south of Canberra because of latitudinal differences. The snowline is also about 200 metres lower on the western slopes than on the eastern slopes at the same latitude due to warming air on the eastern side (Brown and Millner 1989).

Specific site characteristics such as aspect, wind and sun exposure, slope, and vegetative cover also effect the distribution and duration of snow cover at particular sites through their impact on the processes of accumulation and ablation of snow. Temperature influences the proportion of precipitation that falls as snow, whereas wind speed and direction plus topography and vegetative cover affect retention of snow. Once on the ground, the ablation of snow results from the interaction of a number of processes which contribute to melting or erosion by wind. Contributors to melting include heat conduction from the ground; the absorption of solar radiation related to aspect, surface slope and albedo of the snow cover; radiation exchange from the surroundings particularly boulders; the heat content of rain and melt waters; and heat transfer from the air by convection (Ruddell *et al.* 1990, see also Chapter 2). Erosion by wind is related to the velocity of the wind, exposure of the surface and the density of the snow as this affects heat transfer by convection and transport by drifting.

Rainfall events can have a major impact on the duration of snow lie. Heat transmitted to the snow from warm rain speeds up the melting process. Moist air associated with rain is even more effective in accelerating snow melt; moisture condensing directly onto the snow surface, releases enough heat to melt seven or eight times its own weight in snow (Geiger 1950). Rainwater flowing down tree trunks contributes to the areas around the base of trees becoming snow-free early in the thaw (Green and Osborne 1994). Areas of snow exposed to rain can show a dramatic loss compared to protected snow. We documented a loss of 15 cm of snow in one night of rain at Headley Tarn in the Snowy Mountains (by measuring the difference in height between snow protected beneath our tents and exposed snow).

Snowpatch duration

Snowpatches accumulate on leeward slopes by the transportation of snow by winds into relatively calm air and the subsequent deposition of the snow (Green and Osborne 1994). This leads to the formation of areas of deep winter snow which often have sufficient thermal mass to resist melting well past the time of the loss of the general snow cover. Not surprisingly, considering its elevation and the fact that it was the only glaciated area on the mainland in the last glacial (Chapter 4), snowpatches are best developed in the Snowy Mountains where they tend to persist longest in glacial cirques where snow can sometimes metamorphose to the density of firn ice (Chapter 4). For example, in 1996, snowpatches on the Etheridge Range (cirques 2 or 3 in Chapter 4 Fig. 1), Upper Blue Lake Cirque (12) and Twynam Cirque (13) were the longest lasting, remaining until March. However, snowpatches from the winter of 1996 at Wilkinsons Cirque under Mt. Townsend (4), to the right of Mawsons Cirque (7), Club Lake (9) and Upper Blue Lake (12) lasted right through to the winter of 1997, whereas the snowpatches on Etheridge Range were gone by March and the one at Twynam Cirque (13) lasted only until May (K. Green unpubl. data).

The accumulation of snowpatches will vary from year to year depending upon the factors mentioned above. However, the observed interannual variation in the duration of these snowpatches is indicative of more subtle microclimatic effects than are perceived through observations of the depth and duration of the snowpack. Global climate modelling is unable to predict the localised impact of changes resulting from the diverse mountain topography (Whetton *et al.* 1996) and these snowpatches may prove valuable in examining how differences in microtopography and microclimate modify general trends.

Long-term trends in winter precipitation and snow cover

Interannual variations in precipitation, which can be directly related to changes in the occurrence of synoptic circulation systems (Budin 1985) are a major feature of the climate of the Australian Alps, and occur in both annual and seasonal precipitation records. Interannual variations and periods of drought are evident in the winter (June, July and August) precipitation record for Kiandra (1395 m) between 1887 and 1995, Tooma Dam (1220 m) between 1959 and 1995 and Guthega Power Station (1340 m) between 1953 and 1995 (Figs 2 a, b and c) . Of these three stations, Tooma Dam in the north-western Snowy Mountains has the highest annual and winter precipitation; Kiandra in the northern Snowy Mountains has the lowest, and Guthega Power Station in the south-eastern Snowy Mountains is intermediate.

Winter precipitation at Kiandra has declined since the 1920's (Fig. 1a). Although the linear trend was not significant ($r^2=0.018$, $F_{1,107}=1.98$, N.S.), mean winter precipitation during the period 1887 to 1923 was significantly higher than for the remaining period (1924 to 1995) (F test, $p<0.01$). After 1924, the five year moving average trend in winter precipitation at Kiandra, despite some fluctuations, remained generally low until the late 1950's to early 1960's. The trend declined again in the following decades with a small improvement in the early 1970's, and above-average precipitation being recorded in the early 1990's. At both Tooma and Guthega, where records are only available since the 1950's (Figs 1b and c), the increase in winter precipitation in recent years is obvious. However, at Guthega, depression of winter precipitation during the 1980's, and to a lesser extent in the 1960's, was more pronounced. Extremely low winter precipitation was recorded in 1982 (an El Niño year) at all three stations.

Winter snow accumulation is an important factor influencing hydrology in the following spring and summer. Long term trends in snow cover can be demonstrated by referring to snow depth measurements which have been recorded by hydro-electric generating authorities over a number of years. The longest and most reliable snow depth record in the Victorian alpine region has been maintained by the State Electricity Commission of Victoria (SECV) at the Rocky Valley snowcourse (1600–1660 m), Bogong High Plains. From 1935 to 1946 snow depth was measured after each snowfall; from 1954 to 1956 depths were measured from 2 to 5 times each season; and from 1957 to 1972 and 1974 to the present, depths were measured at the middle and end of each month. In the New South Wales alpine region, the longest and most reliable snow depth record has been maintained by the Snowy Mountains Hydro-electric Authority (SMHEA) at the Spencers Creek snowcourse (1830 m) in the Snowy Mountains. Observations at this site have been made approximately every 3–7 days throughout the snow season each year since 1954.

1a

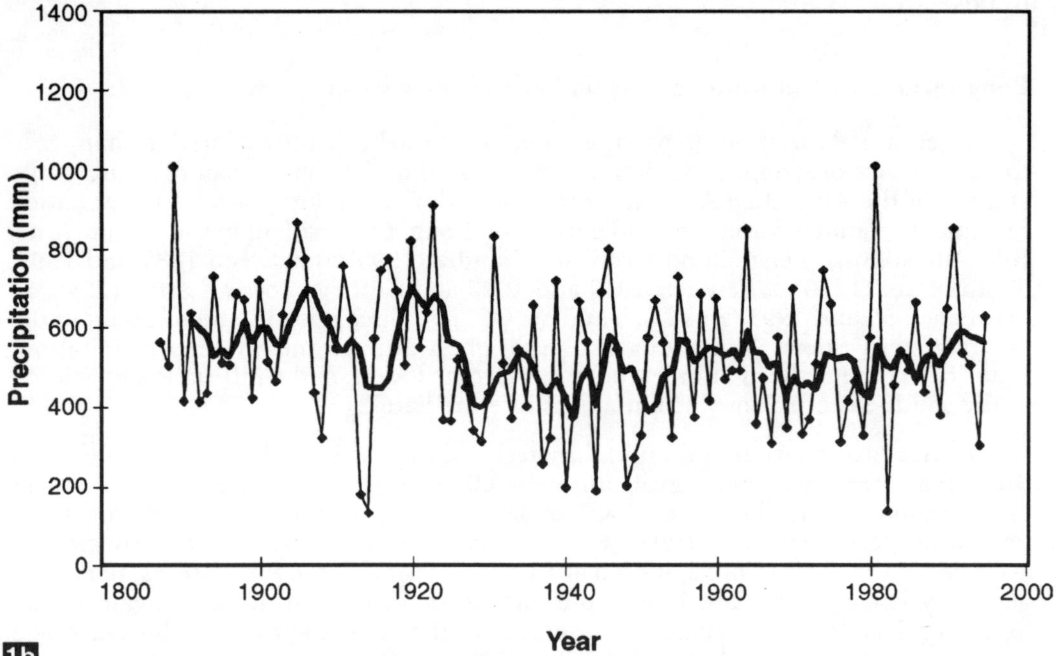
1400
1200
1000
800
600
400
200
0
Precipitation (mm)
1800
1900
1920
1940
1960
1980
2000
Year

1b

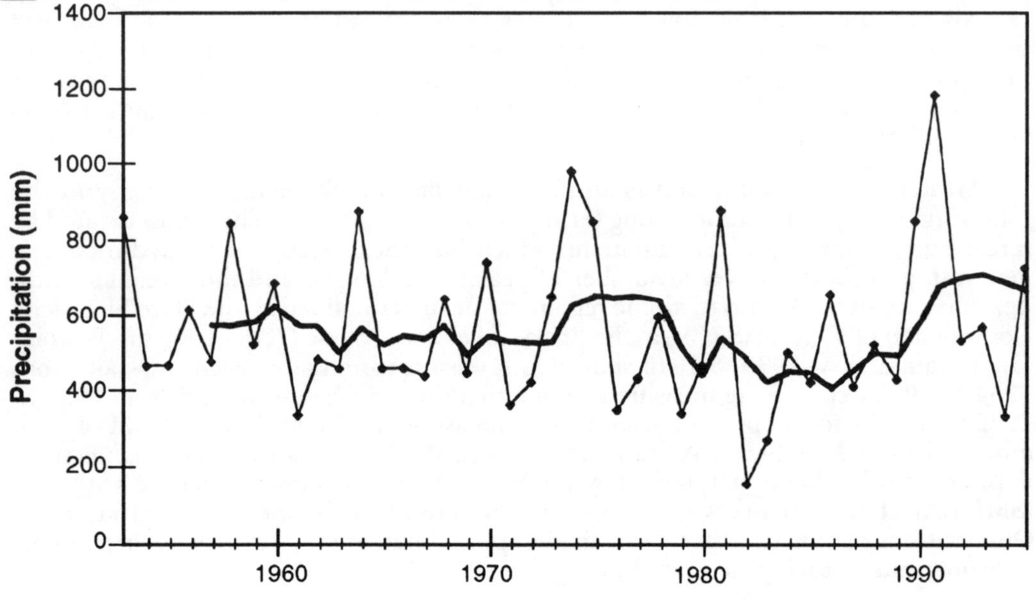
1400
1200
1000
800
600
400
200
0
Precipitation (mm)
1960
1970
1980
1990
Year

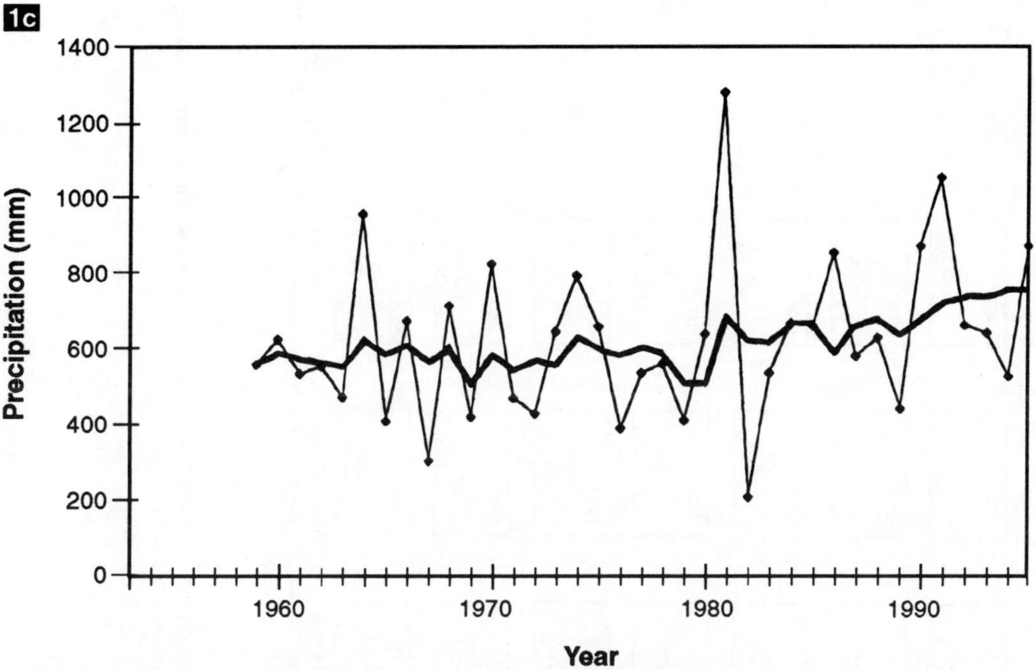

Fig. 1. Combined winter precipitation (June, July and August) for three stations in the Snowy Mountains. Totals and five year running means (heavy line) are shown. (a) Kiandra (1395 m) 1887–1995. (b) Guthega Power Station (1340 m) 1953–1995. (c) Tooma Dam (1220 m) 1959–1995.

The snow depth profiles for 1954 –1996 and maximum depth recorded at Rocky Valley and Spencers Creek snowcourses are compared in Figures 2 and 3. Interannual variability in start and end dates and duration and maximum depth at both sites is evident (see also Figure 4). Maximum depth (in cm of snow) at Spencers Creek ranged between 90 cm and 361 cm, a four-fold variation during the record period. At Rocky Valley, the range in maximum depth was from 39 to 270 cm, a six and a half-fold increase. The depth of snow at Rocky Valley is proportionally less than at Spencers Creek (generally about 60%) which is not unexpected given the difference in altitude, however, there are similarities in the pattern of snow accumulation each season. The maximum snow depth at the two sites is highly correlated ($r^2 = 0.81$). However, whereas Rocky Valley had a slightly positive trend, Spencers Creek had a slightly negative linear trend, neither of which was statistically significant.

Previous analysis of maximum snow depths (in water equivalents) by Ruddell *et al.* (1990) for Rocky Valley in the Bogong High Plains and Spencers Creek and two other snowcourses in the Snowy Mountains snowcourses; Deep Creek (1615 m) and Three

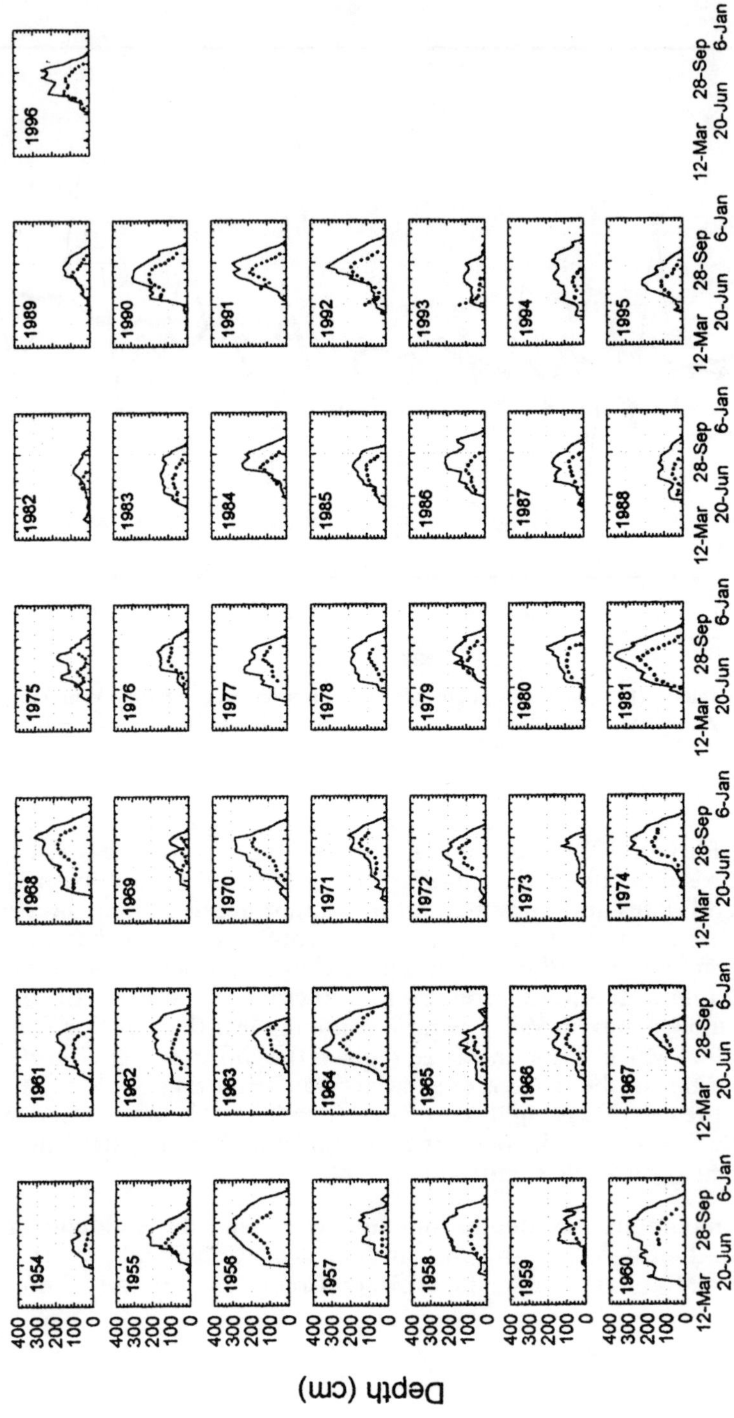

Fig. 2. Variation in snow depth profiles for 37 years (1954–1996) at Spencers Creek (1830 m) SMHEA snowcourse in the Snowy Mountains (solid line) and at Rocky Valley (1610–1660 m) SECV snowcourse in the Bogong High Plains (dotted line).

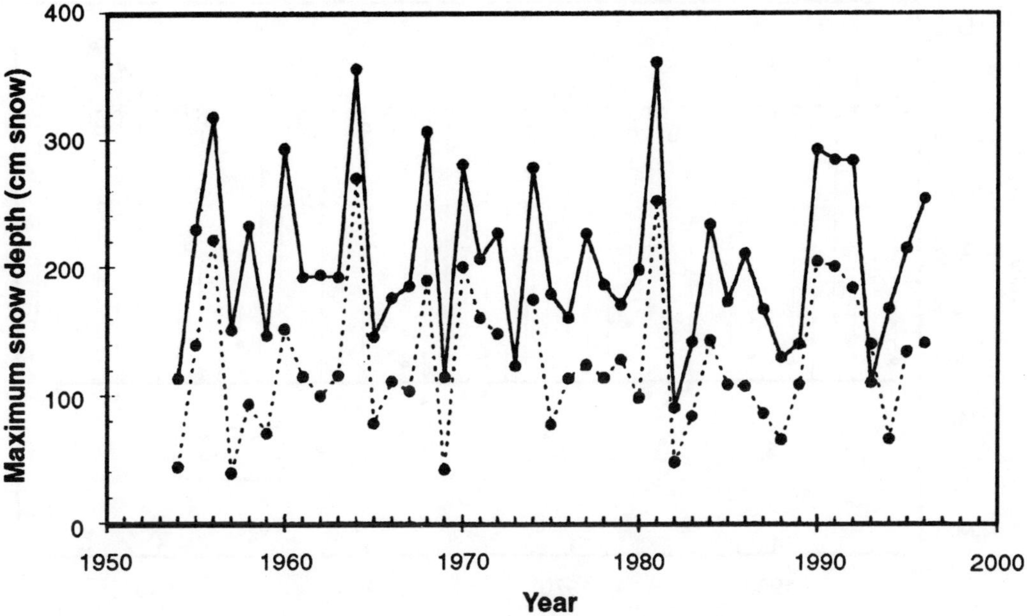

Fig. 3. Maximum snow depth (1954–1996) at Spencers Creek (1830 m) SMHEA snowcourse in the Snowy Mountains (solid line) and at Rocky Valley (1610–1660 m) SECV snowcourse in the Bogong High Plains (dashed line).

Mile Dam (1460 m), between 1957 and 1989, found all snowcourses had slight but not statistically significant negative trends. The number of below-average years during the 1980's was considered by Russell *et al.* (1990) to contribute considerably to the negative trend. They felt a sequence of above average seasons, which could not be ruled out given past variability (and which actually did occur in the early 1990's), would offset the apparent trend.

The Spencers Creek profiles which have observations made at shorter time intervals and more consistently each season than at Rocky Valley (Fig. 2), also demonstrate variations in the depth of snow through individual seasons. Winter snow cover at Spencers Creek usually commences in early June, rarely in April and occasionally not until July. Snow accumulation usually continues, with the biggest build up in July and August, leading to peak snow depths in August or September at an average of about two metres. Snow depth may increase and decrease through the months of July to October, as fresh snow falls and ablation processes take effect. But it is usually in late September or October that the thaw begins in earnest. The ground becomes free of measurable snow in late October or early November, occasionally persisting until December.

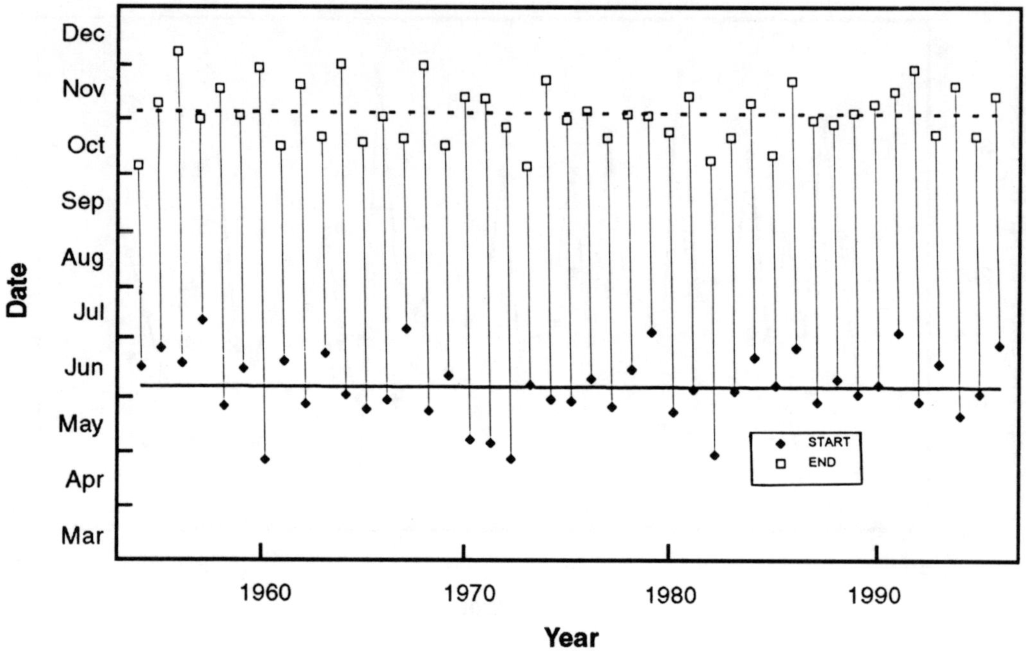

Fig. 4. Duration of continuous snow cover at Spencers Creek (1830 m) 1954–1996. Start and end dates are indicated, connected by a narrow solid line within each year. Dashed line and heavy solid line indicate linear trends during the period.

Although there is considerable variability in the start and end dates of continuous snow cover at Spencers Creek from year to year during 1954–1996 (Fig. 4), the duration of persistent snow cover, the date of commencement and completion did not change significantly from 1954 to 1996 (mean duration = 22.5 weeks; $F_{1,41}$= 0.012, p = 0.91, N.S.: mean start date = June 5; $F_{1,41}$ = 0.045, p = 0.83, N.S.; mean end date = November 2; $F_{1,41}$ = 0.179, p = 0.67, N.S). When depth and duration are expressed jointly as snow cover, calculated by multiplying mean snow depth for each year by the number of days of continuous snow cover (Fig. 5), it is apparent that there was a gradual decline in cover (despite the presence of broad fluctuations) in the three decades prior to 1990, with an upturn in more recent years. The five year trend reveals a reduction in snow cover of about 100 metre-days from the early 1960's to late 1980's (Fig. 5). Whilst snow cover appears to decline over the entire period 1954 to 1996, the linear trend was not significant ($F_{1,41}$ = 1.87, p = 0.18, N.S.). However, when mean snow cover in the two 15 year periods 1960 to 1974 and 1975 to 1989 are compared, there was significantly less snow cover in the second period (t= 1.83, p = 0.04, df = 22) equivalent to a 27% reduction.

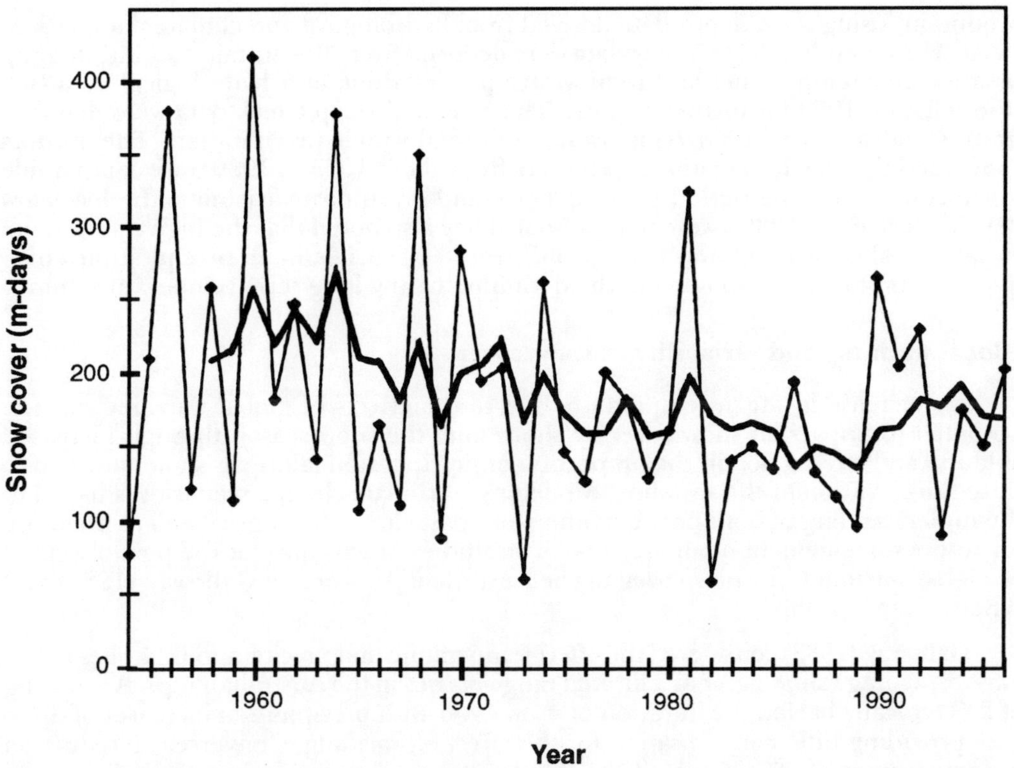

Fig. 5. Estimated snow cover at Spencers Creek (1830 m) 1954–1996. Annual totals and five year running mean (heavy line) are shown.

The difficulty in claiming evidence for climate change based on limited data is discussed by Ruddell *et al.* (1990). They note that a period of 30 years is regarded as suitable for establishing means of precipitation and that evidence of significant trends would require in the order of 50 years of data. Retrospective modelling of snow depth data has been used to provide a longer term perspective to the available data since the 1950's. The Spencers Creek snowcourse data were used to model snow cover quality (incorporating depth and duration) from the turn of the century to 1991 by Duus (1992). According to SMHEA (1993), this modelling indicates that the snow season "quality" in the late 1980's was about half that at the turn of the century. The five year moving average trend indicated an apparent decline in snow cover since about 1960 which was preceded by an increase since the mid 1920's (see Fig. 8.7, SMHEA 1993, p.8–25).

Ruddell *et al.* (1990) extended the snow depth records for Rocky Valley and Deep Creek (another SMHEA snowcourse in the northern Snowy Mountains near Round

Mountain) using snow depth data derived from hydrological and climate data back to 1910. The extended Rocky Valley data were deduced from Mt. Buffalo and Mt. Bogong mean winter temperature and total winter precipitation, and hydrological data (see Ruddell *et al.* 1990 for further details). The extended Deep Creek data were deduced from Kiandra mean winter temperature and total winter precipitation. The authors concluded that the lower snow depth years from about 1975 to 1989 were comparable to the 1930's and the period prior to 1916, and given past variability, the low snow depth years in the 1980's were not unusual. They also noted that the high interannual variability of snow depth was strongly influenced by variability in precipitation which had sufficient capacity to obscure the definition of any long term temperature trend.

Global warming and variability in snow cover

An increase in mean temperature due to the effects of global warming has the potential to impact on snow cover by shortening the snow season through increased ablation and a reduction in the proportion of precipitation falling as snow and, also to raise the elevation of the snowline. Modelling of climate change scenarios is useful to the understanding of both possible future and past climate change as well as planning for future management of alpine areas. Two studies which consider the possible effect of global warming on snow cover in the Australian Alps are by Galloway (1988) and Whetton *et al.* (1996).

Galloway (1988) considered the effect temperature and precipitation changes might have on average snow cover at a limited range of sites in the Australian Alps. A warming of 2°C typically halved the duration of snow cover with precipitation increases of up to 20% providing little compensation for the effect of warming. Conversely, a reduction in temperature of 2°C increased the duration of snow cover and expanded the area subject to snow cover considerably. Sites at lower elevations were most affected. Whetton *et al.* (1996) used climate data from a considerable number of weather stations to simulate snow cover duration under a range of climate change scenarios for the years 2030 and 2070. The simulated average snow cover and frequency of years with more than 60 days cover declined at all sites considered in the best-case scenario. Sensitivity to increasing temperature was found to be very high, particularly at low elevations. Large percentage increases in precipitation were required in the model to compensate for relatively small increases in temperature. For the worst-case scenario a dramatic decline in snow cover was simulated which resulted in snow cover by 2070 being restricted to a very small portion of the highest parts of the Snowy Mountains.

An understanding of the potential of global warming to alter interannual variability — the frequency of low and high precipitation (rain and snow) years – would be helpful in both short and long term planning for industries in the alps such as skiing (Galloway 1988) water resources and hydro-electricity (Whetton *et al.* 1996) as well as for conservation of flora and fauna. Exceptionally high snow depth years are often followed by a low snow depth year, which tend to occur during an El Niño such as occurred in 1981–82 (Budin 1985). Models of global warming are currently only able to take into account areas of uncertainty that are readily quantifiable, and are unable to incorporate future interannual changes that might result from features such as the El Niño-Southern Oscillation (Whetton *et al.* 1996). In order to examine the potential impacts of global warming on the relative frequency of years of high and low snow cover over time on Mt. Baw Baw (1564 m), Mt. Buller (1805 m) and Mt. Bogong (1986 m), Whetton *et al.* (1996) imposed a time series incorporating interannual variations in temperature and precipitation as observed during the 20-year period 1966–85 over best- and worst-case increasing temperature trends. This assumed the interannual variability 1966–85 would be representative of the variability of the next century.

The best case scenario assumed a rise in temperature and precipitation of 0.5°C and +10% by 2030 and 1.0°C and + 20% by 2070, whereas the worst-case scenario assumed a rise of 2°C and 20% reduction in precipitation by 2030 and rise of 5°C and 20% reduction in precipitation by 2070. The worst-case scenario showed a rapid decrease in simulated snow cover at all three sites. Simulated zero snow cover duration commenced at Mt. Baw Baw from around 2020, 2050 at Mt. Buller and 2070 at Mt. Bogong. For the best-case scenario, a declining trend in snow cover duration was evident at the three sites, although it was small compared to the interannual variability. The model predicted some good years, better than occurred in the base data for 1966–1985 occurring through to the end of the next century but with the frequency of good years declining with time, and the frequency of bad years increasing. Whetton *et al.* (1996) draw attention to the best-case scenario of +0.5°C in the forty years to 2030 potentially being an analogue for the current period and the capacity of interannual variability to dominate in trends and to mask small but still significant trends in snow conditions related to temperature.

ACKNOWLEDGEMENTS

We thank the State Electricity Commission of Victoria, the Snowy Mountains Hydro-Electric Authority and the Bureau of Meteorology for access to data.

REFERENCES

Basist, A., Bell, G. D. and Meentmeyer, V., 1994. Statistical relationships between topography and precipitation patterns. *J. Climate*, **7**: 1305–1315.

Brown, J. A. H. and Millner, F. C., 1989. Aspects of the meteorology and hydrology of the Australian Alps. Pp. 297–332 *in* "The Scientific Significance of the Australian Alps: The Proceedings of the First Fenner Conference" ed. by R. B. Good. The Australian Alps National Parks Liaison Committee/Australian Academy of Science.

Budin, R., 1985. Interannual variability of Australian snowfall. *Aust. Met. Mag.* **33**: 145–159.

Duus, A. L., 1992. Determination and analysis of the snow cover in the Snowy Mountains between 1910 and 1991. *Aust. Met. Mag.* **40**: 195–204.

Galloway, R. W., 1988. The potential impact of climate changes on Australian ski fields. Pp. 428–437 *in* "Greenhouse: Planning for Climate Change" ed. by G. I. Pearman. CSIRO/Brill, Melbourne.

Geiger, R., 1950. "The Climate Near the Ground." Harvard University Press, Cambridge, Mass.

Green, K. and Osborne, W. S., 1994. "Ecology of Snow", Pp. 16–29 *in* "Wildlife of the Australian Snow Country." Reed Books, Chatswood NSW.

Ruddell, A. R., Budd, W. F., Smith, I. N., Keage, P. L. and Jones, R., 1990. "The South East Australian Alpine Climate Study." Department of Meteorology, University of Melbourne Report.

Slatyer, R. O., Cochrane, P. M., and Galloway, R. W., 1984. Duration and extent of snow cover in the Snowy Mountains and a comparison with Switzerland. *Search* **15:** 327–331.

Snowy Mountains Hydro-electric Authority, 1993. "Snowy Precipitation Enhancement Project: A proposal to Evaluate the Feasibility of increasing Snow Precipitation over the Snowy Mountains Area". Draft Environmental Impact Statement. SMHEA, Cooma.

Whetton, P. H., Haylock, M. R. and Galloway, R., 1996. Climate change and snow-cover duration in the Australian Alps. *Climatic Change* **32:** 447–479.

CHAPTER 4

Effects of snow on the landscape

ROBERT W. GALLOWAY, KEVIN KIERNAN and JIM A. PETERSON

INTRODUCTION

ONLY a tiny proportion of mainland Australia (<0.1%) carries snow in winter but its significance for conservation, water supply and recreation is considerable, and substantial economic interests are involved. Many of the effects of snow on the landscape are expressed through the vegetation and hydrology, but as these topics are discussed by other contributors we shall confine our attention to the effects of snow on the landforms.

There is a great difference between the effects of snow on the landforms during the cold and rather dry glacial epochs of the Pleistocene and during the relatively warm and humid last 10 000 years (the Holocene). During the glacial epochs, permanent snowfields developed in both Tasmania and New South Wales and resulted in glaciers and nivation processes which had major effects on mountain landforms; in Victoria the impact was confined to nivation effects which were still substantial. In the Holocene, the effects of snow on the highland landforms of south east Australia have been confined to modest nivation effects in and around long-lasting snow patches and there were no glaciers. However, the generally mild conditions were interrupted by periods of minor climatic cooling known as "neoglacials"; the last was at its peak about 150 years ago (Grove 1988). There also seems to have been one or more "altithermal" periods rather warmer than today.

GLACIAL PROCESSES

When annual accumulation of snow exceeds ablation the snowpack thickens and densifies through firn (density >*c*.0.4g. cm^{-3}) to ice (density >*c*.0.8 g. cm^{-3}). On ascending a mountain, air temperature falls and precipitation, including snowfall, usually rises. At a level, variously known as the equilibrium line or the permanent snow line, average annual snowfall exceeds ablation and consequently permanent snow cover and ice masses can develop. The altitude of the equilibrium line on any mountain varies substantially according to the topography being lower in sites favouring snow entrapment: i.e. hollows facing east or southeast protected from direct solar radiation and shielded from the prevailing westerly winds. Thus the height of the equilibrium line on a mountain can vary by hundreds of metres between sites. Its height also varies from year to year and it rises if climatic change increases ablation and/or reduces precipitation. During cold glacial phases of the Pleistocene, the regional snowline was low enough to intersect limited summit areas in the Snowy Mountains that were especially favourable for snow accumulation resulting in small cirque glaciers. Much more extensive areas in Tasmania were also above the regional snowline resulting in much larger glaciation. Today the calculated position of the climatic snowline lies in the atmosphere well above these mountains, and glaciers do not exist.

Once the accumulated snow turns to ice thick enough to flow down slopes under gravity, glaciers form. They can have great erosive effects, scooping out armchair-shaped cirques and depositing some of the resulting rock debris around their lower margins as terminal and lateral moraines. In rather less favourable sites below steep slopes, permanent snow patches may be too thin to form glaciers but provide a smooth surface down which rocks weathered from steep slopes above slide and tumble to form psuedo-moraines or protalus ramparts at their lower edge.

NIVATION PROCESSES

Nivation is a complex of processes occurring in and around long-lasting snow patches (Evans 1969). It includes frost-splitting by freeze/thaw around the patch margins, soil creep induced by the high water content of the ground under melting snow, pressure effects of snow, both static and dynamic, and transport of eroded material and minor gullying by melt water down the slope below the patches. Under snow patches, temperature fluctuations and refreezing of melt water can be minimal and the regolith is saturated for long periods of time. This situation may persist long enough, after thawing commences, to encourage slope instability if the regolith is weathered to sufficient depth. The results are obvious where the slopes are steep and vegetation cover incomplete, but for the most part the emerging slopes during spring are stable under a cover of alpine herbfield. Today, only very long-lasting snow patches can inhibit the development of such stable vegetation cover in Australian mountains.

The extent of snow patches is highly variable from year to year in accord with the variable precipitation on Australian mountains. The sites of snow patches can be readily identified on air photos if the snow patches last sufficiently long to inhibit the normal vegetation cover. Sites with active nivation are more apparent on air photographs taken in the summer after a heavy snow winter than those taken in the summer after a light snow winter.

Snow patches with active nivation are generally situated above the tree line where the mean temperature of the warmest month is less than 10°C. As snow patches with nivation occupy only a small fraction of the areas with such low summer temperatures, heavy snow accumulation made possible by drifting into favourable sites is clearly critical for active nivation. Because of this requirement for high snow accumulation it is probable that nivation was not a widespread process at lower levels even during the coldest part of the relatively dry Pleistocene glacial periods.

A mantle of snow greatly lessens the range of diurnal temperature, and its melting in spring or early summer affects many seasonal cycles such as soil moisture, and habitat availability. Very long-lasting snow cover can inhibit the underlying vegetation so that when the snow finally melts the exposed ground is bare, wet, uninsulated, and subject to freeze/thaw processes such as the growth of needle ice which can lead to soil erosion. Snow cover traps air-borne dust and affects its ultimate distribution and addition to the regolith.

Over long periods nivation creates spoon-shaped hollows with steep backwalls a few metres high fronted by an apron of transported debris which is usually soft and boggy in summer (Derbyshire and Peterson 1977). In winter and spring, they are usually the site of snow cornices or unusually thick snowdrifts. There is a strong positive feed-back effect; as nivation hollows deepen they become ever-better catchment sites for snow which in turn increases nivation. Cirques almost certainly started as nivation hollows forming in valley heads and then grew larger and deeper by this positive feed-back as glaciers developed.

Factors favouring snow patch development are the same as those which favoured cirque development in the Pleistocene. Consequently, the back walls of the Pleistocene glacial cirques (in the Snowy Mountains rather than in Tasmania) form especially favourable sites for modern snow patches. In such sites with heavy snowdrifting, snow patches can survive through spring and well into summer or even autumn. In a few very favourable sites after some heavy snow winters, patches have been known to survive until the following winter and thus attain a semi-permanent status. Permanent status would be the first step towards the formation of glaciers.

Stone pavements associated with snow patches commonly take the form of tightly packed clasts with their flattest face upwards. Tchey may be due to the combined effects of frost action, meltwater saturation and the removal of fine sediment, and the rotation and flattening of blocks by rock weight, snow pressure and possible snow creep (White 1972). Stone pavements can also form by melt-water washing out fines from bouldery solifluction rubble leaving a lag of blocks that cannot be transported. More spectacular pressure effects such as moving of heavy blocks and scratching of rock surfaces can occur in sites where the snowpack is exceptionally thick and dense (Jennings and Costin 1978). The pressure effect of masses of wet snow has caused some concern for ski lodges and other structures situated at the foot of fairly steep slopes. The wet, dense nature of the snow might favour avalanching (Foehn 1992) but in fact avalanches are very rare in Australia presumably because long, steep slopes with thick snow cover are not common. They do, however, occur and vegetation in *Eucalyptus coccifera* forest at Mt. Field (Tasmania) shows multiple avalanche tracks.

Freezing and thawing in the immediate vicinity of the cold snow mass causes small-scale granular disintegration of the rocks to produce sand and silt that contribute to the sedimentary apron below the snow patch. The high water content of the resulting debris favours slow mass movement which may be further accentuated by freezing of the abundant interstitial water at night causing expansion followed by contraction when the water thaws next day.

PERIGLACIAL PROCESSES

Periglacial geomorphic processes are associated with freezing. They can have major landforming effects such as frost-shattering of rocks and creation of extensive solifluction deposits. Most of these processes are not directly related to snow and so fall outside the scope of this book. However, snow and ice between boulders formed by frost-splitting of bedrock probably assisted their movement downslope to form block fields.

Snow meltwaters can percolate the regolith and form permanent ground ice below the surface. Caine (1983) has described ice-thrust dolerite columns that stand up to 50cm above glaciated surfaces at about 1400 m altitude and extensive associated block fields and block streams that raise the possibility of past permafrost conditions on the Ben Lomond Plateau in northeastern Tasmania. Rock glaciers on Mt. Olympus and Mt. Gell in central western Tasmania (Derbyshire 1973) and block streams and block glacis elsewhere in Tasmania, may also have involved permafrost conditions and snow meltwaters though not necessarily so.

NEW SOUTH WALES AND VICTORIA

Pleistocene glaciation

The only indubitable evidence for Quaternary glaciation in the highest part of the Snowy Mountains dates from the last glaciation. Thirteen clear cirques with moraines testify to the presence of ice at a geologically recent time (Fig. 1). Only the Blue Lake cirque is a definite rock basin, but several other lakes have formed behind morainic dams. Previous claims of extensive earlier glaciations in the Snowy Mountains have not stood up to critical analysis (Galloway 1963) and the estimated area that was ice-covered is considered here to have been 30–40 km². Nevertheless, the possibility of much earlier glaciations, comparable with some late Tertiary or early Quaternary evidence in Tasmania, cannot be entirely excluded. The volume of the large Blue Lake and Mawson cirques far exceeds the volume of the associated moraines suggesting that the cirques were initiated long ago and the concomitant moraines were subsequently removed by erosion before the present moraines were deposited.

Very recently, T. Barrows (pers comm.) has begun to re-examine the evidence of glaciation. The glaciated area of the Main Range has been re-mapped and found to be very similar to the "probable" limits of Galloway (1963). Dating the glacial limits has begun using cosmogenic nuclides on the Blue Lake end moraines. Initial results from

^{36}Cl dating suggests that the two outer limits noted by Galloway are actually separate glaciations distinct in age. The inner limit corresponds to the Last Glacial Maximum (about 20 000 years ago) and the outer limit appears to date from 60 to 70 000 years ago. Moraines corresponding to the Marine Isotope Stage 6, the prominent Henty glaciation of Tasmania, have yet to be recognised. In the Victorian Alps, many of the features initially described as glacial are now thought to have a periglacial origin (Peterson 1971).

Pleistocene nivation

Block fields which owe their origin in part at least to snow turning to ice between the boulders are known in several localities in Victoria, in the Snowy Mountains and in the Australian Capital Territory (Rosengren and Peterson 1989). Pro-talus ramparts, marking the lower edge of former permanent snow patches too limited to develop into moving ice, are found on the east ridge of Mount Tate in the Snowy Mountains. Other examples may exist on steep, southeast-facing slopes overlooking the Thredbo Valley. A Quaternary history of regolith adjustment to varying amounts of ground ice and snow cover in sympathy with the Quaternary succession of glacial and interglacial stages is strongly implied by studies of the distribution of block fields in Victoria (McRae-Williams *et al.* 1981).

When the cirques were occupied by ice, nivation was presumably active on adjacent sites slightly less favourable for snow accumulation and indeed such nivation hollows have been recognised on the Main Range and on the Grey Mare Range (Ritchie and Jennings, 1955). They may be quite common in the higher parts of the mountains. In the Victorian Alps, the legacy of Pleistocene nival processes is similar to that in the Snowy Mountains.

Given the requirement of very heavy snowfall, active nivation was not likely at lower altitudes even at the coldest times in the Pleistocene when thousands of square kilometres lay above the tree line but the climate was relatively dry.

Present-day nivation

Today, nivation in New South Wales is confined to narrow belts on the lee side of ridges and plateaus. The best examples are on the eastern side of the crest ridge that runs from the southern spur of Mt. Kosciuszko north-north-east to Mt. Twynam - a distance of some 10 km (Fig. 1). Only sections of this strip particularly favourable for snow accumulation show the characteristic bare surfaces of nivation; elsewhere the underlying vegetation cover is intact. This modern nivation does little more than retouch pre-existing glacial and nivation landforms. A portion of this belt of deep snow, in Mawson's cirque (Fig. 1) was the source of one of the rare avalanches that have caused damage: a young woman was killed when a ski lodge was swept away.

Another notable snowpatch is on the eastern side of Mt. Twynam on the upper backslope of Twynam Cirque. This is usually the longest-lasting snowpatch in the Snowy

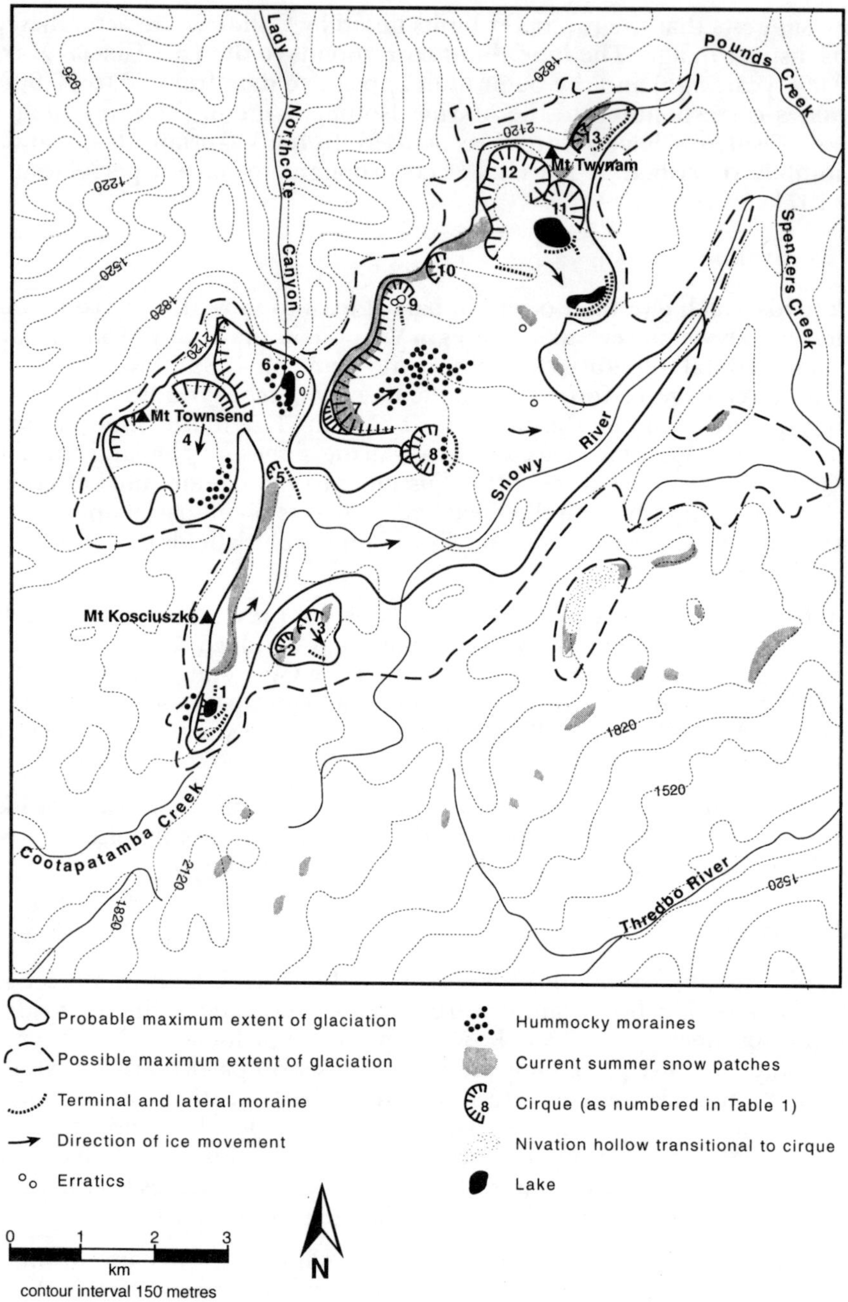

Fig. 1. Glacial and nivational features of the Snowy Mountains, New South Wales.

TABLE 1.
Cirques in the Snowy Mountains

Number on Figure 1	Name	Approx altitude of floor (m)	Remarks
1	Cootapatamba	2010	In Murray headwaters, cf. 6
2	–	2010	Poorly defined
3	–	1950	Poorly defined
4	Wilkinsons	1950–1980	Valley form only slightly modified by glacial action; small cirque at head
5	–	1890	Small cirque, relatively large and distinct lateral moraine; site of age determination by David *et al.* (1901)
6	Albina	1875	Retreat moraine partially drowned in lake
7	Mawsons	1850	Largest cirque, minor cirque perched on headwall
8	Clarke	1860	Minor cirque on headwall
9	Club Lake	1950	Unusual moraines
10	Carruthers	1950	Poorly defined
11	Blue Lake	1850	Best developed cirque, cut in granite; glacially over-deepened basin
12	Upper Blue Lake	1950	Cut in phyllites, form considerably modified by post-glacial frost weathering
13	Twynam	1950	Site of longest-lasting snowpatch in the area

Mountains and sometimes lasts from one winter to the next. Up to 40 m of snow accumulates here and it becomes sufficiently dense to exert major downslope pressure sufficient to move large blocks and cause extensive surface scratchings (Costin *et al.* 1964, 1973, Jennings and Costin 1978). The same phenomenon but of lesser magnitude has been observed on bare rock surfaces near a semi-permanent snow patch at Mt. Howitt (Peterson 1971). Here the close cover of woody shrubs indicates that there has not been significant movement across the floor of the hollow for at least decades.

The amount of snow required to form a long-lasting snow patch today in the Snowy Mountains can be estimated by an empirical model initially developed to predict possible "greenhouse" effects on Australian ski fields (Galloway, 1988, Whetton *et al.* 1996). Inputs to the model for this exercise are the mean monthly precipitation and temperature and standard deviation of monthly temperature at Spencers Creek climate station (altitude 1770 m) that operated from 1952 to 1967. These data are extrapolated upwards to about 2000 m (at which altitude long-lasting snow patches are common) through reducing monthly temperatures by 1.5°C and increasing monthly precipitation by 20%. The results (Fig. 2) show that for a snow patch to last from the beginning of May to the end of December (245 days) in an average year, the normal snowfall on open sites must be multiplied about four times by drifting. Sites where such multiplication can occur are extremely restricted in area. Thus, for nivation to become more widespread would require a major climate change. Field experience in the mountains suggests that 240–250 days is about the minimum duration for a snow patch to generate appreciable nivation effects.

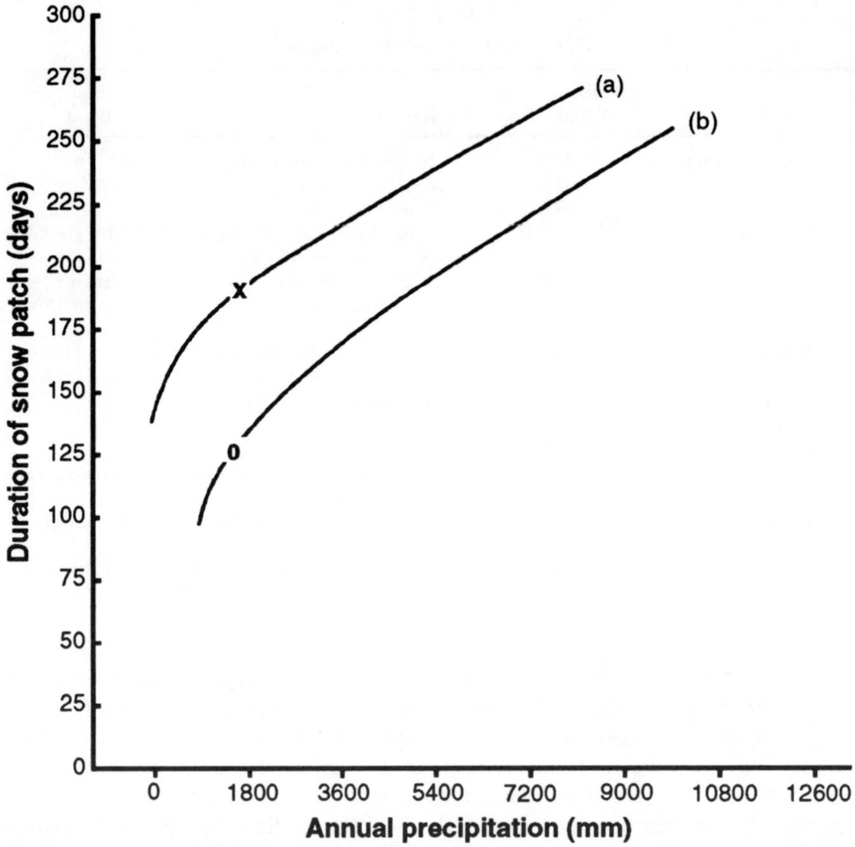

Fig. 2. Relationship between snowfall and duration of a snow patch. (a) For a site at ~2000 m altitude with present mean monthly temperatures. X = approximate present conditions. (b) For a site at ~2000 m with monthly temperatures 1.5°C higher than today. O = approximate present conditions at Spencers Creek climate station, 1770 m.

During the Holocene altithermal a few thousand years ago, mean temperatures were one or two degrees higher than they are now, thus indicating conditions that could be associated with possible "greenhouse" warming over coming decades. When temperatures in the model are adjusted for such warmer conditions, assuming precipitation and standard deviation of monthly temperatures remain the same, the model indicates that precipitation would have to be multiplied by about six times through drifting to form a snowpatch that would last about eight months (Fig. 2). Thus snow patches and nivation, already a minor feature of the landscape, would become even less significant under climatic warming. During the slightly colder Holocene neoglacials, snowpatches presumably lasted rather longer and nivation processes were slightly more effective and widespread.

TASMANIA

Pleistocene glaciation

In Tasmania there is a complex and long glacial record much of which is not well dated. The best evidence for Pleistocene glaciation is from the West Coast Range and Central Plateau, including the valleys to the north and south where glaciation evidence over a period of one million years has been found.

Extensive areas of central and western Tasmania were glaciated during the late Cainozoic. When the glaciers were most extensive in, or before the early Pleistocene, at least 7000 km^2 of ice covered the Tasmanian Central Highlands. Major valley glaciers such as the 70 km long Derwent glacier radiated from the Central Plateau. Many divides were buried beneath ice which accumulated especially in the lee of natural snow fences such as the Du Cane Range. The area between the Central Plateau and the West Coast Range was largely inundated by ice. Smaller glaciers developed in the mountains of southwest Tasmania and on Ben Lomond in the northeast.

Ice cover during the Last Glacial Maximum was less and extended only 10–14 km eastwards from the western rim of the Central Plateau although substantial valley glaciers extended north down the Mersey Valley and south down the Derwent. Diffluence of ice passing from one drainage basin to another was much less common than during more extensive preceding phases. Various ranges in the south and southwest carried small valley and cirque glaciers.

The glacial sediments and landforms indicate that the ice masses were generally of temperate maritime type with high rates of accumulation, rapid mass throughput and ablation in the west but lower activity in the interior. Cirque floors, broadly corresponding to the climatic snowline, rise northeastwards across the Island from 600 m to 1350 m roughly parallel to the modern precipitation gradient (Davies 1967). Palaeosnowline reconstructions suggest temperatures were 6.2°C lower than now at the last glacial maximum and perhaps 7.0°C lower when the glaciers were at their maximum extent (Colhoun 1985).

Glacial landforms were a product of the nature and intensity of glaciation and the rocks on which it operated. The ice-scoured dolerite of the western part of the Central Plateau has thousands of small lakes and ice-abraded knolls but the major pre-glacial relief has not been seriously modified. Cirques are best developed on the eastern and southern slopes of mountains but they are relatively shallow, especially towards the east. The largest cirques formed in the quartzite and schist mountains of the west and southwest. The strong columnar jointing of Jurassic dolerite overlying Carboniferous-Triassic age sedimentary rocks favoured the development of very steep cirque headwalls especially where erosion has cut into the sediments below the dolerite. Glacial troughs are most nearly parabolic in cross-section where they are cut into sedimentary rocks; they are narrower where the valley walls are dolerite and where the long-profile is steep. Some preglacial drainage patterns were deranged by glacial erosion or deposition.

Tasmanian moraines vary in size and form. They tend to be largest in the west and southwest e.g. a multiphase moraine 300 m high that spills west from Lake Margaret and a 200 m high lateral moraine in the Mt. Anne massif. A series of small, steep and narrow moraines lie south of Lake St Clair while hummocky moraine in some tributary valleys implies that the ice decayed *in situ*. Only small moraines formed on the Central Plateau during the last glacial maximum; signs of earlier moraines are indefinite. Proglacial outwash sand and gravel sediments floor many Tasmanian valleys.

Pleistocene nivation

Evidence of former snow effects beyond the glaciers include small nivation cirques, snow patch hollows and protalus ramparts of coarse rock debris that slid or rolled down the surface of semi-permanent snowbanks at the foot of steep slopes. During the cold climatic conditions of glacial periods, snowmelt may well have aided the development of such features as boulder fields, bedded screes and extensive solifluction deposits. On horizontally layered dolerite nivation has produced flat benches. Major valley fills probably owe much to seasonal snow melt waters. Gravel fill in some caves may be of niveofluvial origin.

Present day nivation

There are no glaciers today in Tasmania but nivation continues in sites around long-lasting snow patches notably in some high-altitude cirques and nivation hollows. As long ago as 1925, Lewis recognised active nivation at Mt. La Perouse where a snowbank "moves a few yards down the gentle slope of the ledge. As it passes over the corrugations of the various . . . strata . . . it has scratched the rock with fragments . . . torn from the adjacent cliff face . . . apparently the scratching is repeated each winter."

A circular snow-patch hollow occurs near the summit of Frenchmans Cap on quartzite (Derbyshire and Peterson 1977), with others on Mt. Anne, Mt. Campbell and Cradle Mountain (Colhoun pers. comm.). Other examples of active snowpatch erosion have been deduced from the absence of vegetation on sites where snow accumulates. As on the mainland, present processes seem merely to retouch pre-existing forms. Small nivation pavements have been recorded e.g. on top of the King William range, while on Mt. Rufus small stone-banked terraces are intermittently advancing, probably aided by snowmelt effects. Snowmelt waters affect stream bank and bed deposits to a minor extent.

DISCUSSION

It is evident that effects of snow on the landforms today are miniscule whereas during past glacial periods they were profound. Today snow merely retouches and occasionally freshens landforms created under much colder conditions in the past. No conceivable climate change over the coming decades will seriously amplify the effects of snow but they may well become even more exiguous than they are today. The natural snow cover is already so variable from year to year that additional variation caused by possible climate change will not directly affect the present miniscule landform response.

In fact, the major effects of snow on landforms are indirect through skiing which of course depends on the presence of snow. Even carefully designed ski developments and tracks inevitably involve some effects on snow cover and hence on landforms through slope grooming, installation of tracks and the effects of clearing trees to make ski runs. Some of these developments could locally reduce the insulating snow cover and vegetation thus expanding the area of bare ground subject to freeze/thaw processes and the growth of needle ice. Once needle ice starts to form the area affected tends to increase by a feed-back effect leading to soil erosion.

On the other hand, snow "farming", which locally increases the snow pack by mechanically moving snow to points where the natural cover is inadequate for skiing, is most unlikely to involve quantities sufficient to cause nivation where it did not previously occur. Likewise artificial snow making will never be sufficient to turn ordinary snow covers into the much thicker pack required to initiate nivation.

REFERENCES

Caine, N., 1983. "The Mountains of Northeastern Tasmania." Balkema, Rotterdam.

Colhoun, E. A., 1985. Glaciations of the West Coast Range, Tasmania. *Quat. Res.* **24:** 39–59.

Costin, A. B., Jennings, J. N., Black, H. P. and Thom, B. A., 1964. Snow action on Mount Twynam, Snowy Mountains, Australia. *J. Glaciology* **5:** 219–228.

Costin, A. B., Jennings, J. N., Bautovich, B. C. and Wimbush, D. J., 1973. Forces developed by snowpatch action, Mt. Twynam, Snowy Mountains, Australia. *Arct. Alp. Res.* **5:** 121–26.

David, T. W. E., Helms, R. and Pittman, E. F., 1901. Geological notes on Kosciusko with special reference to evidences of glacial action. *Proc. Linn. Soc. NSW* **26:** 26–74.

Davies, J. L., 1967. Tasmanian landforms and Quaternary climates. Pp. 1–25 *in* "Landform Studies from Australia and New Guinea" ed by J. N. Jennings and J. A. Mabbutt. Australian National University Press, Canberra.

Derbyshire, E., 1973. Periglacial phenomena in Tasmania. *Biuletyn Peryglacjalny* **22:** 131–148.

Derbyshire, E., 1979. Cryonival and Glacial Processes. Pp. 180–198 *in* "Geomorphological Processes" ed by E. Derbyshire, K. J. Gregory and J. R. Hails. Dawson.

Derbyshire, E. and Peterson, J. A., 1977. Australian landform example No. 31 Nivation cirque. *Aust. Geographer* **13:** 416–419.

Evans, I. S., 1969. The geomorphology and morphometry of glacial and nival areas. Pp. 369–80 *in* "Water, Earth and Man" ed by R. J. Chorley.

Foehn, P. M. B., 1992. Climatic change, snow cover and avalanches. Pp. 11–21 *in* "Greenhouse — Impact on Cold-climate Ecosystems and Landscapes" ed by M. Boer and E. A. Koster. Catena Supplement 22. Cremlingen-Destedt, Germany.

Galloway, R. W., 1963. Glaciation in the Snowy Mountains: a reappraisal. *Proc. Linn. Soc. N.S.W.* **88:** 180–198.

Galloway, R. W., 1988. The potential impact of climate changes on Australian ski fields. Pp. 428–37 *in* "Greenhouse: Planning for Climatic Change" ed by G. I. Pearman. CSIRO/Brill, Melbourne.

Grove, J. M., 1988. "The Little Ice Ages". Methuen, London.

Jennings, J. N. and Costin, A. B., 1978. Stone movement through snow creep, 1963–1975, Mount Twynam, Snowy Mountains, Australia. *Earth Surface Processes* **3:** 3–22.

Lewis, A. N., 1925. Notes on a geological reconnaissance of the Mt. La Perouse Range. *Pap. Proc. R. Soc. Tasm.* **1924:** 10–44.

McRae-Williams, M. S., Rosengren, N. S. and Kraemers, S. M., 1981. "Sites of geological and geomorphological significance in East Gippsland, Victoria". Ministry for Conservation, Melbourne.

Peterson, J. A., 1971. The equivocal extent of glaciation in the southeastern uplands of Australia. *Proc. Roy. Soc. Vic.* **84:** 207–212.

Ritchie, A. and Jennings, J. N., 1955. Pleistocene glaciation in the Grey Mare Range. *J. Proc. Roy. Soc. N.S.W.* **89:** 127–130.

Rosengren, N. J. and Peterson, J. A., 1989. Heritage values and the geological and geomorphological significance of the Australian Alpine Zone. Pp. 187–204 *in* "The Scientific Significance of the Australian Alps" ed by R. Good. Australian Alps National Parks Liaison Committee.

Whetton, P. H., Haylock, M. R. and Galloway, R. W., 1996. Climate change and snow-cover duration in the Australian Alps. *Climatic Change* **32:** 447–79.

White, S. E., 1972. Alpine subnival boulder pavements in Colorado Front Range. *Geol. Soc. America Bull.* **83:** 195–200.

PART 2

Life in the snow

CHAPTER 5

Life in the snow: algae and other microorganisms

HARVEY J. MARCHANT

INTRODUCTION

SEASONAL or persistent snow is a feature of polar and alpine areas of the Earth. In spring and summer, the snow can have brightly coloured, usually red, patches. Such coloured snow has been reported from every continent, with the exception of Africa. However, the lack of reports of coloured snow from African alpine regions is more likely to reflect a lack of investigators than the absence of coloured snow. As well as being red, the snow can be orange, green, grey and even black. This colouring of the snow is caused by concentrations of microorganisms, mostly algae - single celled plants. In addition to the green pigment, chlorophyll a, which all plant cells possess, different algae have other pigments of different colours, which give the characteristic colours to the snow.

The Australian Alps are no exception. The red and orange blotches of colour on the snowbanks used to be dismissed as dust blown from inland areas. On occasions dust storms do deposit sufficient material on the snow to colour it (Walker and Costin 1971), but as has been found in other parts of the world, the colours seen in the snowbanks in the Australian high country are principally due to concentrations of microorganisms, mostly algae (Plate 1 and Fig. 1) (Marchant 1982).

In addition to algae, other microorganisms are present. Bacteria and fungi almost invariably accompany the algae as well as single celled animals — protozoa, and metazoa including nematodes, rotifers and insects. These organisms live in extreme conditions of temperature, solar irradiance and desiccation (both when the snow has melted or the temperature is below freezing), as well as high acidity and low nutrients. Some of these organisms are of interest to NASA as one of four life systems on Earth as analogs of possible life in the past or at present on the planet Mars.

Here I give an overview of snow algae and other organisms that inhabit snowfields and their biochemical and physiological strategies for living in snow before considering how global change can influence them and their environment.

MICROORGANISMS THAT LIVE IN THE SNOW

Essentially wherever there is snow on Earth, from the equatorial glaciers to the Antarctic ice cap, it contains microorganisms. That algae are the principal component of coloured snow was reported in the scientific literature early in the 19th century. Since then, there have been monographs and papers published dealing principally with the global distribution, taxonomy and life cycles of the organisms, especially the algae. The reviews by Kol (1968a) and Hoham (1980) provide comprehensive accounts of the biology of organisms living in snow, especially snow algae. The maximum concentrations of snow algae are typically in the range of 50 000 to 1 000 000 cells per ml of snow meltwater in North American snow (Hoham 1992, Hoham *et al.* 1993).

The red snow in alpine and polar regions appears to be due to the presence of one or more of seven species, *Chlamydomonas nivalis*, *Chlamydomonas sanguinea*, *Chloromonas bolyaiana*, *Chlainomonas rubra*, *Chlainomonas kolii* and *Smithsonimonas abbotii* and *Chloromonas rubroleosa*. In the Snowy Mountains *Chlamydomonas nivalis* and *Chloromonas bolyaiana* (formerly *Chlamydomonas bolyaiana*) are the organisms principally responsible for giving the snow a red colour (Plate 2). I found that *Chloromonas brevispina* colours the snow yellow-orange. The green algae *Koliella*, *Raphidonema* and *Chodatia* also occurred in the snow but not in sufficiently high concentration to colour it (Marchant 1982).

The taxonomy of snow algae is based primarily on their morphology ascertained by light and scanning electron microscopy. However, studies of these organisms in culture indicate that they can have several stages in their life cycle. Broady (1996) discusses the problems arising where only a small proportion of the algal flora can be reliably identified by microscopical examination of field-collected material. Several snow algae, especially those that colour the snow, have complex life cycles including motile and resistant resting stages and involving both sexual and asexual reproduction (Fig. 2). Much of the confusion in the taxonomy of these organisms stems from identifying the various life cycle stages. For example, (Hoham *et al.* 1979) reported that seven previously described snow algae were in fact the zygotes of one alga, *Chloromonas brevispina*. Another misidentification was that of *Chlamydomonas nivalis*, also known as *Protoccocus nivalis*, collected near the Australian Antarctic station of Casey which subsequent cultural studies by Ling and Seppelt (1993) found to be a new species, *Chloromonas rubroleosa*. While some of the taxonomic confusion has been resolved by painstaking studies of the organisms in culture and mating studies, it is likely that molecular techniques will be required to clarify the biodiversity of the snow algal flora. Molecular approaches are being increasingly used to resolve systematic and phylogenetic questions. Pertinent to the taxonomy of snow algae is the investigation by Buchheim *et al.* (1990) on *Chlamydomonas* where it was found that the present genus is not monophyletic but includes organisms that may derive from different families or even orders.

late 1. Snow algae colouring a snow bank in the Snowy
[ountains.

late 2. Red and green cells of *Chlamydomonas nivalis* surrounded
bacteria from the Snowy Mountains.

late 3. The snow fungus *Chionaster nivalis* from the Snowy
ountains.

age 4. Red snow algae and windblown debris on the surface of
e snow in the Snowy Mountains.

age 5. A snowbank in the Snowy Mountains with well developed
epressions "suncups" caused by, and leading to further
ncentration of the snow algae.

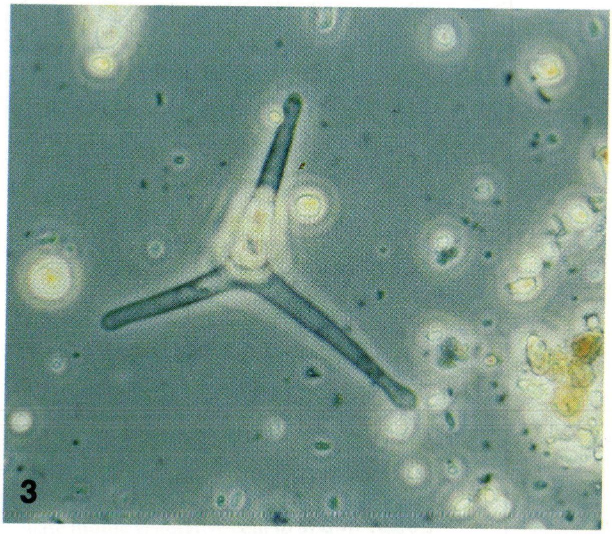

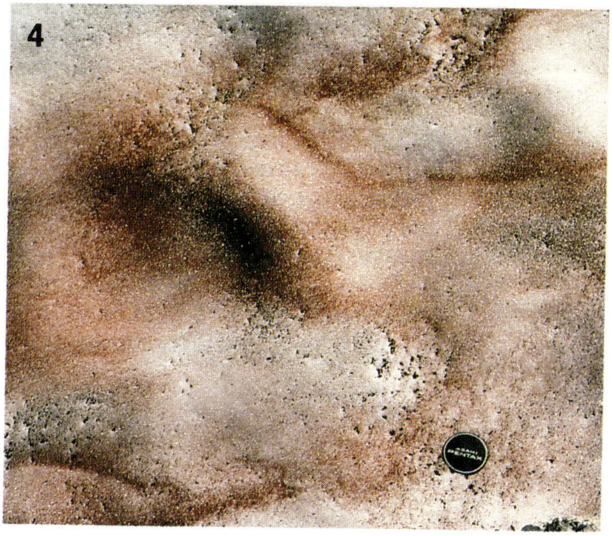

and nine taxa in snow on the South Orkney Islands (Fogg 1967). However, 24 species were found by Ling (1996) in the Windmill Islands region, near the Australian Antarctic station of Casey. Considering the major taxonomic uncertainties with snow algae discussed above, it is clearly premature to speculate on the biogeography of these organisms.

The life cycles of the algae are closely tied to the physical and chemical properties of the snow. Hoham (1980) reviewed the environmental conditions that trigger the development of the various stages in the life cycles (Fig. 3). Germination of zygospores at the soil-snow interface (or the old snow-new snow interface in persistent snow fields) requires light and follows several consecutive days of temperatures above freezing which produce meltwater carrying nutrients and dissolved gasses. The zoospores released from the zygospores swim upwards through the snowbank in the meltwater. They accumulate at a depth in the snowbank determined by light intensity and wavelength where they increase in number by asexual reproduction (Grinde 1983). Nutrient depletion triggers the development of motile gametes and their fusion. Zygospores resulting from this sexual reproduction contain carotenoids giving rise to the coloured snow. Following the melting of the snow, these zygospores accumulate in the soil, or in persistent snowbanks and are covered with the following winter's snow. In the soil or old snow the zygospores apparently remain dormant. Germination of the zygospores and sexual reproduction and growth of the algae probably occurs within the snow banks early in the season. The formation of new zygotes takes place over a period of only a few days late in the season. As a consequence, the process has been observed only a few times in the wild and explains why so many zygotes have been described as new species of Chlorococcales.

In addition to the algae that inhabit the snow, there are also bacteria, fungi, lichen fragments, protozoa and micrometazoa. None of these groups have received the attention that has been given to the algae. Bacteria are usually found associated with snow algae. Little is known about their species composition, and investigations on their temperature responses are confined to a few studies (Thomas and Duval 1995). The concentrations of bacteria and snow algae were found to be directly correlated in studies undertaken on Tioga Pass, California (Thomas 1994). Weiss (1983) reported bacteria within a fibrillar mesh surrounding red cells of *Chlamydomonas nivalis*. Other cases of bacteria and fungi being associated with mucilage surrounding the resting spores of snow algae have been reported (e.g. Marchant 1982, Hoham *et al.* 1993). Several fungi inhabit snow; Kol (1968a) listed more than 70 different taxa including flagellates, yeasts and filamentous forms. The most commonly reported fungi are *Chionaster nivalis*, *C. bicornis* and *Selenotila nivalis* (Plage 3). Bacteria and fungi require an external source of carbon for their nutrition and it is most likely that snow algae provide that source as they excrete >10% of their photosynthetically fixed carbon (Fogg 1967). However, it has not been ascertained whether there is a symbiotic relationship between the organisms with the algae being beneficiaries of some metabolite of the bacteria and fungi. Numerous protozoa and micrometazoa also have been reported to be inhabitants of the snow. Among the protozoa are amoebae, ciliates, and

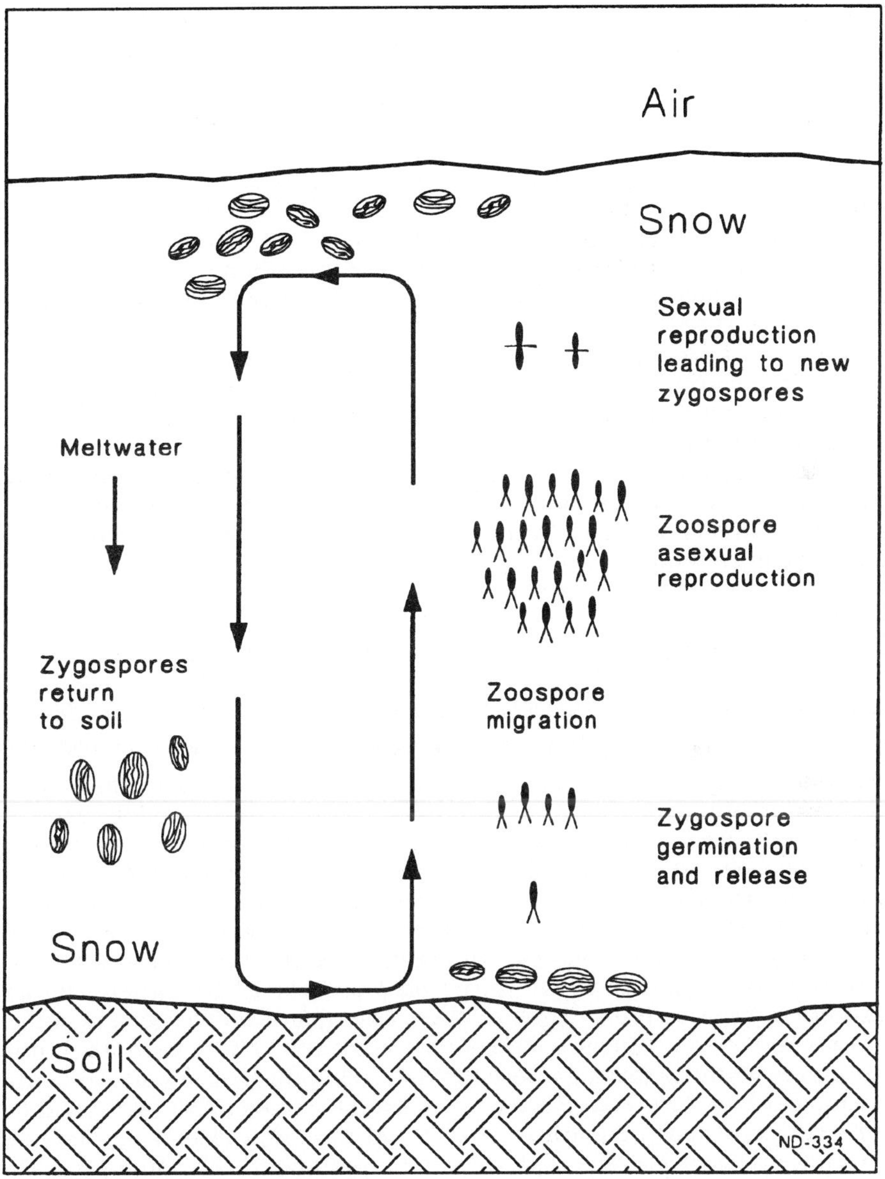

Fig. 3. Seasonal life cycle of snow algae. In the case of persistent snow, the snow-soil interface is the new snow-old snow interface.

flagellates including a colourless euglenoid (Hoham *et al.* 1993). The micrometazoa include rotifers, tardigrades, nematodes, insects and arachnids (Kohshima 1984).

These organisms interact as a food web. Quite complex interactions have been reported from eastern USA (Hoham *et al.* 1993). Snow algae are grazed by protozoa as well as the micrometazoa which would also consume the protozoa. Pollock (1970) reported that the protozoa preferentially graze green cells rather than those coloured red or orange. He speculated that this was probably because the green cells are generally zoospores which lack a cell wall, whereas the orange and red cells are thick walled. While this is a plausible explanation, I suspect that the differences in the size of cells and the extent to which the differently coloured cells secrete organic material could also determine whether they are grazed or not. Grazing of resting cells may also be deterred because of protuberances on the cell wall or mucilage accumulation around individual cells or clusters of cells (Hoham pers. comm). Bacteria and fungi utilize the dissolved organic material that is secreted by all the organisms, particularly the algae. The bacteria and fungi are in turn grazed by the protozoa and some of the micrometazoa. Spiders were found to feed on the micrometazoa, especially the Collembola. In North America, shrews and other small mammals feed on the spiders. The extent to which snow algae in the Australian high country is utilized by other organisms remains unknown.

STRATEGIES FOR LIVING IN THE SNOW

Organisms which live in the snow must be tolerant of low temperature, desiccation, high solar irradiance, large fluctuations in nutrient concentrations and high acidity.

Several studies have investigated the temperature tolerances of snow algae and optimum temperatures for their growth (Hoham 1975, Hoham 1989). Most of the organisms investigated, including *Chloromonas* spp. and *Chlamydomonas nivalis* have growth optima of around 5°C and did not survive at temperatures >10°C. As freshwater organisms, snow algae exist at the lower extreme of temperature. In contrast, polar marine phytoplankton are active at –2°C, the temperature at which seawater freezes, and *Dunaliella* sp. and *Pyramimonas gelidicola* remain motile at temperatures as low as –14°C in hypersaline Antarctic lakes (Burch and Marchant 1983). Differences in fatty acid composition between red and green cells of *Chlamydomonas* has been hypothesized to have a cryoprotectant function (Bidigare *et al.* 1993). These authors found that when samples containing this alga were thawed, following freezing at –70°C, the red cells were intact while the green cells were extensively damaged. The fatty acid pool in the green cells consisted predominantly (72%) of saturated fatty acids. In contrast, 80% of the fatty acids of the red cells were mono-unsaturated. The high proportion of unsaturated fatty acids in cells results in a decrease in the temperature of the fatty acids' melting point and is therefore likely to have a cryoprotective function by making membranes and other cellular components which contain these fatty acids fluid at low temperatures.

As convincingly argued by Kennedy (1993) with regard to Antarctic terrestrial environments, it is the lack of available water rather than low temperature that is the prime determinant of the distribution and abundance of organisms. Organisms that inhabit snow, whether it be in polar or alpine environments also require free water. At low temperatures when the snow is "dry" there is no free water and the organisms, especially those that swim, are inactive. At temperatures above zero the snow crystals are surrounded by free water and it is in this water film that the snow algae and associated organisms are able to move. In summer, when the snow melts and the soil on which the zygospores rest dries out, these organisms are again forced into inactivity. In addition to the role of moisture in determining the distribution and abundance of organisms, the form of intracellular water also determines the activity of organisms. Eukaryotic organisms commonly use glycerol, mannitol or other "compatible solutes" in osmoregulation and in modifying the activity of intracellular water. This influences the structure of macromolecules such as structural proteins and enzymes (Kirst 1990). Antifreeze glycoproteins have been found in several organisms which survive freezing. It is thought that these molecules inhibit the formation of ice crystals or depress the freezing point of intracellular water or partially substitute for intracellular water (Burch and Marchant 1983).

Red pigmented snow algae are usually found in areas of high solar exposure and relatively low nutrient availability while green snow is more commonly found in shaded, more nutrient replete areas. Bidigare *et al.* (1993) reported that the red pigmentation of the snow alga *Chlamydomonas nivalis* is principally due to secondary carotenoids, principally astaxanthin esters, produced in response to the availability of nitrogenous compounds. Twelve pigments were found, in addition to chlorophylls a and b, in a preliminary investigation of the pigments of snow algae of the Snowy Mountains (S. Jeffrey pers. comm.). As I demonstrated in electron micrographs of *Chlamydomonas nivalis* from the Snowy Mountains, these pigments are located in the periphery of the cells, surrounding the chloroplasts (Marchant 1982). The carotenoids have a photoprotective function by reducing the amount of light available to the light harvesting pigments in the chloroplasts.

Organisms living in snow in alpine and polar regions are exposed to high levels of incident ultraviolet light. The reduction in concentration of stratospheric ozone by around 5% per decade in temperate regions has led to an increase in solar UV-B (280–320 nm) irradiance reaching the surface of the Earth. Numerous investigations have demonstrated that growth and other activities of organisms are inhibited by UV-B (Tevini 1993). Thomas and Duval (1995) reported that ambient solar irradiance reduced photosynthetic primary production of red-coloured snow algae (*Chlamydomonas nivalis* spores) by 25% of the rate of cells screened from UV-B, while photosynthesis was inhibited by 85% in green cells (principally *Chloromonas*) exposed to solar UV-B. In addition to the substantially higher concentration of carotenoid pigments in the red cells compared with those that were green, the red cells also had much higher concentrations of compounds that absorb UV wavelengths (Fig. 4). Thomas and Duval (1995) also found that the colour of cells exposed to solar UV for one hour was bleached.

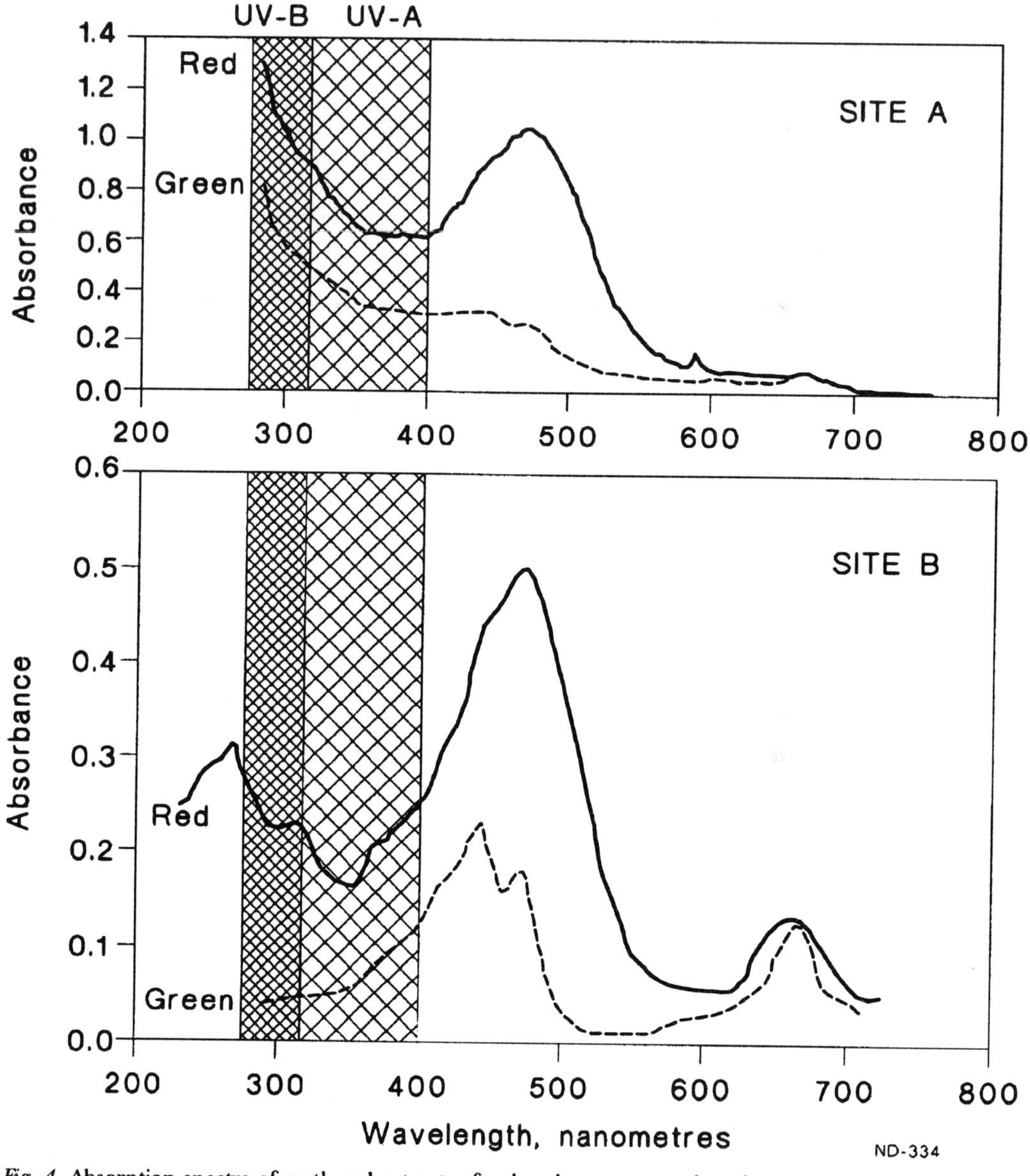

Fig. 4. Absorption spectra of methanol extracts of red and green snow algae from two sites on Tioga Pass, California. The absorption curves for the red and green algae at both sites were normalized to the chlorophyll a peak at 660 nm. The main peak around 450 nm is due to the presence of carotenoid pigments. Note the substantial absorbance in the UV-B by the red algae at both sites and green cells at site A. (redrawn from Thomas and Duval 1995).

There was less bleaching of cells in controls from which UV was screened. However, they did not report whether UV exposure affected the concentration of the UV absorbing compounds. Some organisms have been demonstrated to produce more UV absorbing compounds when exposed to UV irradiation (e.g. Davidson and Marchant 1994). Highest concentrations of red snow algae are often found below the surface of the snow. Hader and Hader (1989) reported that UV exposure impaired the motility of green zoospores of *Chlamydomonas nivalis*. Clearly, the responses of snow algae to UV exposure need to be explored further. One obvious question is whether UV exposure plays a role in triggering the development of resting stages.

Research conducted principally in the USA indicates that many taxa of snow algae have a preference for acidic conditions. For example, strains of *Chloromonas* were found to exhibit optimum growth at pH 4.0 to 5.0 (Hoham and Mohn 1985). Growth of *Chloromonas* at pH 3.5 was double that at pH 7.0 and it has been suggested that the increased acidic precipitation in eastern USA may be leading to the selection of such strains (Hoham 1989).

Snow microbes, because of their ability to live under extreme conditions of low temperature, high acidity, high irradiance, highly variable nutrient conditions and desiccation, are under investigation as one of the four life systems on Earth that could possibly have existed in the past or presently exists on the planet Mars (Rothschild 1990). The other life systems of interest to NASA are endoevaporites (organisms living within salt crystals), endoliths (organisms living within translucent rocks) and chemoautotrophs (organisms in volcanoes, deep sea vents and other habitats which utilize energy sources other than carbon). It is thought that organisms from the polar ice caps of Mars may possibly have analogs in prokaryotic snow dwelling organisms, bacteria and cyanobacteria (blue-green algae) which live in similar conditions on Earth.

ENVIRONMENTAL INFLUENCES ON AND BY SNOW ALGAE

Organisms that live in the snow are obviously influenced by localized as well as global changes in their environment.

The very reason that the algae that colour the mountain snow red or orange or yellow is due to the production of carotenoid pigments at the onset of dormancy in their life cycles. The production of these pigments is stimulated by high light as protection against photoinhibition of photosynthesis at high irradiance (Bidigare *et al.* 1993). However, these authors found that in areas of high nutrient availability, adjacent to bird rookeries, under high light the algae did not produce the carotenoids which colour the snow. Thus, production of these pigments is not simply a response to high light but is also influenced by their nutrient status. Snow algae act as dark bodies absorbing more heat than the surrounding snow leading to an acceleration in the melting of the snow and an increase in the availability of meltwater. Melting snow can cause depressions "sun-cups" in the surface of the snow which can act as traps for windblown debris including insects and other organic and inorganic matter (Plates 4 and 5). Elevated levels of nutrients further promote the growth of microorganisms and the metazoa that feed on them.

The presence of organisms in the snow directly affect its albedo. Studies in Antarctica, Greenland, the Himalayas and on the equatorial glaciers of Irian Jaya have demonstrated that organisms promoted snow melting resulting in surface irregularities. The colonies of black snow algae found on equatorial glaciers were 1°C to 7°C warmer than their surroundings leading to the development of holes and channels (Kol and Peterson 1976). These authors suggest that the snow algae determine the surface morphology of the glaciers and may be causing melting rates several times greater than if a clean surface was present. In the Snowy Mountains, melting of the snow associated with patches of snow algae contribute to the fine scale (cm–m) morphology of the surface of the snow with the formation of depressions "suncups", and other irregularities (Plate 5).

In addition to snow algae, the snow of the Australian Alps is occasionally discoloured by dust blown from inland regions, ash from bushfires, insects and wind-blown plant material, particularly pollen. The importance of this windblown material compared to snow algae in affecting the albedo of the snow banks has not been quantified. Our estimates, based on analysis of photographs of snow banks, suggest that windblown debris and snow algae have similar effects on the albedo except when the snow algae are concentrated (e.g. Plage 4). Similar results have been reported by Thomas and Duval (1995) in the Tioga Pass area of the Sierra Nevada, California. They found that although there was a highly significant negative correlation between snow albedo and the concentration of algal cells on a small scale, snow algae did not decrease the mean albedos of snow fields because of algal patchiness. They also found that dirt on the snow contributed to decreased albedo. Thus while snow algae can have fine scale effects it apparently does not affect the hydrological relationships between snow pack and water supply in the watershed.

The high concentrations of organisms found in the snowbanks accumulate by both active and passive processes. As previously discussed, most of the snow algae have motile stages in their life cycle and swim in the melt water component of the snow. In addition, as snow banks melt, water percolates through them carrying organisms in addition to debris. Thus, a variety of material, including living organisms can accumulate in the lower parts of snow banks. Organic and inorganic materials provide substrates for some of the living organisms. Yoshimura *et al.* (1997) suggest that the biomass and species diversity of snow algae from annual accumulation zones in glaciers could provide a detailed record of regional past climatic change.

Considering how prominent snow algae can be in alpine and polar regions, they have received little attention. Their taxonomy, life histories and geographical distribution have been investigated but as discussed earlier, their taxonomy is continually being reassessed and there are only few studies on their biochemistry and physiology. I am not aware of any work having been done on them in the Australian high country since my report of their presence in the Snowy Mountains some 15 years ago (Marchant 1982). It is becoming clear that our understanding of snow physical and chemical processes is not complete without knowledge of the microbial activity.

Global warming is possibly leading to an increase in elevation of the snow-line. It has been predicted that the area receiving regular snowfalls in Australia could be reduced to a small area around Mt. Kosciuszko within 100 years (Whetton 1996 and Chapter 11). With no snow, or a markedly reduced snow pack, obviously there will be no snow algae or associated organisms. Some of these organisms, especially the heterotrophs, will continue to live in the alpine soils. It is expected that a decrease in the extent and duration of our snow fields will lead to a loss of biodiversity of microorganisms that depend on the snow for their existence or expression of some phase of their life cycle. At this stage we cannot say which of these organisms are likely to be lost. Considering the relatively small part of the snowfields in which snow algae reach a concentration that is visible, it is unlikely that the loss of snow algae and associated microorganisms as a result of the predicted loss of snow will have discernible ecological consequences.

However, microorganisms, especially those from extreme environments, are coming under increased scrutiny as interest accelerates in tapping their biotechnological potential (Roessler 1990, Margesin and Schinner 1994). The commercial prospects of microorganisms should not be underestimated. These organisms, or genes from them, are presently being utilized as a source of antibiotics and other drugs used in human and veterinary pharmaceuticals, as additives for human and stock food, and a host of other applications ranging from sunscreening compounds to industrial cleaning compounds. The increasing consumer demand for natural products will further drive this search. Organisms inhabiting snow are living in an extreme environment and it is quite probable that they will yield compounds or genes of commercial value. It would be a tragedy if this was recognized in Australia after snow algae were utilized overseas or after the snow had vanished from a warmed Australia.

REFERENCES

Bidigare, R. R., Ondrusek, M. E., Kennicutt II, H. C., Iturriaga, R., Harvey, H. R., Hoham, R. W. and Macko, S. A., 1993. Evidence for a photoprotective function for secondary carotenoids of snow algae. *J. Phycol.* **29:** 427–434.

Broady, P. A., 1996. Diversity, distribution and dispersal of Antarctic terrestrial algae. *Biodiversity and Conservation.* **5:** 1307–1335.

Buchheim, M. A., Turmel, M., Zimmer, E. A. and Chapman, R. L., 1990. Phylogeny of *Chlamydomonas* (Chlorophyta) based on cladistic analysis of nuclear 18s rRNA sequence data. *J. Phycol.* **26:** 689–699.

Burch, M. and Marchant, H. J., 1983. Low temperature motility and microtubule stability in Antarctic algae. *Protoplasma* **115:** 240–242.

Davidson, A. T. and Marchant, H. J., 1994. The impact of UV-B radiation on *Phaeocystis* and selected species of Antarctic marine diatoms. *in* "Ultraviolet Radiation and Biological Research in Antarctica" ed by C. S. Weiler and P. A. Penhale. *Antarctic Research Series, American Geophysical Union* **62:** 187–205.

Fogg, G. E., 1967. Observations on the snow algae of the South Orkney Islands. *Phil. Trans. Roy. Soc. Lond., Ser. B* **40:** 293–338.

Grinde, B., 1983. Vertical distribution of the snow alga *Chlamydomonas nivalis* (Chlorophyta, Volvocales). *Polar Biol.* **2:** 159–162.

Hader, D.-P. and Hader, H. A., 1989. Effects of solar UV-B irradiation on photomovement and motility in photosynthetic and colorless flagellates. *Environment. Experiment. Bot.* **29:** 273–282.

Hoham, R. W., 1975. Optimum temperatures and temperature ranges for growth of snow algae. *Arct. Alp. Res.* **7:** 13–24.

Hoham, R. W., 1980. Unicellular chlorophytes — snow algae. Pp. 61–84 *in* "Phytoflagellates" ed by E. R. Cox. Elsevier North Holland, Inc., New York.

Hoham, R. W., 1989. Snow microorganisms and their interaction with the environment. Pp. 31–35 *in* "Proceedings of the Fifty-seventh Annual Western Snow Conference" ed by B. Shafer. Ft. Collins, Colorado.

Hoham, R. W., 1992. Environmental influences on snow algal microbes. Pp. 78–83 *in* "Proceedings of the Sixtieth Annual Western Snow Conference" ed by B. Shafer. Jackson Hole, Wyoming.

Hoham, R. W. and Blinn, D. W., 1979. Distribution of cryophilic algae in an arid region, the American Southwest. *Phycologia* **14:** 133–145.

Hoham, R. W., Laursen, A. E., Clive, S. O. and Duval, B., 1993. Snow algae and other microbes in several alpine areas in New England. Pp. 165–173 *in* "Proceedings of the Fiftieth Annual Eastern Snow Conference" ed by H. Ferrick and T. Pangburn. Québec City, Québec, Canada.

Hoham, R. W. and Mohn, W. W., 1985. The optimum pH of four strains of acidophilic snow algae in the genus *Chloromonas* (Chlorophyta) and possible effects of acid precipitation. *J. Phycol.* **21:** 603–609.

Hoham, R. W., Mullet, J. E. and Roemer, S. C., 1983. The life history and ecology of the snow alga *Chloromonas polyptera* comb. nov. (Chlorophyta, Volvocales). *Can. J. Bot.* **61:** 2416–2429.

Hoham, R. W., Roemer, S. C. and Mullet, J. E., 1979. The life history and ecology of the snow alga *Chloromonas brevispina* comb. nov. (Chlorophyta, Volvocales). *Phycologia* **18:** 55–70.

Kennedy, A. D., 1993. Water as a limiting factor in the Antarctic terrestrial environment: A biogeographical synthesis. *Arct. Alp. Res.* **25:** 308–315.

Kirst, G. O., 1990. Salinity tolerance of eukaryotic marine algae. *Ann. Rev. Plant Physiol. Plant Mol. Biol.* **41:** 21–53.

Kohshima, S., 1984. A novel cold-tolerant insect found in a Himalayan glacier. *Nature* **310:** 225–227.

Kol, E., 1968a. Kryobiologie. Biologie und Limnologie des Schnees und Eises I. Kryovegetation. *in* "Die Binnengewasser, Vol. 24" ed by H.-J. Elster and W. Ohle. E. Schweizerbart'sche Verlagsbuchhandlung, Stuttgart, 216 pp.

Kol, E., 1968b. A note on red snow from New Zealand. *N. Z. J. Bot.* **6:** 243–244.

Kol, E. and Peterson, J. A., 1976. Cryobiology. Pp. 81–91 *in* "The Equatorial Glaciers of New Guinea" ed by G. S. Hope, J. A. Peterson, I. Allison and U. Radok. Balkema, Rotterdam.

Ling, H. U. and Seppelt, R. D., 1993. Snow algae of the Windmill Islands, continental Antarctica. 2. *Chloromonas rubroleosa* sp. nov. (Volvocales, Chlorophyta), an alga of red snow. *Eur. J. Phycol.* **28:** 77–84.

Ling, H. U., 1996. Snow algae of the Windmill Islands region, Antarctica. *Hydrobiologia* **336:** 99–106.

Marchant, H. J. 1982. Snow algae from the Australian Snowy Mountains. *Phycologia* **21:** 178–184.

Margesin, R. and Schinner, F., 1994. Properties of cold-adapted microorganisms and their potential role in biotechnology. *J. Biotech.* **33:** 1–14.

Pollock, R., 1970. What colors the mountain snow? *Sierra Club Bull.* **55:** 18–20.

Roessler, P. C., 1990. Environmental control of glycerolipid metabolism in microalgae: commercial implications and future research directions. *J. Phycol.* **26:** 393–399.

Rothschild, L. J., 1990. Earth analogs for Martian life. Microbes in evaporites, a new model system for life on Mars. *Icarus* **88:** 246–260.

Stein, J. R. and Brooke, R. C., 1964. Red snow from Mt. Seymour, British Columbia. *Can. J. Bot.* **42:** 1183–1188.

Tevini, M., 1993. "UV-Bradiation and ozone depletion: effects on humans, animals, plants, micro-organisms and materials". Lewis Publishers, Boca Raton.

Thomas, W. H., 1994. Tioga Pass revisited: Interrelationships between snow algae and bacteria. Pp. 56–62 *in* "Proceedings of the Sixty-second Annual Western Snow Conference" ed by C. Troendle. Santa Fe, New Mexico.

Thomas, W. H. and Duval, B., 1995. Sierra Nevada, California, USA, snow algae: snow albedo changes, algal-bacterial relationships and ultraviolet effects. *Arct. Alp. Res.* **27:** 389–399.

Walker, P. H. and Costin, A. B., 1971. Atmospheric dust accession in South-eastern Australia. *Aust. J. Soil Res.* **9:** 1–5.

Weiss, R. L., 1983. Fine structure of the snow alga (*Chlamydomonas nivalis*) and associated bacteria. *J. Phycol.* **19:** 200–204.

Whetton, P. H., Haylock, M. R. and Galloway, R., 1996. Climate change and snow-cover duration in the Australian Alps. *Climatic Change* **32:** 447–479.

Yoshimura, Y., Kohshima, S. and Ohtani, S., 1997. A community of snow algae on a Himalayan glacier: changes of algal biomass and community structure with altitude. *Arct. Alp. Res.* **29:** 126–137.

CHAPTER 6

Changing snow regimes and the distribution of alpine vegetation

ROGER GOOD

INTRODUCTION

SNOW contributes in excess of 60 percent of the total precipitation of approximately 2800mm per annum in the alpine zone, but snowfalls are irregular and the bulk of the snow occurs in three or four large falls each winter, often separated by weeks of relatively mild winter conditions (Snowy Mountains Authority 1970, Brown and Milner 1989). The snowpack may accumulate to a depth in excess of two metres and, in the leeward side of slopes, snowdrifts accumulate to very much greater depth. Zwinger and Willard (1972) regarded snow as "the most important factor controlling plant communities in the alpine region". Persistent snow cover, for example, favours herbaceous plants even in rocky areas (Kirkpatrick 1989). Beneath snow, plants are protected from the harsh cold and desiccating winds of winter, but the persistence of the snowpack reduces the time necessary for a plant to grow, flower and set seed. In such locations plants with vegetative means of reproduction may be favoured (Zwinger and Willard 1972). In some areas snowpack duration (as snowdrifts) may be too long for any plant growth and in these situations, boulders normally covered by lichens will be bare. Snow also restricts plant growth in certain areas of deep accumulation where the weight of the snow compresses the plants beneath it (Zwinger and Willard 1972). Additionally, summer soil temperatures are kept lower below melting snowdrifts and the soil may be waterlogged and subject to creep or needle ice formation, all conditions suboptimal for plant growth (Zwinger and Willard 1972). In the Australian Alps plant species and communities have also had to adapt to the erratic pattern of snowfalls and differing snowpack depths and duration. On average, snow cover lasts for approximately 130 days a year but varies widely between a minimum duration of about 80 days and a maximum of about 160 days (Snowy Mountains Authority 1970). At the highest elevation (Mt. Kosciuszko) the snow duration is about 180 days while at the lowest elevation of the alpine zone at 1850 metres the duration is in the order of

100 days (Slatyer *et al.* 1985). The variability in the snow regime has long been a feature that the alpine plants have adapted to and which has provided a capacity within the alpine flora to adapt to further changes in the alpine climates.

The plant species which now constitute the alpine vegetation have invaded, adapted and evolved within this climatically harsh environment such that a characteristic alpine vegetation now exists. The vegetation is diverse in origin, species number and structural forms, and contains a high number of endemic species. The structural forms of the various plant groups and communities, together with the edaphic features in which they exist, will determine to a large degree the influence snow will have on defining the occurrence and distribution of vegetation, as much as the direct influence of snow itself.

The existing alpine vegetation is composed of species from two main centres of origin, one being adjacent or nearby lower elevation vegetation communities. Many of these autochthonous species evolved in cold climates and migrated upwards to the current alpine zone, as a direct response to the altitudinal rise of the climatic conditions to which they had adapted over thousands of years. Billings (1979) noted that climatic fluctuations are a common feature of small and isolated mountain ranges such as the Australian Alps and this encourages enrichment of alpine communities by speciation and upward migration of biota from the lower slopes and adjacent lowlands (subalpine and montane zones). Approximately 60 percent of the alpine flora is of this group but many species have also adapted to the changes to warmer and drier weather conditions at the lower elevations and so the group is now represented in a very wide range of climatic conditions from coastal environments to the alpine zone. Species in the genera *Scleranthus, Celmisia, Aciphylla, Gingidia, Euphrasia*, and *Ranunculus* and the grass genera of *Poa, Danthonia (Rytidosperma), Erythanthera, Chionochloa* and *Australopyrum* are common examples of this group (Barlow 1989, Good 1992). Interestingly, the majority of the genera are represented by only a single species (Smith 1986, Barlow 1989).

Many other species invaded and colonised the alpine area following distance dispersal from vegetation communities predominantly occurring distant from the Alps, particularly on other continents or islands. These "peregrine" species also evolved in cold climates, with many migrating to the Australian Alps, particularly when sea levels were lower and colder climates prevailed. This group of species arguably has been the most adaptable to past climatic drifts with a number of the species still occurring on other continents but with some species now existing at lower altitudes and in milder temperate climatic zones. Species of the genera *Isolepis, Carpha, Uncinia, Oreobolus* and *Deschampsia* are examples of this group but the two groups are not mutually exclusive as there are many species which indicate much earlier distance dispersal before establishment as austral species.

The Australian alpine flora has been infiltrated more by lowland genera and species than any other alpine area, which is reflected in the low number of alpine "exclusives", there being only some 16 obligate alpines in a total species number of approximately 220 (Good 1992). Barlow (1989) suggested that this reflects the long history of response in the lowlands flora to increasing dryness, an adaptation which also provides physiological adaptation and tolerance to frosts, desiccation and extremes of maximum and minimum temperatures.

Only species having the capacity for physiological adaptation to wide ranges in temperatures and moisture regimes have therefore been able to colonise and survive in the alpine zone and the present alpine vegetation is one which has retained the capacity to survive in very variable climatic conditions. As a consequence of these and the prevailing high wind conditions, the alpine vegetation has also been able to endure very variable soil moisture regimes ranging between wilting point and saturated soil conditions. Moisture stress in summer exists for up to a third of the growing season, while in late autumn, winter and early spring, physiological drought occurs, where cold conditions prevail against active uptake-of any available soil moisture (Soil Conservation Service — summit area works program records 1959–74). It is in protection from the wind that the snowpack plays an important part in plant growth; below the snow, plants are protected, above it they are pruned back (Zwinger and Willard 1972). In addition, some plants are dependent upon the snow cover for their survival as their leaves cannot tolerate winter ambient temperatures and are killed by the cold (Zwinger and Willard 1972).

PLANT LIFE FORMS

The life forms of the alpine vegetation world-wide reflect adaptations to the severe environmental conditions of mountainous areas, and those of the Australian Alps are no exception. Dormancy is a dominant part of the life-cycle of the alpine species, with dormancy extending up to four or five months a year. Early onset of dormancy once severe climatic conditions set in late autumn is a life form feature of many alpine species, but due to the variability of the onset of these conditions in the Australian Alps, particularly the occurrence of semi-permanent snow, this feature is not an obvious survival mechanism in Australian alpines. The onset of dormancy in the Australian setting is more a response to a much shorter photoperiod, in the order of six to seven hours. Billings (1978) noted that northern hemisphere arctic and alpine species exhibit onset of dormancy at photoperiods as long as 14 to 15 hours in arctic areas while, alpine ecotypes under the same temperature conditions do not do this until the photoperiod is reduced to 12 to 13 hours.

As a response to the long periods of unfavourable growing conditions, particularly beneath snow cover, the perenneating structures of the herbaceous plants are predominantly near-soil surface or subsurface buds, roots systems and rhizomatous root structures. These living structures provide for early break of dormancy after snow thaw or rapid regrowth when favourable climatic growth conditions arrive after winter. Many alpine plants have adapted to enable rapid growth of above ground stems and foliage, and early flowering from primordal buds (often formed the previous summer), when mean ambient temperatures increase to about 0°C. Root system growth generally commences a month or more in advance of the above ground structures, as the soils in the Australian Alps seldom freeze to any great depth. Frost crystal formation in the soil surface does occur, but soil permafrost is not a winter phenomenon as it is in high latitude and higher elevation mountains. Snow cover is important here in contributing to the maintenance of soil temperatures at or above 0°C during winter (SCS records 1959–74).

THE PLANT COMMUNITIES

Within the broad gradients of topographic environments (macrogradients) numerous microenvironments exist which influence the amount of solar radiation to which plants are exposed, temperatures, prevailing wind conditions and the amount of precipitation captured as mists, rain and as snowpack. These microenvironmental features and gradients exert a major control over the occurrence and distribution of the alpine communities, and the species that can survive in any locality. Any change in one or more of the climatic conditions will be reflected in changes in the occurrence and distribution of species and communities. These changes will be most obvious at or above the treeline.

Subalpine Woodland — treeline

The treeline is the upper limit of tree growth and occurs at about 1750 to 1900m in Victoria and 1850 to 2050m elevation in New South Wales. At the treeline, tree-dominated communities give way to herbaceous plant and woody shrub communities. The lowest levels of the treelines occur on southern and eastern slopes, with higher treelines being on the warmer south-westerley to north-westerly aspects. Slatyer (1989) noted that this is accentuated by the prevailing south-westerley snow-bearing winds because snow is blown off the windwards slopes onto leeward slopes thus extending the growing season to windward. Generally the treelines around the world occur at an elevation where the mean summer temperature for the warmest month is 10°C and this is also true for the Australian Alps. This isotherm is where trees are no longer able to maintain a positive carbon balance. Environmental factors, particularly temperature, control the rate and duration of photosynthesis and hence the carbon gain by plants. Below 10°C mean temperature, growth of trees cannot be sustained because there is a net carbon loss and they are replaced by the herbs and shrubs of the higher elevation alpine zone. Tree islands on rocky outcrops occur above the general treeline where the growing season is prolonged by early snowmelt in spring and later accumulation in autumn (Slatyer 1989) creating locally warmer habitats (Wimbush and Costin 1973).

Slatyer (1989) suggested that treelines have been relatively stable in the Australian Alps since 8000 to 4000 years BP, with little evidence of treeline shifts since that time. He suggested that this has been due to the high degree of inertia provided to the treeline communities by the presence of trees themselves and the associated dense understorey of shrubs. The existing treeline community would require a significant change in climate and/or vegetation structure for any major shift in the treeline. Slatyer (1989) speculated on the rate of upslope movement of the treeline if global warming was to occur. Slatyer (1978) showed that the slope of mean annual temperature in the Snowy Mountains is 5.9°C km^{-1}, so that an increase of 1°C would represent an increase of about 160 metres in equivalent elevation in the 10°C isotherm but he asserts that the movement of the treeline by this amount is a very remote possibility because natural recruitment at these elevations would only occur with significant introductions of large quantities of seed to disturbed sites. Slatyer (1989) further suggested that any response of the treeline to global warming will lag well behind the temperature potential for upward movement and has calculated from seedling establishment trials, that in favourable conditions this would be in the order of 3 metres altitudinally in 100 years and 120 metres in 4000 years, based on current predictions of increases in mean temperatures in the Alps (Slatyer 1989).

However, this only takes into account one controlling factor in tree growth, that of the average temperature in the warmest month. There are other factors that affect trees growing in areas subject to seasonal snow. Mechanical damage to trees often results from a heavy load of snow, particularly after heavy snowfalls early in the season, leading to loss of shoots and resultant coppicing (Ashton and Williams 1989). Such damage is quite common because of the umbrella-shape of the tree with large leaves that hold snow, in contrast to northern hemisphere conifers which are better able to shed snow with their needles and general conical shape (Slatyer 1989). The lag period between warming alpine climatic regimes and the upward elevational movement of the treeline may be extended by the increasing length of exposure of seedling trees through winter. A series of years with little snow at and below the treeline would also place stresses on the trees. This would not be greatly manifest in mature or semi-mature trees but would be very obvious in seedling trees up to six or seven years old. In years such as 1982 when there was a very shallow cover of snow, the canopies of many seedling and juvenile trees extended above the snowpack and were exposed for several weeks to very low temperatures and the full impact of desiccating and mechanically damaging winds. The importance of the latter in causing damage can be observed in small trees emerging from the snow cover where the browning of leaves is more apparent on the windward side of the canopy. A high number of seedlings and many juvenile trees, as well as the understorey shrubs were killed in that year as a result of desiccation and the physical impacts of very low temperature frosts. The one ameliorating factor in years of generally shallow snow might be the tendency for snowgums to accumulate snow. Ashton and Williams (1989) analysed the depth and persistence of snowbanks and found that both were greater where trees were not trapping snow upwind. Grazing by sheep on coppice regeneration snowgum after the 1939 fires in the Snowy Mountains also had the effect of reducing the amount of snow intercepted in woodland (Costin 1954).

Cochrane and Slatyer (1988) noted that direct temperature damage by radiation frosts occurring in autumn or spring, when the leaf tissue is not fully hardened, may be a major cause of dieback. Ferrar *et al.* (1988) also showed, through trial plantings of seedlings above and below the treeline, that the most critical periods for seedling establishment following germination were the first growing season and the first winter. They noted the importance of the interaction of snow protection and the length of the snow-free period on the success or failure of seedling establishment. They also noted that the probability of survival was enhanced as biomass increased, with seedlings up to 30cm high being readily killed by frosts while seedlings in the order of 50–100 cm usually survived a single adverse season. To survive several consecutive adverse autumn and winter seasons then by extrapolation it could be estimated that only seedlings in the order of 100–200 cm in height with a basal diameter of approximately 5cm would survive. In this case, the upward movement of the treeline in response to declining snow cover and depth would be extremely slow as upslope seedling recruitment would be very low.

The alpine herbfields

The alpine vegetation is characterised by tall alpine herbfields in which perennial grasses are dominant but in which perennial herbs fill the inter-tussock spaces and give the alpine flora its reputation as one of the most diverse and colourful. The tall alpine herbfields produce a large above-ground biomass during the summer growth period. While the above-ground biomass is obvious, it is the even greater below-ground living biomass that provides for regeneration and survival following the spring snow-thaw. Many of the inter-tussock species have taproot or bulbous root systems which provide for starch (carbohydrate) storage and hence the capacity for rapid growth as soon as milder growth conditions occur in spring.

The alpine herbfields occur in sites where snowpack lasts for varying lengths of time each year, a regime to which the species of the tall alpine herbfield are well adapted. However, heavy snow cover may occasionally kill off large areas of *Poa* which allows a massive growth of composite herbs (Costin *et al.* 1979). Winters with a heavy snowpack are also likely to provide conditions suitable for the infestation of the grassland with the larvae of swift moths and case moths resulting in the death of tens or hundreds of square metres of tussock (Carr and Turner 1959). Many species of the tall alpine herbfield posses the capacity to migrate up or down the elevation gradient, this being well exhibited in the genera *Ranunculus, Craspedia* and *Celmisia* which appear to have evolved phenotypic forms and possibly genotypically, as a response to relatively recent environment changes.

The occurrence of short alpine herbfields is determined by the presence of snowdrifts. These communities exist around the downslope edges of long-lasting semi-permanent snowdrifts where cold thaw water flows over shallow gravelly soils. These sites support a specialized ground-hugging community which is characteristic of short herbfields world-wide. The community is dominated by *Plantago muelleri* and *Neopaxia australasica* but lacks the mosses and liverworts which dominate the leeward snowpatch vegetation in the northern hemisphere. These communities may be exposed for only a short period each summer after the late thaw of the snowdrifts and the plant species have evolved adaptations to short summer growing conditions with *Caltha introloba* for example, being able to initiate flowering as soon as light filters through the melting snow (see Chapter 7).

Neopaxia australasica is arguably the most adaptable of all alpine species, having over the past 20 to 30 years been observed to be colonising sites in almost every environmental gradient from the snow-free exposed and dry feldmarks, to groundwater communities and previously severely eroded areas. With this capacity to adapt to a range of sites and environmental conditions, the survival of the majority of short alpine herbfield species, including the botanically interesting *Diplaspis hydrocotyle, Dichosciadium ranunculaceum* and *Caltha introloba,* is not greatly influenced by the snowpack regime.

Bog and Fen

The occurrence and distribution of valley-side and raised bog communities is little influenced directly by occurrence of snow. These groundwater communities are essentially maintained by summer runoff regimes and much less by winter climatic regimes and runoff during the spring thaw.

The *Epacris paludosa-Sphagnum cymbifolium* raised bog is a physiographic climax community which exists where acid and base-deficient conditions are produced by the existence of permanent free-standing water at and below ground level. The raised bogs are dynamic communities which oscillate through a cycle of dominance from herbaceous species to shrubs and have a capacity to accommodate changes in their climatic and microenvironments. Costin (1957) noted the influence of relative humidity and cloudiness in the maintenance of these communities latitudinally from north to south in the alpine areas.

In areas of acid soil conditions under free-standing water in valley bottom sites, a characteristic bog community dominated by *Carex gaudichaudiana* and *Sphagnum cristatum* occurs. This community is considered to be transitional between the raised bog and the acid fen communities. The bog communities are underlain by peat and have a significant water holding capacity. These communities are important natural regulators of water release to the streams and rivers during summer. These communities were extensively damaged by past stock grazing and much of the total bog area is still to regenerate to an active hygrophilous community.

The acid fen community of *Carex gauchaudiana* occurs commonly along watercourses in small basins where surface water accumulates. Peatbeds have accumulated to considerable depth in many sites but their continued development is dependent on the continued input of base material or mineral matter from surrounding areas which may be affected to a small degree by snow cover and runoff from snowmelt.

Feldmark communities

At higher elevations, along the upper edges of snowpatches and in the windswept cols above the alpine herbfields, the feldmark communities exist in sites of little or no winter snow cover. The feldmarks, particularly the exposed cold feldmark of the upper cols, exist on very shallow gravelly soils as a result of the continual removal of soil fines and the movement of gravels downwind by saltation. The feldmark communities cover the smallest area of any alpine plant community and have long been considered as one of the most fragile of alpine plant communities. Under stable natural environments the feldmarks are actually very resilient communities, able to exist in the most severe of edaphic conditions and alpine climates.

The snowpatch feldmarks are dominated by *Coprosma pumila* and *Colobanthus benthamianus* and are in a moisture-stressed state for the greater part of the snow-free months in most years. This community only occurs in the mainland Australian Alps, while in the Tasmania alpine area and elsewhere in the southern hemisphere, cushion plant communities occur in similar feldmark-like areas. The cushion plants are all but absent from the mainland Australian Alps suggesting that post-glacial climatic changes towards drier conditions may have eliminated many of them from the snowpatch feldmark.

In the cols between the highest peaks, strong wind conditions prevail for much of the year including the winter months and, as with the soil fines, snow is continually blown from these feldmark areas. The occurrence of vegetation on these feldmark areas is therefore more influenced by the lack of a snow cover and is little affected by annual differences in snow occurrence. The windswept feldmarks are dominated by the very low growing shrubs, *Epacris microphylla* and *Veronica densifolia*. The plants growing on the gravelly lithosols are continually pruned and buried by the movement downwind of the surface gravels. Over many years the community migrates across the feldmark with new growth and foliage being produced at the furthermost downwind extensions of the plants. With this creeping migration the feldmark communities exist in various stages of exposure and vegetative cover. This cyclic pattern of growth, burial, foliage destruction and regeneration provides a capacity for the feldmark species to adapt to other environmental perturbations including that of climatic change.

The capacity of many alpine herbaceous plants to colonise and adapt to changes in microenvironments is evident in many areas of artificially created, feldmark-like gravel pavements. These have developed following many years of alpine grazing and extensive erosion of organic alpine humus soils down to the underlying gravel basement. These areas now support a sparse but stable herbaceous cover of *Leucochrysum albicans, Ranunculus muelleri, Drapetes tasmanicus, Epacris microphylla, Veronica densifolia, Colobanthus benthamianus, Ewartia nubigena* and *Euphrasia* species.

Heath communities

Extensive heaths of woody shrubs dominate the alpine landscape in sites of very free drainage, such as boulderfields and glacial moraines, and in areas of poor drainage where inadequate aeration limits the growth of herbaceous alpine herbfields. The composition of heath communities will also be affected by exposure to wind and snow cover (Zwinger and Willard 1972). The *Oxylobium ellipticum* and *Podocarpus lawrencei* heath occurs on the boulderfields and rock screes. This community grades into and intersperses the tall heath community dominated by *Phebalium ovatifolium*, but growing in association with a number of other heath species, including *Grevillea australis*, *Prostanthera cuneata*, *Olearia* and *Ozothamnus* spp., which are more commonly subalpine heath and woodland understorey species.

The tall heath is replaced or grades into the short heath dominated by *Epacris seryllifolia* and *Kunzea muelleri*, which occurs over large areas of shallow soils, particularly around feldmarks, and the *Epacris glacialis* heaths in damp soils bordering bogs and fens. The transition from feldmarks to heathlands and herbfields reflects the degree of exposure and desiccation from exposed, very dry, shallow soils to the protected, well developed, alpine humus soils of the herbfield areas. The capacity of the heaths to adapt to climatic and edaphic changes is exhibited in many sites where dry heaths have replaced wet heaths following disturbance and drying out of fen and bog areas as a result of past alpine grazing. *Kunzea muelleri* is very much an opportunistic species which has colonised large areas resulting from edaphic changes during and after alpine grazing.

Heath plants are pruned by heavy snowfalls each winter and are considerably flattened by the accumulating snowpack. The snow-pruning of the shrubs tends to keep the spring and summer growth compact and better able to support the weight of snowpack in the following winter.

Lichens

Lichens cover many small rocks in the feldmarks and the sheltered surfaces of large granite boulders in the herbfields. The alpine zone lichen flora includes the very conspicuous cosmopolitan *Thamnolia vermicularis*, the bright yellow *Rhizocarpon geographicum*, the black *Umbilicaria decussata* and the alpine endemic species, *Centraria australiensis*. As with all lichens, they survive under extremes of exposure and throughout the world, feldmark lichens have been found to be intolerant of snowcover (Zwinger and Willard 1972). Prolonged snow cover can result in death of some lichens as a result of depletion of thallus carbon reserves (Benedict 1990). As a result, they are often absent from boulders occurring in areas where snowpatches persist long into the summer. Thus areas of long snow duration can be identified even when the snow cover has gone, by the transition from lichen-covered boulders to bare rock. The future occurrences of lichen populations will be little changed by putative changes associated with global warming such as a decline in snow cover, but may be influenced by increasing lengths of exposure, generally to the benefit of the species.

Weed species

Costin (1954) recorded six introduced plant species in the alpine zone of the Snowy Mountains with 44 in the subalpine zone. A checklist compiled by Thompson and Gray (1981) recorded the presence of 27 species in the alpine zone and 67 in the subalpine zone, indicating an increased diversity of introduced species over that time period (Mallen-Cooper 1990). The main reasons for this increase were trampling and grazing by stock (Costin 1958, Bryant 1971) and increased disturbance, particularly around ski resorts, associated with increased numbers of visitors (Mallen-Cooper 1990). Despite this high number of species, most plants are restricted to disturbed sites and many are represented by only a single individual (Mallen-Cooper 1990). The decreasing number of weed species with altitude was ascribed to a number of factors including the severity of the conditions, the short growing season and the period of snow cover. The lower rate of weed invasion in the alpine vegetation of Tasmania was, however, put down to the lack of a protective snow cover during winter by Kirkpatrick (1989) who considered that disturbance was not a factor as the Tasmanian alpine area has large areas of sheet erosion. In the Snowy Mountains, many weed species were found to have their upper altitudinal limit at about the treeline (Mallen-Cooper 1990). Although germination of weed seeds was possible in harsh environments such as feldmark and snowpatch, high seedling mortality and low overwinter survival of the plants was found as a result of desications, frost heave, low temperatures retarding growth, early snowfalls and predation (Mallen-Cooper 1990). An aspect of the failure of seedlings of two experimentally grown weed (*Rumex crispus* and *Verbascum virgatum*) to survive highlighted by (Mallen-Cooper 1990) was their slow growth rates in the growing season, and she concluded that "their failure to overwinter precludes their successful establishment in the alpine zone." It would seem, therefore, that along with lack of disturbance, snow cover, by reducing the length of the growing seasons excludes annual weeds and in fact only 7 of 200 native species in the alpine zone recorded by Costin *et al.* (1979) were annuals.

POSSIBLE IMPACTS OF GLOBAL WARMING

Whetton *et al.* (1996) suggest that under a scenario of global warming by 0.5 to 2.0°C, by the year 2030, the duration of snow in the alpine zone at Mt. Kosciuszko could decline to between 135 and 178 days, with commensurate declines in the number of snow days with decreasing elevation to the subalpine area. The effects of a decline in the total amount of snow, the seasonal distribution of snowfalls and the length of snow cover will vary between vegetation communities and will be very much influenced by topographic features.

With global warming and the possible consequent rise of the semi-permanent snowline in the next 30 to 50 years, snowfalls and the total area of snow cover will influence a redistribution of plant communities. This may be significant for the survival of many species if the predicted rise in temperatures results in an altitudinal rise in the subalpine zone and a decline in the area of the alpine zone. Unless the changes in snow conditions are dramatic and outside the existing natural range of variability the opportunism of the alpine plants to colonise new microenvironments could result in the establishment and formation of some small areas of new plant associations and even community types, possibly increasing the number and diversity of alpine plant communities.

The influence of snow on individual plant species in such a scenario would be manifest in how alpine plant species respond to the lack of a protective and insulating cover through winter. This will depend on their adaptability to extremes of cold, moisture and desiccation and their capacity to acclimate to changes in the precipitation regime and groundwater and soil moisture status, as future precipitation tends more to high intensity rainfall and away from snowfalls (Whetton *et al.* 1996). The desiccating influence of wind on plant communities not protected by snow cover may increase to the extent that wind alone could be the major influence in determining future vegetation communities and species occurrences. With a decrease in snow cover, the photoperiod influencing growth in the Australia Alps may exceed six to seven hours for a few extra days per year, but possibly would not to any great extent influence the onset of dormancy. Hence the plants would be little affected by any extra exposure beyond the normal range of conditions which influence growth and dormancy.

A reduction in snow cover could potentially lead to a greater intensity of frost crystal formation in the soil surface and "frost heave", but this impact is prevented or greatly ameliorated where a dense stable vegetative cover is maintained.

Many and varied responses by alpine plants to changes in the temporal and spatial occurrence and distribution of snow are expected even if they are not possible to predict accurately. These responses will be most noticeable at the community level where the smaller and more sensitive communities may decline in area or possibly disappear altogether from some sites.

Should the extent of snowpack decline as a result of lower snow precipitation, the extent of the short alpine herbfield will decline as a direct response, because it is a community dependent upon, and adapted to, the continuous flow of snowmelt water well into the summer months. An analogue of the possible changes under global warming was documented by Ashton and Williams (1989). Snowpatches on the Bogong High Plains, existing annually since the 1939 fires in both grazed and ungrazed areas developed a typical snowpatch vegetation. These communities were gradually invaded by shrubs (on grazed and ungrazed areas) and Ashton and Williams (1989) postulated that the growth of trees upwind of the snowpatches intercepted much of the wind-blown snow, reducing the extent of snow accumulation and therefore resulting in a gradual lengthening of the growing season favouring shrub invasion. While the total extent of the short alpine herbfield community may decline to a very small area, or even disappear as an alpine plant community, the individual species will in all likelihood survive, because both of the dominant species also occur as inter-tussock herbs in the tall alpine herbfield, sod-tussock grasslands and groundwater communities. As with the tall alpine herbfield, disturbance will remain the major threat to the community as the extent of the short alpine snowpatch community has already been very much reduced by past domestic stock grazing. The detrimental impacts of declining snowfall and snowpack regimes on the tall alpine herbfields should be relatively minor and in fact may be to the benefit of the community as the spatial extent of the short alpine herbfield decline and are replaced by tall alpine herbfield.

Many bogs and fens are currently regenerating to active hygrophilous communities after the many years of grazing and, as noted earlier, the impact of predicted declining snow and groundwater regimes may lead to a break in the regenerating processes and a loss of these groundwater communities. The extent of the impact will depend on the total precipitation regime which is predicted to include sporadic high intensity storms. Any change in the accumulation of material in peat beds is not likely to be obvious if total precipitation and runoff conditions remain similar to that which presently exist.

The impacts of changing snow regimes on the heath communities should be minimal and may actually result in additional subalpine shrubs invading the alpine heath communities and some alpine herbfield sites. The existing alpine heath communities may move into sites which become more exposed as a result of the predicted decline in the protective and insulating snow cover in a similar manner to the invasion by shrubs of declining snowpatches on the Bogong High Plains (Ashton and Williams 1989).

Similarly, changes in snowfall regimes will have little or no impact on the survival of the existing feldmarks. With larger areas in nearby alpine herbfields being exposed to increased severity of wind conditions, feldmark communities may also actually extend their spatial extent if the organic soils supporting herbfields are blasted away to gravel lithosols.

At the species level, some loss of obligate alpine species may occur. These species will be threatened by changes in the climatic environment in the Alps, because they do not have the capacity to adapt readily to relatively rapid changes in climatic conditions. Similarly they have narrow habitat tolerance ranges and can only tolerate very small changes even over long periods. Species such as *Drapetes tasmanicus, Diplaspis hydrocotyle, Schizeilma fragoseum, Oreomyrrhis pulvinifica, Uncinia compacta, Carex cephalotes* and *Plantago glacialis* are in this category. Not all would be threatened on a southern hemisphere scale as several also occur in New Zealand and other southern alpine areas. Conversely, there may be greater opportunities for the invasion of weeds species which are presently excluded by the duration of the snowpack reducing the length of the growing season to the extent that invasion by these species is not possible.

Many species may be affected indirectly by the decline in snow cover. Increased exposure to solar radiation and high levels of ultra-violet light will affect the onset and break of dormancy, flowering times and seed set. The occurrence and distribution of species such as *Herpolirion novae-zealandiae* in which flowering is initiated by a very specific combination of temperature, humidity and length of day and which flower for very short periods (usually less than a day), may also be influenced and possibly threatened by minor changes in these conditions. Whether the reduction in snow cover will be beneficial or detrimental to the growth and spread of species such as *Caltha introloba* which have evolved adaptations to short summer growing conditions will be an interesting indicator of the possible long term effects of global warming and warming alpine climatic conditions.

Generally, the detrimental impacts of declining total precipitation as snow and a reduction in the length and extent of snow cover should not be as great as has been expressed in many forums and publications and will be mainly expressed in a spatial redistribution of plant communities. Alternatively, it can be predicted that several communities, such as tall alpine herbfield, windswept feldmark and sod tussock grasslands may extend their distribution and area of occurrence, albeit at the expense of the smaller and more climatically sensitive communities.

A number of species can be taken as early indicator species of the early changes in climatic regimes, but the impacts of global warming will not necessarily be expressed first in the alpine vegetation as has often been suggested by a number of writers such as Krockenberger and Kincade (1997).

CONCLUSION

Snow has been a dominant weather feature which has influenced the occurrence and distribution of plant communities in the Australian Alps. Snowfalls have many influences on the form of erect branched structures and can open up clearings, snap tops off trees and tall shrubs, break off limbs etc. However, it is, and has been, the responses of plant species to the depth and length of snowpack, and the temperature and moisture regimes which have been associated with the snowfalls over many years, which have given the alpine area its characteristic plant communities and floral landscape. Long term weather records indicate that the Australian climate is characteristically very variable, and vegetation in any biophysical zone has to endure great ranges in climatic conditions. The climatic conditions of the alpine zone are no exception, with very wide temperature ranges in summer and extreme minimum temperatures in winter. This variability is also apparent in the amount, distribution and the number of days each year on which snow falls.

Moisture stress is experienced by the vegetation in both winter and summer, while severe wind conditions are a regular feature at any time during the year, particularly in late autumn, winter and spring. As a consequence, the vegetation has evolved and survives under the harshest of Australian cold climatic conditions.

Global warming has been predicted by many scientists to lead to major changes in global climates and hence have a considerable impact on the occurrence and distribution of snow in the Australian Alps (Whetton *et al.* 1996). Research suggests that with the predicted temperature rises and precipitation increasingly occurring as storm rains, temperature and moisture regimes conducive to snowfall may, in the future, occur only in the vicinity of the highest peaks of the Alps (Whetton *et al.* 1996). The impact that this rise in snowline may have on the vegetation is difficult to assess but a consideration of the origin and structural forms of the alpine plant species enables a prediction to be made, that the number of species will change little, although the occurrence, distribution and extent of several plant communities will decline significantly such that more species will be of very limited distribution or be considered rare and threatened. These species will predicably be the present obligate alpine species and several of the alpine endemics. eg. *Dichosciadium ranunculaceum, Colobanthus nivicola, C. pulvinatus* and *Euphrasia alsa*.

A consequent feature of the predicted reduction in the spatial and temporal snow cover will be the increased length of exposure of the plant species and communities to the high levels of ultra-violet light which occurs at high elevations. The influence of this in terms of plant distribution is difficult to ascertain but it is expected that the increased radiation will result in mutations and changes in plant morphology in some genera and species. Over time, increased speciation will possibly occur and endemism increase from the current level of approximately eleven percent of native species. Hybridisation will similarly increase in several genera such as, *Ranunculus*, *Craspedia*, *Luzula* and *Oreomyrrhis* which all exhibited increased hybridisation during the many years of alpine grazing. This is still apparent in Victoria where some areas of alpine grazing still exist.

To predict the full extent of the impacts of declining snow cover on alpine plant species will require a greater knowledge and understanding of the ecology and floral biology of many species. Several species for example are known to exhibit cleistogamy (the setting of seed without the flowers opening to full blooming), which may be induced by changes in environmental factors such as moisture stress or flooding. Several other alpine species are androdioecious, andromonoecious or gynomonoecious as an adaptation to moisture stress and temperature extremes during summer and winter.

Other species exhibit specific responses to environmental conditions such as *Chionogentias muelleriana* subsp. *alpestris* which exhibits flower closure with decreased light intensity during the day, such as may occur on misty or overcast days. An understanding of the impacts of declining snow cover on these species will require much research before any definitive statement can be made as to the long-term viability of the species.

REFERENCES

Ashton, D. H. and Williams, R. J., 1989 Dynamics of the sub-alpine vegetation in the Victorian region. Pp. 143–168 *in* "The Scientific Significance of the Australian Alps: The Proceedings of the First Fenner Conference" ed by R. B. Good. The Australian Alps National Parks Liaison Committee/Australian Academy of Science.

Barlow, B. A., 1989. The alpine flora: autochthones and peregrines. Pp. 69–78 *in* "The Scientific Significance of the Australian Alps: The Proceedings of the First Fenner Conference" ed by R. B. Good. The Australian Alps National Parks Liaison Committee/Australian Academy of Science.

Benedict, J. B., 1990. Lichen mortality due to late-lying snow : results of a transplant study. *Arct. Alp. Res.* **22:** 81–89.

Billings, W. D., 1979. High Altitude Geoecology. Pp. 97–125 *in* Proceedings of the American Association for the Advancement of Science. Selected Symposium 12. ed by P. J. Webber. Westview Press.

Brown, J. A. H. and Millner, F. C., 1989. Aspects of the meteorology and hydrology of the Australian Alps. Pp. 297–332 *in* "The Scientific Significance of the Australian Alps: The Proceedings of the First Fenner Conference" ed by R. B. Good. The Australian Alps National Parks Liaison Committee/Australian Academy of Science.

Bryant, W. G., 1971. Deterioration of vegetation and erosion in the Guthega catchment area, Snowy Mountains, New South Wales. *Soil Con. J.* **27:** 62–81.

Carr, S. G. M. and Turner, J. S., 1959. The ecology of the Bogong High Plains. I. The environmental factors and the grassland communities. *Aust. J. Bot.* **7:** 12–33.

Cochrane, P. M. and Slatyer, R. O., 1988. Water relations of *Eucalyptus pauciflora* near the alpine treeline in winter. *Tree physiol.* **4:** 45–52.

Costin, A. B., 1954. "A Study of the Ecosystems of the Monaro Region of New South Wales." NSW Government Printer, Sydney.

Costin, A. B., 1957. The high mountain vegetation of Australia. *Aust. J. Bot.* **5:** 173–189.

Costin, A. B., 1958. The grazing factor and the maintenance of catchment values in the Australian Alps. CSIRO Div. Plant. Ind. Tech. Pap. No. 10, 1–14.

Costin, A. B., Gray, M., Totterdell, C. J. and Wimbush, D. J., 1979. "Kosciusko Alpine Flora." CSIRO Collins Melbourne/Sydney.

Ferrar, P. J., Cochrane, P. M. and Slatyer, R. O., 1988. Factors influencing germination and establishment of *Eucalyptus pauciflora* near the treeline. *Tree Physiol.* **4:** 27–43.

Good, R. B., 1992. "Kosciusko Heritage." NSW National Parks and Wildlife Service, Sydney. Surrey Beatty & Sons, Chipping Norton.

Kirkpatrick, J. B., 1989. The comparative ecology of mainland Australian and Tasmanian alpine vegetation. Pp. 127–142 *in* "The Scientific Significance of the Australian Alps" ed by R. B. Good. Australian Alps Liaison Committee and Australian Academy of Science, Canberra.

Krockenberger, M. and Kinkade, P., 1997. The Impacts of Global Warming. *Habitat* **25:** 13–20.

Mallen-Cooper, J. 1990. "Introduced plants in the high altitude environments of Kosciusko National Park, south eastern Australia." Unpublished Ph.D. thesis, Australian National University.

Slatyer, R. A., 1978. Altitudinal variation in the photosynthetic characteristics of snow gum, *Eucalyptus pauciflora* Sieb. ex Spreng. VII Relationships between gradients of field temperature and photosynthetic temperature optima in the Snowy Mountains Area. *Aust. J. Bot* **26:** 111–121.

Slatyer, R. A., 1989. Alpine and Valley Bottom Treelines. Pp. 169–184 *in* "The Scientific Significance of the Australian Alps" ed by R. B. Good. Australian Alps Liaison Committee and Australian Academy of Science, Canberra.

Slatyer, R. A., Cochrane, P. M. and Galloway, R. W., 1984. Duration and Extent of Snow Cover in the Snowy Mountains and a Comparison with Switzerland. *Search* **11–12:** 327–331.

Smith, J. M. B., 1973. Evolution of subalpine and alpine plant groups in New Zealand. *N.Z. J. Bot.* **11:** 177–200.

Snowy Mountains Hydro-electric Authority, 1970. Catalogue of meteorological data of the Snowy Mountains Region, Australia. SMHEA. Cooma.

Soil Conservation Service, 1974. Summit area works program records 1959–74. SCS. Cooma.

Thompson, J. and Gray, M., 1981. A check-list of the subalpine and alpine species found in the Kosciusko region of New South Wales. *Telopea* **2:** 299–364.

Whetton, P. H., Haylock, M. R. and Galloway, R. W., 1996. *Climate Change* **32:** 447–479.

Wimbush, D. J. and Costin, A. B., 1973. Vegetation mapping in relation to ecological interpretation and management in the Kosciusko alpine area. CSIRO Div. Plant. Ind. Tech. Pap. No. 32, 1–22.

Zwinger, A. H. and Willard, B. E., 1972. "Land Above the Trees. A Guide to American Alpine Tundra." Harper & Row, New York.

CHAPTER 7

Plant Activity Beneath the Snow

IAN F. WARDLAW

INTRODUCTION

RECORDS that have been collected on snow cover over time are often a product of their end use. Thus information of use to skiers relates in general to snow on south facing downhill slopes, while observations in relation to the Australian Alps as a water catchment includes deep snow accumulated in gullies and depressions, which can last well into the summer (see Slatyer *et al.* 1985). When considering the effect of snow cover on alpine vegetation in Australia a number of factors have to be taken into account, but the most basic are the duration and depth of snow cover which can vary with altitude, aspect and topography as well as from year to year.

Light and temperature conditions under the snow are governed by both ground (rock, soil and moisture) temperatures and the depth of snow. Unlike more severe arctic and alpine regions, the soil in the Australian Alps is not frozen at depth, and the lower layers of the snow often melt resulting in an air space between the ground and the snow cover (see also Chapter 8). In this space plants may exist without being compacted by the snow.

Light does not penetrate readily through fresh snow (Kimball *et al.* 1973, Salisbury 1984, Hamerlynck and Smith 1994) so that at a depth of 25 cm the light level is very low (see Fig. 1) and at 80 cm there is almost complete darkness. Light penetration through the snow varies, not only with depth but also with the degree of compaction, reaching a minimum at a density of 0.5 g cm^{-2}. As compaction increases further, light penetration starts to increase, possibly reflecting some icing of the snow (Marchand 1984) and this may account for light penetrating through nearly two metres of snow in some studies (Richardson and Salisbury 1977). Ice allows a much greater penetration of light than snow with measurable levels (4.5% of incident radiation) at a depth of 5m. Contamination with dirt is another factor that will reduce light penetration through the snow. Light quality is also affected by snow which selectively absorbs the longer

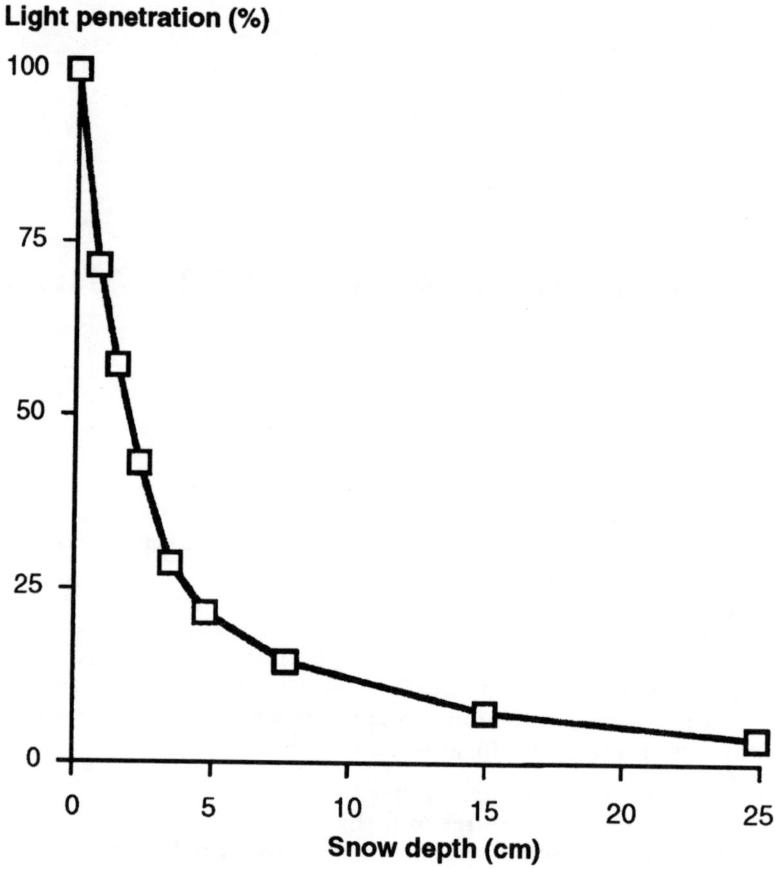

Fig. 1. Light penetration through mildly compacted snow, where only 0.1% of the incident radiation would be expected to penetrate to a depth of 60 cm. (Based on observations by Thomas 1972, Kimball *et al.* 1973, Hamerlynck and Smith 1994).

(red) wavelengths, resulting in a relative enhancement of the blue end of the spectrum (Kimball *et al.* 1973, Richardson and Salisbury 1977, Salisbury 1984).

Air temperatures in the open at Charlottes Pass have fallen as low as 23°C below zero in the winter, but under the snow cover (which acts as a thermal blanket) air temperatures in the space above the soil are still 1°C, or possibly 2°C, above freezing (Billings and Moony 1968, Kimball *et al.* 1973: Salisbury 1984, Inouye and McGuire 1991). Thus the plants under snow are protected from extreme low temperatures that might otherwise cause tissue freezing and severe damage.

Snow is permeable to CO_2 and the concentration of CO_2 at the soil/snow interface is dependent on a number of factors such as snow thickness and compaction, as well as air and soil temperatures and the respiratory activity of plants and micro-organisms. CO_2 levels as high as 0.14% have been measured under thaw crusts (Schmid 1984). Excessive CO_2 can be very damaging to plant tissues and this can occur if plants become encased in ice, because gaseous diffusion through ice is very limited (Andrews 1996). Red algae are often concentrated in the top 10 cm of the snow, but have been observed at depths of up to 30 cm, and this distribution is thought to reflect the attenuation in both CO_2 and light with depth through the snow (Thomas 1972).

In the Australian Alps, melting snow provides a useful source of water for the growth of many plants in the spring and early summer, while during the winter, packed snow can reduce water loss from the soil surface (and plants) by providing protection from cold dry winds (Billings and Mooney 1968).

DEVELOPMENTAL CYCLES

At the higher altitudes (2000m) in the Australian Alps snow cover lasts on average for about 142 days (Slatyer *et al.* 1985), which would suggest a growing period for plants without snow of between seven and eight months. However, low temperatures at the beginning and end of the season may limit the effective duration of this snow-free period. In this time, some species are able to initiate and develop flowers and set and disperse seed. However, it is clear that many other species which open their flowers in spring as the snow melts have another strategy. These plants actually extend their developmental period over two years instead of one and make use of the snow cover as a protecting blanket to provide temperatures that allow some growth and also influence development (Mark 1965, Kimball and Salisbury 1974, Wardlaw *et al.* 1989a).

The pattern of floral development in a snow patch plant is shown in Figure 2 (see Sørensen 1941, Bliss 1971, Kimball and Salisbury 1974). This development pattern is exhibited in Australia by Alpine Marsh Marigolds (*Caltha introloba* F. Muell. I.) (Wardlaw *et al.* 1989a). A vegetative shoot formed during the summer is subject to a number of conditions that may result in the eventual initiation of a floral apex and the development of a flower. Firstly, as the shoot passes through autumn and is covered by snow, the temperature drops to 0–2°C which is in the vernalizing temperature range (temperatures that can shorten the growth period before flowering). The most effective temperatures for vernalization are generally considered to fall in the range from 2°C to 7°C (see Fig. 3), but depending on species, may still have a positive action in the range from several degrees below 0°C to nearly 15°C (see Leopold 1964, Salisbury and Ross 1985). There is some evidence that these chilling temperatures enhance flowering in some alpine species, but this does not appear to be an obligatory cold requirement (see Billings and Mooney 1968). Without the snow cover, ambient temperatures might not only damage the plants but be too low for effective vernalization of some plants (Fig. 3). Unfortunately, the requirement for this low temperature preconditioning to induce flowering has not been directly assessed in many Australian alpine species, although at

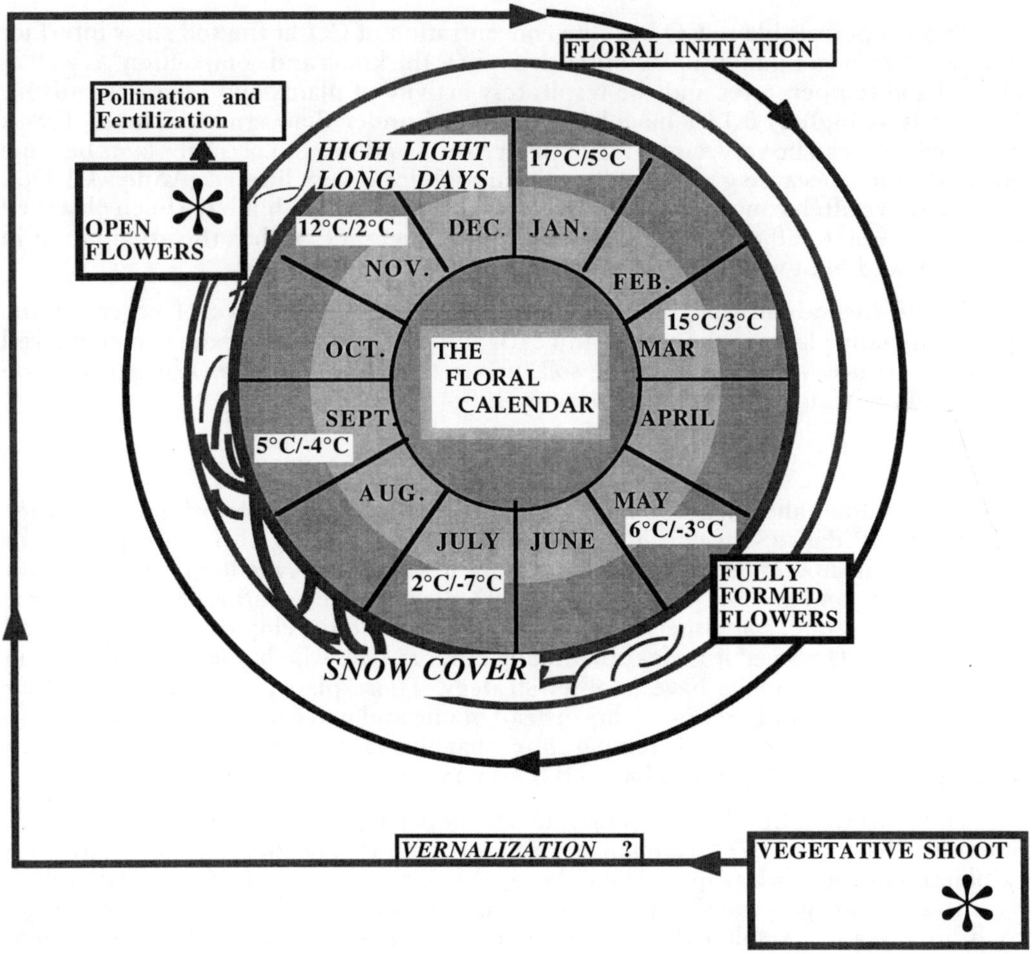

Fig. 2. The floral calendar, illustrating the timing of development of a snow patch plant from a vegetative shoot to an open flower, as this relates to seasonal temperatures, photoperiod and snow cover.

The vegetative shoot is often covered by snow at the start of the winter months (late May or June). In some alpine species this cold (vernalization) period under the snow may enhance the subsequent flowering. When the snow melts, the still vegetative shoot emerges into longer days and higher temperatures which can act as a further stimulus for floral initiation. Floral organs are evident at the shoot apex by mid summer (January). The flowers, which can be fully formed by the end of May (late autumn), do not however open until they have had sufficient cold and the snow melts in the following spring (October /November).

The temperatures shown are mean maximum and minimum values for every second month.

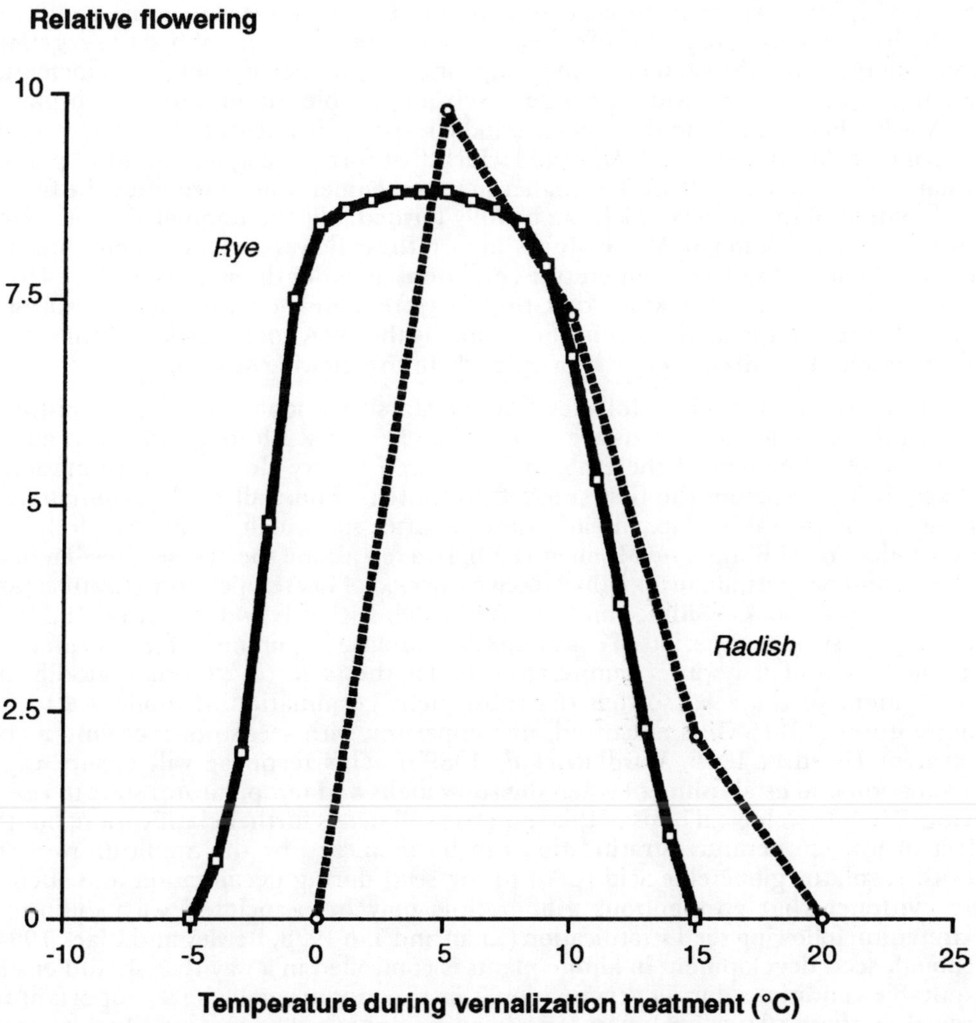

Fig. 3. The effect of vernalizing (low) temperatures on the flowering of rye and radish. The optimum temperature for vernalization is about 5°C and the effective temperatures range from about –4°C to +15°C, depending on the species. (See Leopold 1964, Salisbury and Ross 1985).

least five alpine *Ranunculus* species are known to flower under glasshouse conditions without having received a period of cold (C. Pickering pers. comm.). When the vegetative shoot emerges from the snow cover in spring, or early summer, it is subject to increasing daylength (photoperiod) which provides a second possible stimulus to floral initiation (see Mark 1965). This long day photoperiodic response is known to be a factor in the flowering of the Alpine Marsh Marigold which first forms recognisable floral parts in January (Wardlaw *et al.* 1989a). The higher summer temperatures encourage the further development of the flowers which can be fully formed, but still unopened, by the time of the first snow fall in late May or June. In fact, these flowers will not open until they have several months of low temperature conditioning under the snow (see Mark 1970). This is a built-in safety feature that prevents premature opening in the autumn. Normally the flowers then open in the spring as the snow melts back and this allows pollination and fertilization of the ovary early in the growing season.

Seed development in the follicle of Alpine Marsh Marigolds (see Fig. 4) continues through the summer, and seeds are shed in late February. Although they can absorb water, the development of the embryo in the seed is very slow, and germination is unlikely to occur before the first snowfall in winter. Almost all seed germination in alpine conditions takes place in early summer after snow melt in the year following seed production (Billings and Mooney 1968). In a few alpine species, seeds will remain dormant and not germinate until they receive a period of low temperature (stratification) of four to eight weeks (Billings and Mooney 1968, Richards and Beardsell 1987). In other alpine species, a period of cold is not an absolute requirement for germination, but the period of low winter temperature under the snow (0–2°C) may modify the development of the seed so that the subsequent germination at moderate to low temperatures (10–15°C) is enhanced, in comparison with seeds not receiving a cold treatment (Junttila 1976, Wardlaw *et al.* 1989b). This response will ensure rapid germination and establishment when the snow melts and temperatures start to rise in spring. The physiological basis of this response still needs further clarification, but the effect of low temperature stratification can be mimicked by the application of the growth regulator gibberellic acid (GA_3) to the seed during germination and there is also evidence that endogenous gibberellins may be associated with enhanced germination following seed stratification (Khan and Tao 1978, Bewley and Black 1994). Although seed development in alpine plants is controlled in a way that should ensure favourable conditions during germination, this may not normally be so important for survival as these plants are often long lived perennials that can readily propagate vegetatively (Billings and Mooney 1968, Bliss 1971).

PHYSIOLOGICAL ACTIVITY AND GROWTH UNDER THE SNOW

It is known that some plants, but not all, may grow while still under the snow (see Salisbury 1984, Hamerlynck and Smith 1994) and growth can take place at temperatures below 0°C (Kimball *et al.* 1973). Therefore, it is reasonable to assume that growth may proceed slowly in the air space between the soil and the snow cover in the Australian Alps where temperatures can be several degrees above zero. However, at these low

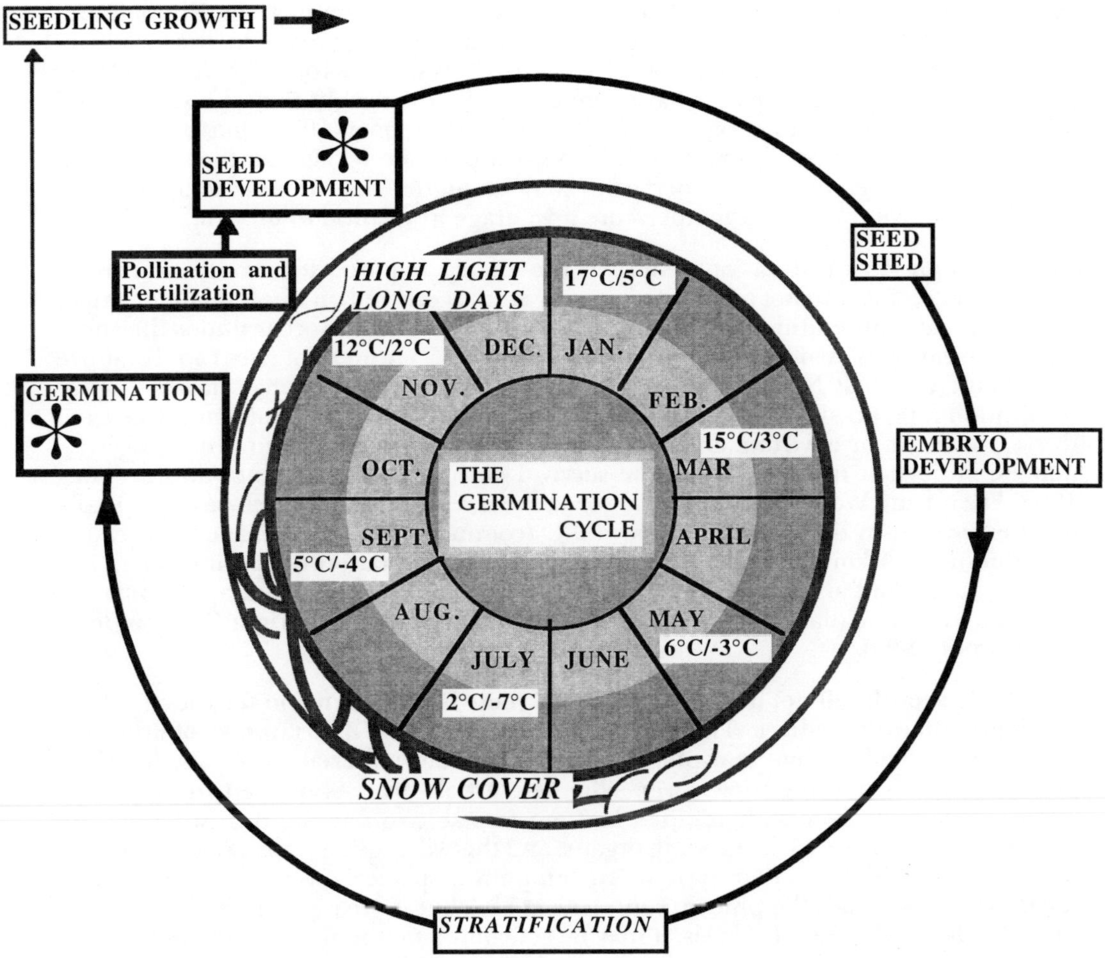

Fig. 4. The germination cycle illustrating for a snow patch plant the timing of development and germination of the seed.

Seeds start developing in the flower following pollination in late spring (November) and are shed from the fruits in late summer or early autumn (February/March). The embryo continues to develop in the seed, but the seed will not be ready to germinate before the first snow falls in winter. A period of low temperature under the snow may be of value as low temperature preconditioning (stratification) can result in more rapid germination at a lower temperature when the snow melts in the spring. This early start in turn provides an opportunity for better seedling establishment.

temperatures organ development may not necessarily be complete as Kimball and Salisbury (1974) have shown in relation to stomatal development in the leaves of *Claytonia lanceolata*. The underground parts (rhizomes and roots) of many alpine perennial dicotyledonous plants have a much greater dry weight than the shoots, and act as carbohydrate storage organs (see Billings and Mooney 1968), but not much is known about the seasonal growth of the roots of alpine species. Under the snow cover, the soil and therefore root temperatures may be higher than the air temperatures, which could give the roots a temperature advantage in relation to growth.

Although dark respiration generates the energy required for growth and tissue maintenance, there do not appear to be many studies centred on whole-plant or organ respiration at temperatures in the range from 0°C to 2°C which occur under the snow. Dark respiration is evident at temperatures as low as –5°C and then rises rapidly above 0°C (Mark 1975). It has been suggests that alpine plant species may have a higher respiration rate, possibly associated with a greater concentration of mitochondria, than those from warmer environments, and the efficiency of respiration in producing ATP may be important in determining the survival of these plants (Billings and Mooney 1968, Friend and Woodward 1990). However, these relationships, which have generally been examined in the range well above 0°C (commonly 10–20°C), have been called into question (Atkin and Day 1990, Collier 1995). Adaptation of respiration to low temperature could also be an important factor, but low temperature adaptation is not restricted to alpine plants and can also occur in plants of lowland origin (Larigauderie and Körner 1995).

While growth will not be rapid at 0 to 2°C there is still a demand for the supply of nutrients and carbohydrate. This must be met, either by the direct uptake of nutrients from the soil and current leaf photosynthesis, or from material stored in the plant prior to the onset of the snow cover. Whether carbohydrate is derived from current photosynthate, or from storage, the plant must be able to relocate this carbohydrate from the supply tissue to the growing organs, and this will require translocation through the phloem of the vascular system. In temperate species active translocation of carbohydrate through the phloem can occur at 1°C, but is greatly inhibited below 0°C (see Wardlaw 1979). On this basis, it may be concluded that alpine species are capable of phloem transport at temperatures close to 0°C (see Wallace and Harrison 1978), but this does not appear to have been evaluated experimentally and little is known about translocation in plants at lower temperatures.

Photosynthesis, or more appropriately net CO_2 exchange (NCE) has been observed at sub-zero temperatures (Mark 1975, Bate and Smith 1983, Wardlaw *et al.* 1989b), but whether this occurs will depend not only on temperature, but also on snow depth and light penetration. Figure 5 combines information on the penetration of light through snow with the effect of light on photosynthesis by the leaves of the Alpine Marsh Marigold at 0°C (Wardlaw *et al.* 1989b). From this analysis, it would appear that net CO_2 exchange would be negative at snow depths greater than 40 cm. It is not only the effect of light on photosynthesis that must be considered, but also the effect of

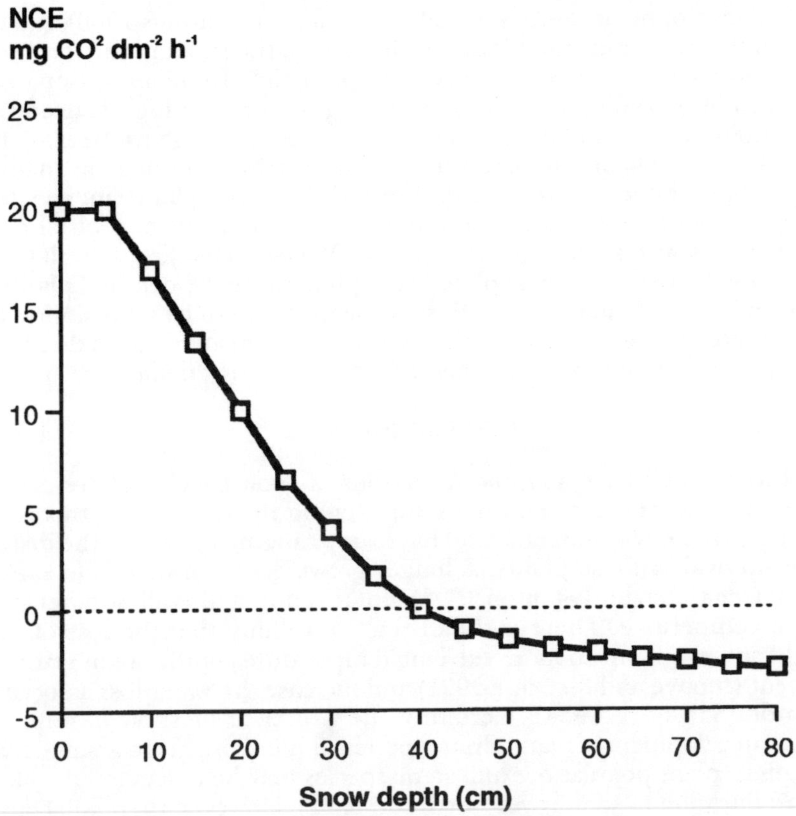

Fig. 5. The effect of the depth of snow cover on net CO_2 exchange (NCE) by the leaves of a snow patch plant. This figures combines the data on the change in light transmission as the snow depth increases and the photosynthetic response of the leaves of Alpine Marsh Marigolds to irradiance. (Wardlaw 1989b).

temperature and light on the photosynthetic apparatus. Thus it has been shown that the chlorophyll content of the leaves of some plants is inversely proportional to the depth of the snow cover (Kimball *et al.* 1973), a response more closely associated with variation in light rather than temperature. More recently, Hamerlynck and Smith (1994) found that the NCE by the leaves of Glacier Lily (*Erythronium grandiflorum* Pursh) was negative at snow depths of 10 cm (although light levels at this depth would be adequate to maintain a positive NCE in many plants), and development of the photosynthetic apparatus was retarded until after emergence from the snow cover.

On the basis of these observations, reserve carbohydrates rather than current photosynthate would need to be utilized to maintain growth under the snow (Mooney and Billings 1960, Salisbury 1984). Peak storage of carbohydrate occurs in autumn following seed development (Sørensen 1941, Mooney and Billings 1960) and much of this is found in the roots and crown tissue at the base of the shoot in the form of starch. However, the form of storage does vary between species and from one season to another (Mooney and Billings 1960), and it has been suggested that a high concentration of sugars rather than starch would provide a degree of protection against freezing (Shibata and Nishida 1993). Probably the main use of carbohydrate reserves in many alpine species is to supply the substrate for rapid growth after the plant emerges from the snow cover, when both light and temperature increase, and this may also be the case for nutrient reserves such as nitrogen (Jaeger and Monson 1992). Although there is no evidence that reserves will ever be depleted enough to cause the death of plants subject to snow cover in the Australian winter, this has been shown to be a possibility in other areas. For example, in some lichens prolonged snow cover can result in death and this is associated with the depletion of thallus carbon reserves (Benedict 1990).

CONCLUSION

Many alpine plants have adapted to periods of snow cover and are competitive with other species under these conditions. Shortening the duration of this cover, due to the warming of the environment, could have wide ranging effects on the distribution and possibly survival of these plants. A longer growing season would be expected to favour growth (see Galen and Stanton 1993), but this might also allow other species to become more competitive. There is the distinct possibility that the loss of the snow blanket would subject plants to more extreme temperature conditions at critical stages of development (Inouye and McGuire 1991) and increase the water loss associated with cold dry winds (Salisbury 1984). Certainly the presence of snow is important in controlling the development of many alpine species. It is reasonable to assume that there will be changes to plant populations and some species may be endangered by a shift in the depth and duration of snow cover (see Kudo 1992). However, there will continue to be speculation on the consequences of a changing snow cover until further experimental data and careful ecological observations are obtained to clarify the situation.

ACKNOWLEDGEMENTS

I would like to acknowledge the help and support of the National Parks and Wildlife Service of New South Wales and my colleagues Colin Totterdell and Mike Moncur for their efforts in getting me into the snow country.

REFERENCES

Andrews, C. J., 1996. How do plants survive ice?. *Ann. Bot.* **78:** 529–536.

Atkin, O. K. and Day, D. A., 1990. A comparison of the respiratory processes and growth rates of selected Australian alpine and related lowland plant species. *Aust. J. Plant Physiol.* **17:** 517–526.

Bate, G. C. and Smith, V. R., 1983. Photosynthesis and respiration in the sub-antarctic tussock grass *Poa cookii. New Phytol.* **95:** 533–543.

Benedict, J. B., 1990. Lichen mortality due to late-lying snow : results of a transplant study. *Arct. Alp. Res.* **22:** 81–89.

Bewley, J. D. and Black, M., 1994. "Seeds: Physiology of Development and Germination. 2nd Edition." Plenum Press, New York and London.

Billings, W. D. and Mooney, H. A., 1968. The ecology of arctic and alpine plants. *Biol. Rev.* **43:** 481–529.

Bliss, L. C., 1971. Arctic and alpine plant life cycles. *Ann. Rev. Ecol. Syst.* **2:** 405–438.

Collier, D. E., 1996. No difference in leaf respiration rates among temperate, subarctic and arctic species grown under controlled conditions. *Can. J. Bot.* **74:** 317–320.

Friend, A. D. and Woodward, F. I., 1990. Evolutionary and ecophysiological responses of mountain plants to the growing season environment. *Adv. Ecol. Res.* **20:** 59–124.

Galen, C. and Stanton, M. L., 1993. Short-term responses of alpine buttercups to experimental manipulations of growing season length. *Ecology* **74:** 1052–1058.

Hamerlynck, E. P. and Smith, W. K., 1994. Subnivean and emergent microclimate, photosynthesis, and growth in *Erythonium grandiflorum* Pursh, a snowbank geophyte. *Arct. Alp. Res.* **26:** 21–28.

Inouye, D. W. and McGuire, A. D., 1991. Effects of snowpack on timing and abundance of flowering in *Delphinium nelsonii* (Ranunculaceae): implications for climate change. *Amer. J. Bot.* **78:** 997–1001.

Jaeger, C. H. and Monson, R. K., 1992. Adaptive significance of nitrogen storage in *Bistorta bistortoides*, an alpine herb. *Oecologia* **92:** 578–585.

Junttila, O., 1976. Seed germination and viability in five *Salix* species. *Astarte* **9:** 19–24.

Khan, A. A. and Tao, K-L., 1978. Phytohormones, seed dormancy and germination. Pp. 371–422 *in* "Phytohormones and Related Compounds — A Comprehensive Treatise. Vol II Phytohormones and the Development of Higher Plants" ed by D. S. Letham, P. B. Goodwin and T. J. Higgins. Elsevier/North-Holland Biomedical Press: Amsterdam, Oxford, New York.

Kimball, S. L., Bennett, B. D. and Salisbury, F. B., 1973. The growth and development of montane species at near-freezing temperatures. *Ecology* **54:** 168–173.

Kimball, S. L. and Salisbury, F. B., 1974. Plant development under snow. *Bot. Gaz.* **135:** 147–149.

Kudo, G., 1992. Effect of snow-free duration on leaf life-span of four alpine plant species. *Can. J. Bot.* **70:** 1684–1688.

Larigauderie, A. and Körner, C., 1995. Acclimation of leaf dark respiration to temperature in alpine and lowland plant species. *Ann. Bot.* **76:** 245–52.

Leopold, C. A., 1964. "Plant growth and development." McGrae-Hill, New York.

Marchand, P. J., 1984. Light extinction under a changing snowcover. Pp. 33–37 *in* "Winter Ecology of Small Mammals" ed by J. F. Merritt. Special Publication of Carnegie Museum of Natural History, Pittsburgh.

Mark, A. F., 1965. Flowering, seeding, and seedling establishment of narrow-leaved snow tussock, *Chionochloa rigida. N. Z. J. Bot* .**3:** 180–193.

Mark, A. F., 1970. Floral initiation and development in New Zealand alpine plants. *N. Z. J. Bot.* **8:** 67–75.

Mark, A. F., 1975. Photosynthesis and dark respiration in three alpine snow tussocks (*Chionochloa* spp.) under controlled environments. *N. Z. J. Bot.* **13:** 93–122.

Moony, H. A. and Billings, W., 1960. The annual carbohydrate cycle of alpine plants as related to growth. *Amer. J. Bot.* **47:** 594–598.

Richards, D. and Beardsell, D., 1987. Seed dormancy. Pp. 1–13 *in* "Germination of Australian Native Plant Seed" ed by P. Langkamp. Inkata Press, Melbourne, Sydney.

Richardson, S. G. and Salisbury, F. B., 1977. Plant responses to the light penetrating snow. *Ecology* **58:** 1152–58.

Salisbury, F. B., 1984. Light conditions and plant growth under snow. Pp. 39-50 *in* "Winter Ecology of Small Mammals" ed by J. F. Merritt. Special Publication of Carnegie Museum of Natural History, Pittsburgh.

Salisbury, F. B. and Ross, C. W., 1985. "Plant Physiology. 3rd Edition." Wadsworth Publ. Co, Belmont, California.

Schmid, W. D., 1984. Materials and methods of subnivean sampling. Pp. 25–32 *in* "Winter Ecology of Small Mammals" ed by J. F. Merritt. Special Publication of Carnegie Museum of Natural History, Pittsburgh.

Shibata, O. and Nishida, T., 1993. Seasonal changes in sugar and starch content of alpine snowbed plants, *Primula cuneifolia* ssp. *hakusanensis* and *Fauria crista- galli*, in Japan. *Arct. Alp. Res.* **25:** 207–210.

Slatyer, R. O., Cochrane, P. M. and Galloway, R. W., 1985. Duration and extent of snow cover in the Snowy Mountains. *Search* **15:** 327–331.

Sørensen, T., 1941. Temperature relations and phenology of the northeast Greenland flowering plants. *Meddr. Grønland* **125:** 1–305.

Thomas, W. H., 1972. Observations on snow algae in California. *J. Phycol.* **8:** 1–9.

Wallace, L. L. and Harrison, T. A., 1978. Carbohydrate mobilization and movement in alpine plants. *Amer. J. Bot.* **65:** 1035–1040.

Wardlaw, I. F., 1979. The physiological effects of temperature on plant growth. *Proc. Agron. Soc N.Z.* **9:** 39–48.

Wardlaw, I. F., Moncur, M. W. and Totterdell, C. J., 1989a. The growth and development of *Caltha introloba* F. Muell. I. The pattern and control of flowering. *Aust. J. Bot.* **37:** 275–289.

Wardlaw, I. F., Moncur, M. W. and Totterdell, C. J., 1989b. The growth and development of *Caltha introloba* F. Muell. II. The regulation of germination, growth and photosynthesis by temperature. *Aust. J. Bot.* **37:** 291–303.

CHAPTER 8

A winter niche: the subnivean space

KEN GREEN

MOUNTAIN regions are renowned for their harsh environments, particularly in winter. However, severe weather conditions encountered in the mountains give little indication as to the environment inhabited by small animals, in fact they can be very misleading (Fuller *et al.* 1969). There exists beneath the snow cover an environment that is relatively moist and warm and, in the average year, remarkably stable (Pruitt 1958) and below a certain depth, dark. Additionally, this subnivean space provides a variety of food types and is protected from most predators (Evernden and Fuller 1972). However, the subnivean space is dependent upon the insulative cover of snow and so, with changing depths of snow cover, the severity of the subnivean microclimate varies from year to year. An understanding of this subnivean environment is important to our understanding of the survival of small animals through the winter.

FORMATION OF THE SUBNIVEAN SPACE

The common perception that snow lies flat on the ground is in error; there is a space, sometimes substantial, that occurs between the ground and the underside of the snow pack. This is called the subnivean space from the Latin *nivis* for snow. The subnivean space may be formed in a number of ways. Snow falling on vegetation may form an interconnecting layer over the vegetation, supported by both adherence to the plants themselves, and by the intercrystaline bonds within the snow. This enables it to bridge over open spaces very much like a blanket. The depth of subnivean space depends upon the height of the vegetation and how dense and how strong it is (Coulianos and Johnels 1962). The strength of vegetation is greater than often imagined. A pasture of clover and grass for example can hold the snow from three to eight centimetres off the ground (Coulianos and Johnels 1962). In Australia, in areas of open spreading shrubs such as Ovate Phebalium (*Phebalium ovatifolium*) or on grassland, the snow will penetrate to ground level, whereas with more thickly growing shrubs such as Mountain Plum Pine (*Podocarpus lawrencei*) snow is more likely to lie on top of the canopy (K. Green pers. obs.).

However, strong winds will carry the snow through any vegetation to fill in the space. It is then, and also in areas with little plant cover, that the other main factor in the formation of the subnivean space comes into play. It is generally true that even the driest and coldest soil is warmer and more moist than the air above the snow and, as a result, there is a temperature and vapour pressure gradient through the snow cover (Pruitt 1970). The thermal gradient existing in the snow from the warmer, lower levels to the cooler, upper layers causes water vapour (formed by sublimation) to be transported upwards through the profile (Coulianos and Johnels 1962). As a result, the snow nearest the ground is eroded away, leaving behind a subnivean space between lattice-like walls and roof (Pruitt 1970).

The subnivean space is widespread, and an almost continuous subnivean space may cover several thousand square metres (Johnson 1957). The subnivean space will usually last through the winter under most normal circumstances. Only 10–15 cm of snow are required to enable a network of tunnels for small mammals to exist through winter (Formozov 1964). However, the subnivean space may collapse under pressure in years of heavy snow. In many average years it will collapse in deep gullies alongside creeks where the snow may build up to a great depth unsupported from beneath as snow is gradually worn away by the flowing water. The subnivean space is also unable to persist where the snow is heavily compacted, particularly by vehicle movement (K. Green pers. obs.).

TEMPERATURES

Snow cover is a mixture of air and water (in its frozen state) and the greater the amount of trapped air, the greater the amount of heat that can be trapped. The thermal conductivity of snow varies approximately as the square of its density (Fuller *et al.* 1969). The insulative value of snow being given by $I = 1/K$ where $K = 0.0068p^2$, K being the thermal conductivity and p the snow density) (Williams and Gold 1953). Generally, new snow has a density of 0.1 to 0.2 $g.cm^{-3}$ and old snow 0.3 to 0.5 $g.cm^{-3}$ (Geiger 1950). The Snowy Mountains receives wetter snow (true powder being rare) and the average density of snow measured at the first snowfall contributing to the winter cover, is about 0.27 $g.cm^{-3}$ (Green and Osborne 1994, see also Chapter 2). However, new snow falling in the middle of the season when the air is colder may well be of lower density.

The depth at which snow achieves effective insulation is thought to be at about 20 cm, with further snow doing little to stabilise the microclimate within the subnivean space (Williams and Gold 1953); this point was called the "hiemal threshold" by Pruitt (1970). Above an unbroken 20 cm of snowpack the temperature may vary by 25°C while beneath it the fluctuation may be no more than 2°C (Coulianos and Johnels 1962). Formozov (1964) reported that the surface of the soil near Leningrad averaged 15°C higher beneath the snow than in the open, with the average temperature beneath the snow in January being 19°C higher than above the snow (Formozov 1964). However, if the snow depth is insufficient to cover the vegetation completely, particularly on a

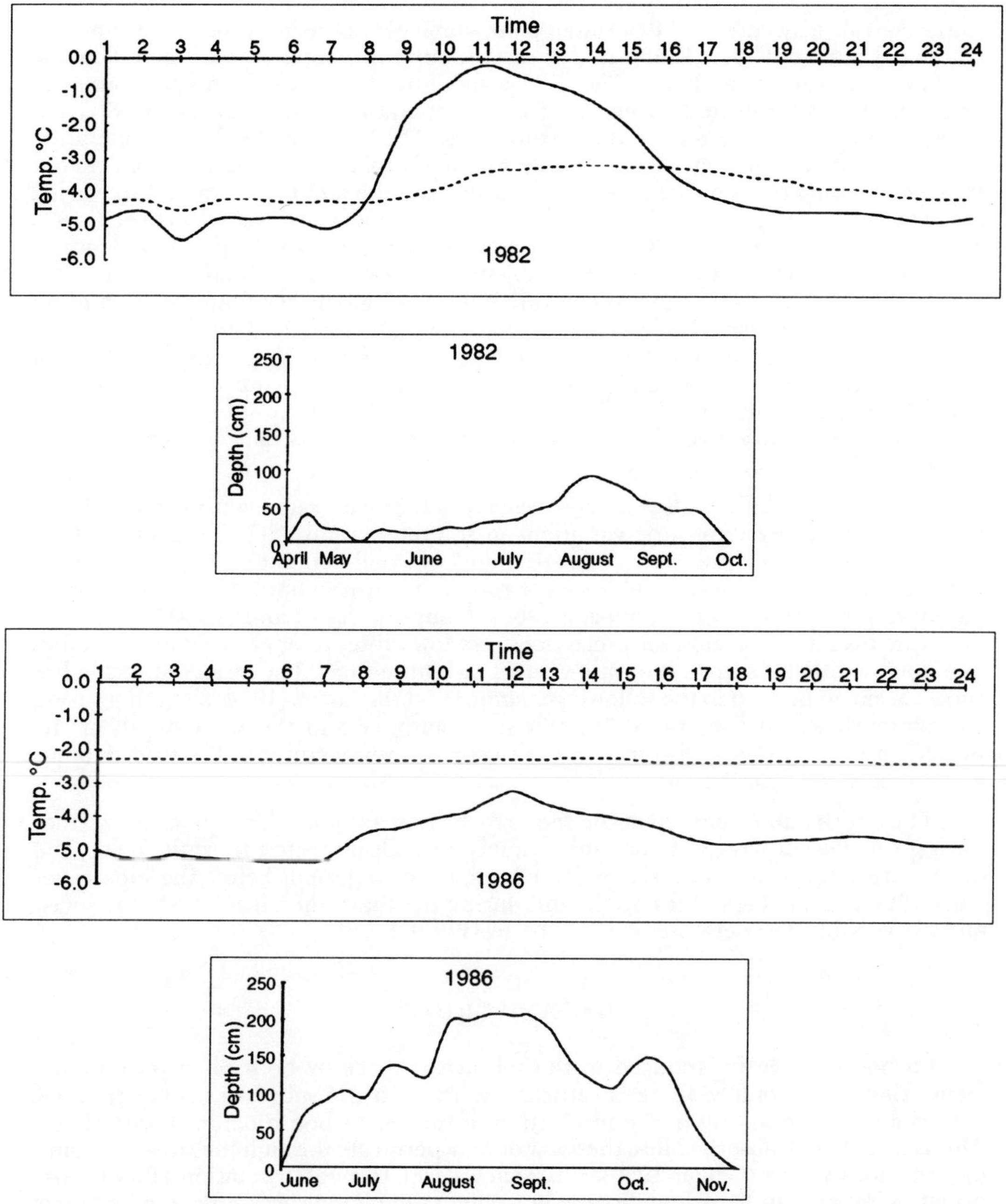

Fig. 1. Average temperature for each of 24 hours of the day beneath the snow (dashed line) and above the snow (solid line) on the South Ramshead in an average snow year (1986) and a low snow year (1982). Smaller graphs show snow depths recorded at the SMA snow course at Spencers Creek.

slope, cold air may enter and flow through the subnivean space in a process of subnivean cold air drainage (Green 1988). In this case, subnivean temperatures may follow the air temperatures more closely (Fig. 1). During July, 1982 in the Snowy Mountains when there was little snow cover, subnivean temperatures fell to as low as –8°C and remained below –5°C for many days at a time (Green 1988). On a site where temperature was being recorded in July, 1982, snow depths at 13 stakes averaged 40.3±6.0 cm but there were many holes in the snow cover. At this site the ambient temperature varied between –10°C and +7°C, while the subnivean temperature varied between –8°C and –1°C. By contrast, in July, 1986 at the same site with an average snow depth of 65.7±7.6 cm and full snow cover, the ambient temperature fluctuated between –10°C and +3°C with the subnivean temperature only varying between 0 and –3°C, and on average remaining remarkably steady at just below –2°C (Fig. 1). This shows that with an incomplete snow cover, winter conditions in Australia can be as extreme for small mammals as those encountered in the Arctic. In the taiga of the northern hemisphere it has been shown that small mammals normally inhabiting the subnivean space are incapable of survival above the snow — or more relevant to Australia, without the snow cover and the protection it affords.

The high insulative value of snow leads to a thermal gradient through its depth (Geiger 1950) preventing wide variations in soil temperature. Thus the snow offers protection to the soil, preventing "freezing out" of small mammals (Formozov 1964). With "freezing out", the soil and litter are completely frozen hard, hindering animals attempting to obtain food (Formozov 1964). Many species of animals which depend upon the insulation of snow for protection from low winter temperatures are therefore in a doubly difficult situation during winters with little snow. The results of such a low snow cover can be seen in the following summer and Formozov (1964) showed a strong correlation between the number of shrews in summer and the snow depth of the preceding winter. This is also true in the Australia snow country of the insectivorous marsupial species (see below).

The depth and insulation of the snow reaches a maximum through winter, so that during the coldest days of winter the animals are well protected from the cold. The most dangerous periods for small mammals are the cold periods before the snow cover has built up to the hiemal threshold and during the thaw when insulation is reduced and yet very low temperatures can still occur (Pruitt 1970).

GAS AND LIGHT

Carbon dioxide is produced in the subnivean space by both plant and animal respiration. Additionally, there is sufficient warmth in the subnivean space to allow bacterial action, resulting in the production of further carbon dioxide (Pruitt 1970). This may either be absorbed into the snow or may permeate through the snow, escaping into the atmosphere (Geiger 1950). If the snow cover, through formation of icy layers, becomes impervious to carbon dioxide movement then it may accumulate in low spots in sufficient quantity to be harmful to small mammals (Pruitt 1970). Schmid (1984), however, found no more than 0.14% CO_2 in samples taken from beneath such ice layers during the thaw compared to a normal 0.03% in the atmosphere.

In areas of homogeneous subnivean microhabitat with respect to temperature and food, the heterogenous distribution of small mammals may be related to CO_2 (Penny and Pruitt 1984). Because small mammals react physiologically to increased CO_2 they avoid such areas of elevated CO_2, moving their home ranges away and returning later when the subnivean level of CO_2 have dropped (Penny and Pruitt 1984). However, if there is any opening to the surface, such as where plant stems protrude or via a small mammal burrow, the cold air will sink to these low levels to replace the warmer carbon dioxide. Northern small mammals such as voles are known to excavate ventilator shafts to the surface (Pruitt 1970) and "popholes" in the snow are common in Australia beneath snowgums.

Sunlight falling on the surface of the snow may be reflected, transmitted or absorbed. Most of the light not reflected by snow is absorbed as heat energy giving rise to melting above 0°C or sublimation below 0°C. New snow has one of the highest albedos of naturally occurring surfaces at 75 to 95% reflectivity, although as it gets older and dirtier it reflects less light and the albedo may drop to as low as 40% (Geiger 1950). A light fall of new snow on old snow has a similar effect on light levels beneath the snow as has the sun going behind a cloud (Fig. 2). Thirty cm of snow transmits about 8% of the incident light and by 60 cm that level is down to only 1% (Pruitt 1960 and Fig. 3). Because the albedo of older snow is less, light will be transmitted to a much greater depth and at 85 cm the amount of light may be equivalent to that found in the corner of a photographic darkroom. This means that for most of winter with a normal cover of snow the small animals of the subnivean space exist in virtual darkness beneath a ceiling of snow. For animals that normally use visual cues for hunting and are dependent upon changes in daylength for the initiation of reproduction, life in the subnivean environment is akin to living in a cave.

Light is not transmitted through snow in its white form; as snow depth builds up there is a shift in the transmitted light to the red end of the spectrum (Evernden and Fuller 1972). As the season progresses, the amount of light entering the subnivean space changes, through increased transmittance, even if the snow depth remains the same. This is due to a higher angle of incidence of the sun towards the equinox, the reduction in the albedo of the snow and changes in the snow itself that aid transmittance (Evernden and Fuller 1972). An increase in activity of mites towards the end of the winter beneath the snow has also been attributed to their sensitivity to increases in light intensity (Aitchison 1979e). As snow metamorphoses over winter, becoming more dense, shorter wavelength light (blue) is able to penetrate further into the snow (Evernden and Fuller 1972). The transmission of light through snow is important, because both the photoperiod and the spectral composition of light have been shown to affect reproduction. Photoperiod is a well known cue for reproduction; additionally, red light can inhibit sexual maturation in, for example, microtine rodents (Evernden

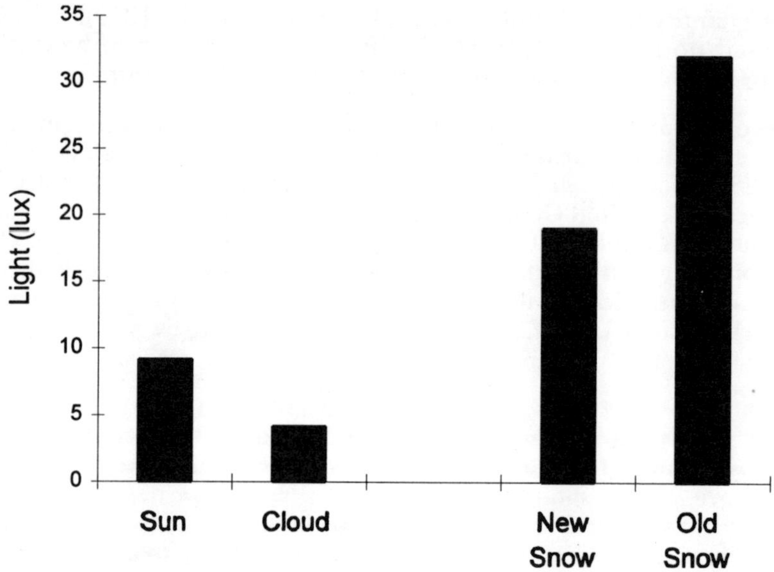

Fig 2. Light levels at Smiggin Holes at the same depth of snow beneath full sun and cloud, and at 36 cm between old snow and the same snow with a light dusting of a few millimetres of fresh snow.

and Fuller 1972). In the subnivean space, four things may help to determine the onset of sexual maturation: photoperiod, spectral composition of light, the condition of the animal and cold. All of these will be determined to a certain extent by snow depth and duration. Early loss of snowcover in Canada resulted in early weight gain and sexual maturation in Red-backed Voles (*Clethrionomys gapperi*) (Fuller 1969). However, related species are capable of subnivean breeding (Evernden and Fuller 1972). Reproduction under the snow is common in a number of species throughout the world. In the Snowy Mountains the Antechinuses are able to breed synchronously in spring with very little change in the time of breeding from year to year. How the cues for breeding are received are poorly known. Some animals may come to the surface of the snow where they receive cues about daylength, but most over-snow excursions by small mammals are in the early rather than mid-winter period immediately before the mating time. Differences in snow depth do not affect the time of breeding of the Dusky Antechinus (*Antechinus swainsonii*) (which appear to be quite active above the snow surface), whereas the Broad-toothed Rat (*Mastacomys fuscus*)(which Osborne (1980) regarded as not greatly active above the snow) seems to breed earlier when the snow season ends earlier (Green and Osborne 1994).

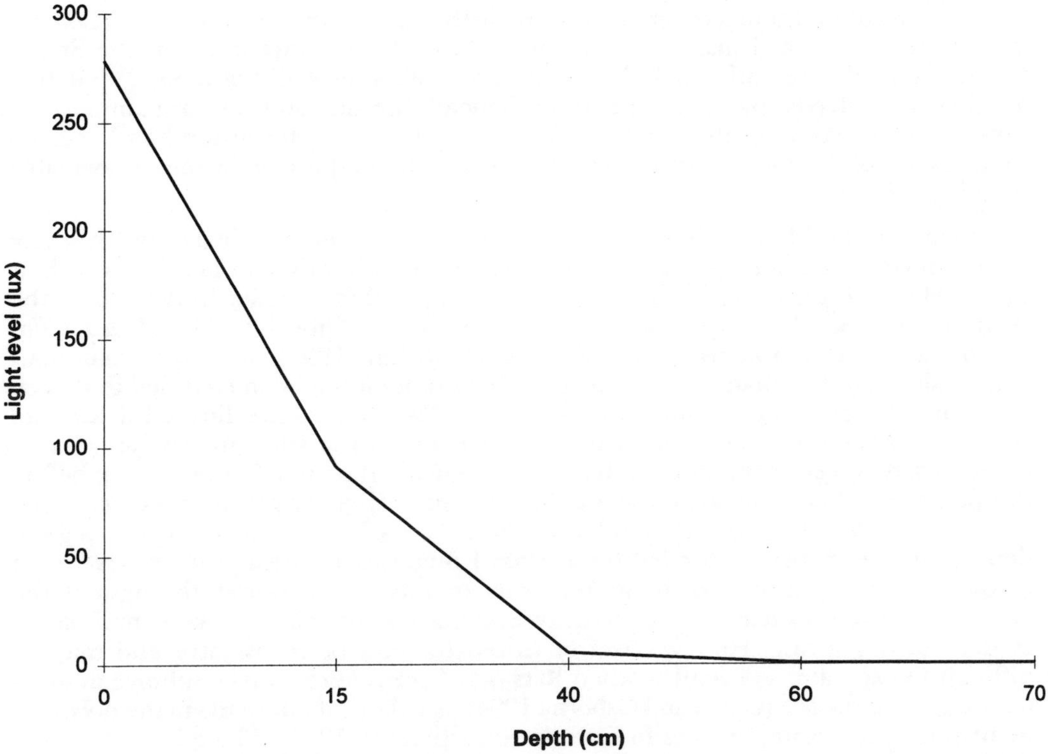

Fig. 3. The loss of light with increasing depth of snow measured at Smiggin Holes.

THE ANIMALS OF THE SUBNIVEAN SPACE

Mammals

The temperature experienced by small mammals on the snow surface is appreciably lower than that recorded in a standard meteorological screen (Coulianos and Johnels 1962). Many small mammals would be unable to survive in this environment, especially such species as the Dusky Antechinus which declines in body mass with increasing altitude and is therefore highly dependent upon the snow cover to protect it from very low temperatures. The life of mammals is, therefore, considerably more comfortable beneath the snow than above it. The subnivean space is so well protected that mid-winter is not a time of great thermal stress and, in fact, it is in both the pre-nival period (Pruitt 1960, Halfpenny and Ozanne 1989) and during and after the thaw (Courtin *et al.* 1991) that the most thermally stressful times for animals occur due to both the exposure to cold without a cover of snow and with rain falling instead of snow (Courtin *et al.* 1991). The small insectivorous mammal, the Dusky Antechinus in the Snowy Mountains is adversely affected by both deep and shallow snow (Green 1988). Population numbers are affected by low temperatures beneath the shallow snow and animals are exposed to greater levels of predation. Deep snow, on the other hand, causes compression of the lower snow levels and of the subnivean space, reducing the available foraging area.

Mammals and birds will often seek shelter beneath a bush or leafy branch where cavities form under the mantle of snow. This is true both for larger mammals such as Brown Hares (*Lepus capensis*) (K. Green pers. obs.) and for smaller birds such as the White-browed Scrubwren (*Sericornis frontalis*) and Brown Thornbills (*Acanthiza pusilla*) which may even roost beneath the snow at night (Schulz 1991). Subnivean mammals almost all weigh less than 250g although in the tundra it has been recorded that even foxes may become subnivean at times (Pruitt 1984) and I have flushed a Ringtail Possum *(Pseudocheirus peregrinus)* from a nest beneath the snow in daylight. Small mammals, however, spend most of their winter beneath the snow. In the autumn before the first snow, there appears to be a large amount of digging going on especially by Bush Rats (*Rattus fuscipes*) as they enlarge their burrows. The burrows can go to a great depth; one I excavated extended for at least 1.8 metres at which point it was 78 cm below the soils surface. This digging activity by rats may continue throughout the winter in some sites leaving the characteristic bolsters of soil and plant material so evident in the spring. However, wet or frozen soil is a poor insulator and may be difficult to excavate, so Broad-toothed Rats *(Mastacomys fuscus)* make subnivean nests on the ground surface (Green and Osborne 1994), like their counterparts in the northern hemisphere, for example voles in Ontario (Courtin *et al.* 1991). There is no need for deep snow before this occurs, and Formozov (1964) recorded voles building winter nests under as little as 20–25 cm of snow. Broad-toothed Rats appear to be the most likely to move out from subterranean nests and construct nests on the surface of the ground. The nests are generally compact bundles of grass (Green and Osborne 1994), and elsewhere Formozov (1964) recorded subnivean nests as differing from subterranean nests in having a greater thickness of wall and more insulative material in the construction. Because of the good insulation of the snow surrounding the well-insulated nest, overwintering above the soil is favoured (Formozov 1964). This, however, makes life easier for Foxes which only have to dig down through soft snow to get to their prey rather than through the soil into nests (Formozov 1964). Mammals beneath the snow will use subnivean tunnels to store waste earth as they dig into the soil or will tunnel through the grass at the snow-earth interface and store the cut vegetation in snow tunnels. These are compacted under the snow and emerge in spring as a cast showing just a small example of the complicated network of tunnels that existed in winter.

Pruitt (1960) found that Red-backed Voles (*Clethrionomys rutilus*) avoided areas of shallow snow through winter while Whitney and Feist (1984) found that their density and survival was higher in forested and shaded aspects than sunny aspects because of the better subnivean temperature conditions. Spencer (1984) found that long-tailed voles (*Microtus longicaudus*) chose habitats which provided easier subnivean access even, where necessary, resorting to less preferred, but more easily obtained foods to reduce the costs of burrowing. Energy is saved both by this behavioural change and by creating a skeleton network of passages while the snow is still at low density early in the season (Spencer 1984). The simple act of mowing a pasture prior to snowfall with an ordinary lawnmower was sufficient to prevent animals moving into it, and runways led through the unmown pasture to the border of the mown/unmown area but not beyond (Coulianos and Johnels 1962). A similar situation can be seen in Australia where groomed cross-country ski trails cut through small mammal habitat creating a barrier to subnivean movement (K. Green pers. obs.). Coulianos and Johnels (1962) concluded that areas with dense ground vegetation attract small mammals in winter, whereas areas with the same vegetation but mown were not so attractive. They concluded that the presence or absence of a subnivean space was the most important factor in their choice.

Small mammals will avoid high density snow and Schmid (1971) found that voles would come to the snow surface to cross over-snow vehicle tracks rather than burrow through dense snow compacted to 0.4 $g.cm^2$. When mammals do tunnel, they tend to pick layers which have lower densities. One advantage of snow as a structure is that it enables rodents to tunnel upwards to parts of plants inaccessible to them in summer so that after the thaw rodent teethmarks can be seen where they have chewed the bark on plants a metre or more above the ground, or have completely denuded the tops of shrubs such as ovate phebalium.

Oversnow movement

Early in the season, with only a few centimetres of snow on the ground there is major activity of small mammals on the snow. This decreases with increasing snow cover until the subnivean space is complete and the numerous access points onto the snow surface are reduced to just a few such as at the base of trees.

Although Carron (1985) found Broad-toothed Rats were often negatively associated with canopy cover, trees are important for animals in the subnivean space. They provide holes for the venting of CO_2, access to the surface to cross areas of little or no subnivean space and more importantly access back beneath the snow to avoid predators. Thus it is not surprising that in a study of distribution of Broad-toothed Rats in the Perisher Valley, Holloway and Osborne (1997) found "that across all sites *Mastacomys* preferred sites close to trees". The importance of breaks in the snow was also shown by Sonerud (1986) who found significantly fewer trails of small mammals in deep snow on a clearcut forest than in surrounding uncut forest. This was despite the fact that with a light snow cover there was more activity in the clearcut area. A similar result has been found in Australia where more small mammal tracks can be found in woodland than in heath (K. Green unpubl. data). Also, tracks in heath tended to be longer because there is usually a greater distance between breaks in the snow surface into which an animal can retreat. Thus, the average length of tracks may be three times as long in heath (56.7±91.0m, n=12) as in woodland (18.1±17.1m, n=49) (K. Green unpubl. data). The reason for this can easily be seen by comparing similar woodland sites in the thaw (with many popholes at the base of trees) and after a snowfall. In 1995, tracks in woodland with old snow and popholes leading to the subnivean space averaged 9.5±13.4m in length (n=13) and were significantly (t=2.719, p=0.01) shorter than eight days later (30.7±24.3m, n=12) after a fall of fresh snow blocked many such popholes (K. Green unpubl. data).

Invertebrates

The insects most commonly seen in relation to snow cover are those found on the snow surface (Edwards 1973, Osborne *et al.* 1978). However, these insects are generally non-resident species, either spring migrants such as Bogong Moths (*Agrotis infusa*) or insects recently metamorphosed at lower altitudes and blown up into the mountains while the winter snowpack still exists. These insects are probably far outnumbered by the invertebrates active beneath the snow. It was generally assumed that invertebrates beneath the snow were inactive until the pioneering subnivean pitfall trapping study by Näsmark (1964), but the activity beneath the snow is still not a widely recognised phenomenon. However, a number of invertebrates have been recorded as active at temperatures close to 0°C (references in Aitchison 1979f) with Collembola recorded as active by Aitchison (1979d) at temperatures down to –6°C, Acari to –10°C (Aitchison 1979e), and spiders to –8°C (Aitchison 1978). The factors that favour winter activity of invertebrates, are higher temperatures (nearer to 0°C) and high relative humidity (Aitchison 1974), these are the very conditions that occur beneath the snow.

The subnivean invertebrate fauna worldwide appears to be dominated by Acari (mites), Araneae (spiders), Collembola (springtails) and Coleoptera (beetles) (Aitchison 1984), a situation that is no different in Australia (Green 1997). Subnivean captures of invertebrates in the Snowy Mountains are dominated numerically by the Collembola, Coleoptera and Araneae with Acarina and Diplopoda occasionally important at high subalpine and alpine sites with the greater prominence of Diptera and Hymenoptera at lower subalpine sites (Green 1997). Aitchison found activity of some groups (Collembola, Coleoptera, and Diptera) to be highest in early winter then lowest in late winter Aitchison (1979d, f and a respectively).

In a study undertaken in the Snowy Mountains Green (1997) found there was a reduction in the activity of invertebrates as measured by their numbers in pitfall traps from snow-free months to winter with an increase in spring (Fig. 4). The lowest numbers of invertebrates in pitfall traps in winter 1982 fell to 12.2% and 4.6% of maximal numbers from the preceding snow-free months in an alpine site and subalpine site respectively (Green 1997). In a Canadian study, the numbers of invertebrates in pitfall traps, and the diversity of species generally, decreased in autumn, were lowest in winter beneath the snow and increased gradually in late spring (Aitchison 1984). The situation in the Snowy Mountains differs with respect to autumnal peaks and the rapid increase in numbers in spring (Fig. 4).

There are marked differences in the activity of subnivean invertebrates in the Snowy Mountains with altitude, season and year presenting an opportunity to study

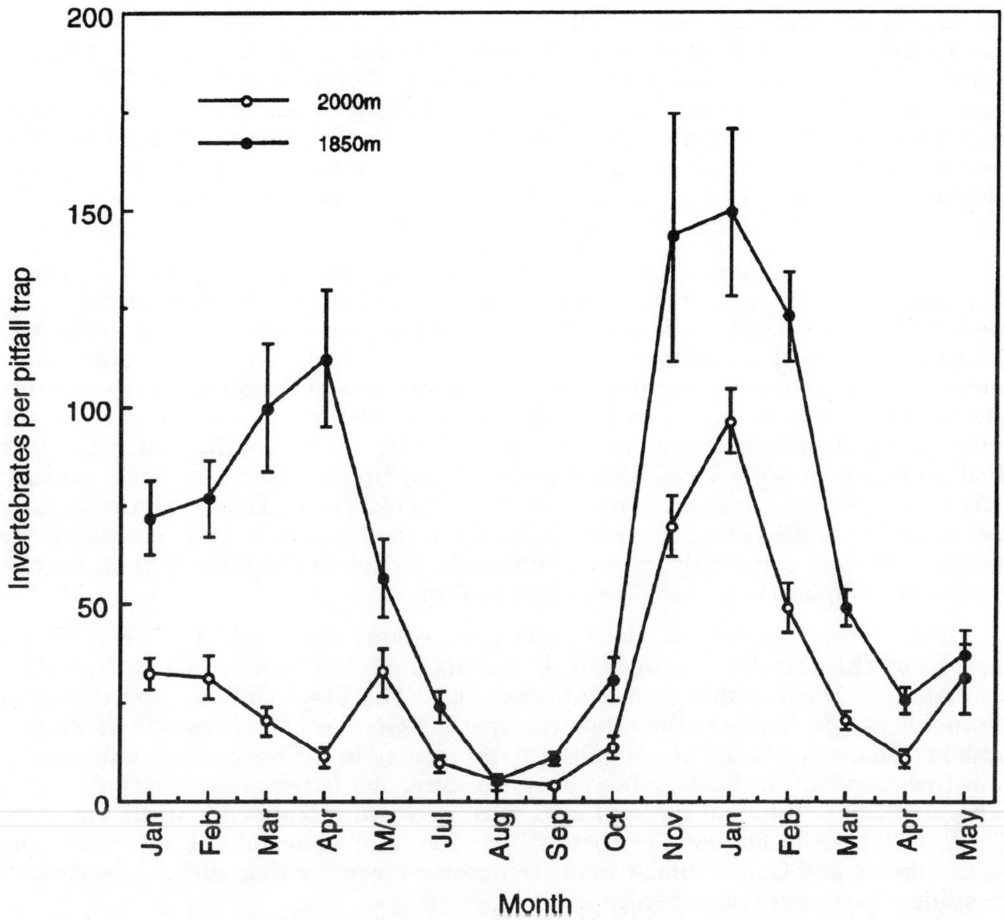

Fig 4. The mean number of individuals per pitfall trap with standard errors at an alpine site (2000m) and high subalpine site (1850m) in the Snowy Mountains for 1982 and 1983.

the long-term impact of climate change, particularly in regard to the effects of differing snow cover (Green 1997). However, Green (1997) found there were few differences attributable to snow depth between numbers of invertebrates caught in pitfall traps on the South Ramshead Range following the two winters (1981 and 1982) having the deepest and shallowest snow covers since records began in the Snowy Mountains (Green 1988). Only two groups suffered significant declines in numbers from February, 1982 to February, 1983. These declines, of amphipods and springtails which are normally associated with moist habitats, were probably attributable more to the dry conditions of 1982/83 (Green 1988). Overall, the number of invertebrates increased.

Among the spiders recorded in the Snowy Mountains there are six families found beneath the snow; in general order of abundance these are: 'Amaurobiidae', Linyphiidae, Lycosidae, Micropholcommatidae, Orsolobidae and Thomisidae (Green 1997). The most common species in summer, '*Rubrius*' sp. and *Storenosoma* sp., are also the most common winter species. This order differs from overseas where the "Amaurobiidae" have not been found at all in studies of the subnivean fauna and only two families (Lycosidae and Linyphiidae) are shared with Europe; and these together with the Thomisidae are also shared with Canada (Green 1997).

Collembola are characteristic of arctic and high alpine environments (Fjellberg 1975, Chernov 1985) and are the most commonly trapped subnivean invertebrates in Sweden (Näsmark 1964) and, after the Acari, the second most common in Canada (Aitchison 1979d, e). In the Snowy Mountains, Green (1997) found that Collembola were the most commonly trapped subnivean invertebrates in alpine and subalpine sites through one year but in another the Araneae were more common early in the winter, and Collembola were more common in the second half of the winter. In order of abundance in subnivean pitfall traps the springtail families were ranked: Sminthuridae, Neanuridae, Entomobryidae, Tomoceridae, Hypogastruridae and Isotomidae. This differs from overseas studies where the Isotomidae and Entomobryidae were the two most commonly trapped families in the subnivean space both in Europe (Näsmark 1964) and in Canada (Aitchison 1979d).

In the Snowy Mountains, the beetles most commonly caught by Green (1997) were from the families Leiodidae, Carabidae, Melandryidae, Curculionidae, Lathridiidae, Scymaenidae and Endomychidae. The Staphylinidae was the most commonly caught family in the subnivean space in Germany (Heydemann 1956) and Canada (Aitchison 1979f) but was only well represented in the Snowy Mountains in the snow-free months. The beetles showed a great difference between summer and winter with only 8.5% of the total species found during the study trapped beneath the snow (Green 1997). Eight families (11 species) were found in winter pitfall traps with only the Carabidae and Curculionidae being represented by more than one genus. As with the spiders there were no specialist winter species.

In the Snowy Mountains, Diptera and Hymenoptera are active beneath the snow throughout the winter with both winged and apterous species being present (Green 1997). Molluscs, oligochaetes, centipedes and pseudoscorpions are generally caught infrequently under snow in the Snowy Mountains (Green 1997), in Sweden (Näsmark 1964) and in Canada (Aitchison 1979b, 1979c) but activity has been found in some species whenever temperatures exceed 0°C (Weygoldt 1969). Interestingly, Green (1997) found that pseudoscorpions were more likely to be caught in the subnivean space than in summer.

There are differing trends in numbers of species that are adapted to the subnivean space in Australia and Canada; among the Collembola, 19 species have been trapped in the Snowy Mountain (Green 1997) compared to only 16 species (mainly isotomids) in Canada (Aitchison 1979d). By contrast, 62 species of Coleoptera, from five families,

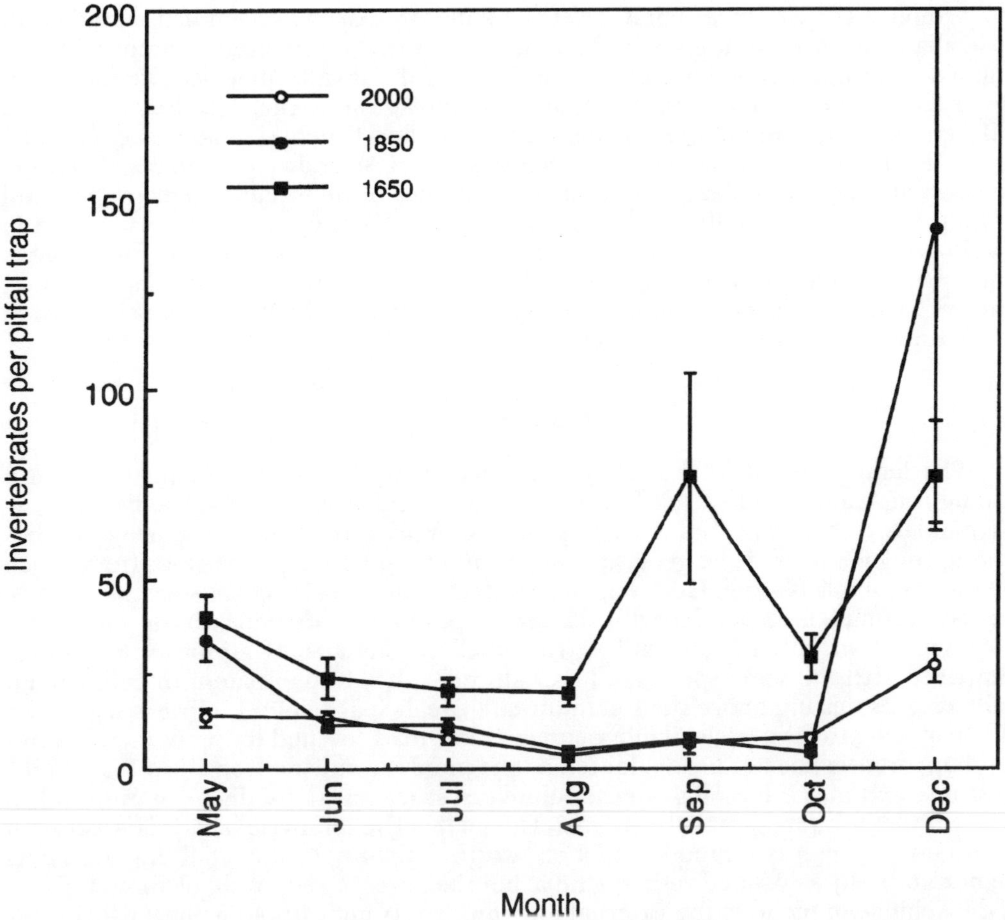

Fig 5. The mean number of individuals per pitfall trap with standard errors in an alpine site (2000m), high subalpine site (1850m) and low subalpine site (1680m) in the Snowy Mountains for May to December, 1986.

are active in winter beneath the snow in Canada (Aitchison 1979f) compared to only 11 in the Snowy Mountains (Green 1997). All of these active species must possess some adaptations to a cover of snow, and invertebrates that have spent the winter beneath the snow in diapause may lack such adaptations and are therefore particularly susceptible to adverse environmental conditions once they have emerged for the summer. Green (1997) documented an increase in the numbers of invertebrates at low subalpine altitudes after the thaw but a subsequent decrease after a heavy October snowfall. However, at higher subalpine and alpine altitudes where the snow cover had not thawed, the invertebrates had not emerged and were therefore unaffected (Fig. 5).

Winter activity of invertebrates in the subnivean space has been better studied in Australia than in mountains elsewhere in the world. The Australian mountains are isolated from other snow-covered mountains, and the taxa from which the mountain fauna has evolved differs from the faunal assemblages elsewhere leading to marked differences in the families of Araneae, Collembola and Coleoptera active beneath the snow. The invertebrate fauna of the subnivean space of Australian mountains, therefore, presents an ideal assemblage of animals for studies of mountain biogeography and evolution. This is true in their own right but additionally they are an important part in a subnivean food chain proposed by Aitchison (1984) consisting of fungivorous collembolans (and herbivorous cicadellids) as primary consumers, predatory mites, spiders and beetles as secondary consumers and the shrew (for Australia read Antechinus) as the tertiary consumer.

CONCLUSION

The fauna present in the subnivean space of Australian mountains represent a unique collection of animals differing from those studied overseas in their taxonomy (marsupials rather than shrews among the mammals and a host of differing families among the insects) and the geographical origin of the source populations from which the mountain forms arose. However, faced with the same challenges as species overseas, the local fauna has responded with similar adaptations to life beneath the snow. This is evident in morphology, including the smaller body size of alpine Antechinuses compared with the same species at lower altitudes. It is also apparent in behavioural traits such as nesting above the soil in the subnivean space, found in overseas species which appears to have evolved independently in Broad-toothed Rats, but on the other hand apparently has not been adopted by the more recently-arrived Bush Rat. Whilst most research into the ecology of the subnivean space has naturally been undertaken overseas these differences and similarities mean that the subnivean space of Australian mountains represents a unique situation, worthy of study. Harsh cold environments are not normally associated with Australia, but the potential exposure of animals fitted to the subnivean niche to the external environment is not without its hazards. In this respect it is important in any studies of global warming and the possible disappearance of snow cover to consider the transition that such animals will face in being exposed to the full brunt of winters without snow.

REFERENCES

Aitchison, C. W., 1974. A sampling technique for active subnivean invertebrates in southern Manitoba. *Manitoba Ent.* **8:** 32–36.

Aitchison, C. W., 1978. Spiders active under snow in southern Canada. *Symp. Zool. Soc. Lond.* **42:** 139–148.

Aitchison, C. W., 1979a. Winter-active subnivean invertebrates in southern Canada. IV. Diptera and Hymenoptera. *Pedobiologia* **19:** 176–182.

Aitchison, C. W., 1979b. Notes on low temperature activity of oligochaetes, gastropods and centipedes in southern Canada. *Amer. Midl. Nat.* **102:** 399–400.

Aitchison, C. W., 1979c. Low temperature activity of pseudoscorpions and phalangids in southern Manitoba. *J. Arachnol.* **7:** 85–86.

Aitchison, C. W., 1979d. Winter-active subnivean invertebrates in southern Canada. 1. Collembola. *Pedobiologia* **19:** 113–120.

Aitchison, C. W., 1979e. Winter-active subnivean invertebrates in southern Canada. 111. Acari. *Pedobiologia* **19:** 153–160.

Aitchison, C. W., 1979f. Winter-active subnivean invertebrates in southern Canada. 11. Coleoptera. *Pedobiologia* **19:** 121–128.

Aitchison, C. W., 1984. A possible subnivean food chain. Pp. 363–372 *in* "Winter Ecology of Small Mammals" ed by J.F. Merritt. Special Publication of Carnegie Museum of Natural History, Pittsburgh.

Carron, P. L., 1985. "The ecology of three species of small mammals in subalpine habitat." Unpublished Ph.D. thesis. Australian National University, Canberra.

Chernov, Y. U. I., 1985. "The Living Tundra" Cambridge University Press, Cambridge.

Coulianos, C. C. and Johnels, A. G., 1962. Note on the subnivean environment of small mammals. *Arkiv For Zool.* 15 nr **24:** 363–370.

Courtin, G. M., Kalliomaki, N. M., Hillis, T. and Robitaille, R. L., 1991. The effect of abiotic factors on the overwintering success in the meadow vole *Microtus pennsylvanicus:* winter redefined. *Arct. Alp. Res.* **23:** 45–52.

Edwards, J. S., 1973. Insect fallout on snow in the Snowy Mountains, New South Wales. *Aust. Ent. Mag.* **1:** 57–59.

Evernden, L. N. and Fuller, W. A., 1972. Light alteration caused by snow and its importance to subnivean rodents. *Can. J. Zool.* **50:** 1023–1032.

Fjellberg, A., 1975. Organisation and dynamics of Collembola populations on Hardangevidda. Pp. 73–79 *in* "Fennoscandinavian Tundra Ecosystems. Part 2: Animals and Systems Analysis" ed by F. E. Wielgolaski. Springer, Berlin.

Formozov, A. N., 1964. "Snow Cover as an Integral Factor of the Environment and its Importance in the Ecology of Mammals and Birds (English translation)." Boreal Institute for Northern Studies, University of Alberta, Edmonton.

Fuller, W. A., 1969. Changes in numbers of three species of small rodent near Great Slave Lake, N.W.T., Canada, 1964–1967, and their significance for general population theory. *Ann. Zool.i Fenn.* **6:** 113–144.

Fuller, W. A., Stebbins, L. L. and Dyke, G. R., 1969. Overwintering of small mammals near Great Slave Lake northern Canada. *Arctic* **22:** 34–55.

Geiger, R., 1950. "The Climate Near the Ground". Harvard University Press, Cambridge Massachusetts.

Green, K., 1988. "A study of *Antechinus swainsonii* and *Antechinus stuartii* and their prey in the Snowy Mountains". Unpublished Ph.D. thesis. Australian National University, Canberra.

Green, K., 1997. Inter-annual, seasonal and altitudinal differences in invertebrate activity in the Snowy Mountains. *Vict. Nat.* **114:** 222–229.

Green, K. and Osborne, W. S., 1994. "Wildlife of the Australian Snow-Country". Reed, Sydney.

Halfpenny, J. C and Ozanne, R. G., 1989. "Winter: An Ecological Handbook". Johnson Books, Boulder, Colorado.

Holloway, S. and Osborne, W. S., 1997. "Distribution and habitat of the Broad-toothed Rat *Mastacomys fuscus* in the Perisher Blue Ski Resort". A report to David Hogg Pty Ltd. on behalf of Perisher Blue Pty Ltd. Perisher Valley.

Johnson, H. M., 1957. Preliminary ecological studies of microclimates inhabited by the smaller Arctic and subarctic mammals. Pp. 125–131 *in* "Proceedings of the Second Alaskan Science Conference."

Näsmark, O., 1964. Vinteractivet nabboss under sno hos landlevande evertebrata. Zool. Revy. **26:** 5–15. cited in : Aitchison C. W. 1984. A possible subnivean food chain. Pp. 363–372 *in* "Winter Ecology of Small Mammals" ed by J.F. Merritt. Special Publication of Carnegie Museum of Natural History, Pittsburgh.

Osborne, W. S., 1980. Movement over the snow by the bush rat *Rattus fuscipes*. *Vict. Nat.* **97:** 163–164.

Osborne, W. S., Preece, M., Green, K. and Green, M., 1978. Gungartan: A winter fauna survey above 1500 metres. *Vict. Nat.* **95:** 226–35.

Penny, C. E. and Pruitt, W. O., 1984. Subnivean accumulation of CO_2 and its effect on winter distribution of small mammals. Pp. 373–380 *in* "Winter Ecology of Small Mammals" ed by J.F. Merritt. Special Publication of Carnegie Museum of Natural History, Pittsburgh.

Pruitt, W. O., 1958. Qali, a taiga snow formation of ecological importance. *Ecology* **39:** 169–172.

Pruitt, W. O., 1960. Animals in the snow. *Sci. Amer.* **202:** 61–68.

Pruitt, W. O., 1970. Some ecological aspects of snow. Pp. 83–99 *in* "Proceedings of the 1960 Helsinki Symposium on Ecology of the Subarctic Regions." UNESCO, Paris.

Pruitt, W. O., 1984. Snow and small mammals. Pp. 1–8 *in* "Winter Ecology of Small Mammals" ed by J.F. Merritt. Special Publication of Carnegie Museum of Natural History, Pittsburgh.

Schmid, W. D., 1971. Modification of the subnivean microclimate by snowmobiles. Pp. 251–257 *in* "Proceedings of the 'Snow and Ice in relation to wildlife and recreation' Symposium." ed by A. O. Haugen. Iowa State University, Ames Iowa.

Schmid, W. D., 1984. Materials and methods of subnivean sampling. Pp. 25–32 *in* "Winter Ecology of Small Mammals" ed by J.F. Merritt. Special Publication of Carnegie Museum of Natural History, Pittsburgh.

Schulz, M., 1991. Observations on birds in snow gum woodland and alpine heathland during mid winter in south-eastern Australia. *Aust. Bird Watch.* **14:** 95–102.

Sonerud, G. A., 1986. Effect of snow cover on seasonal changes in diet, habitat, and regional distribution of raptors that prey on small mammals in boreal zones of Fennoscandia. *Hol. Ecol.* **9:** 33–47.

Spencer, A. W., 1984. Food habits, grazing activities, and reproductive development of long-tailed voles, *Microtus longicaudus* (Merriam) in relation to snow cover in the mountains of Colorado. Pp. 67–90 *in* "Winter Ecology of Small Mammals" ed by J.F. Merritt. Special Publication of Carnegie Museum of Natural History, Pittsburgh.

Weygoldt, P., 1969. "The biology of the pseudoscorpions." Harvard University press, Cambridge Massachusetts.

Whitney, P. and Feist, D., 1984. Abundance and survival of *Clethrionomys rutilus* in relation to snow cover in a forested habitat near College, Alaska. Pp. 113–119 *in* "Winter Ecology of Small Mammals" ed by J.F. Merritt. Special Publication of Carnegie Museum of Natural History, Pittsburgh.

Williams, G. P. and Gold, L. W., 1953. Snow density and climate. *Trans. Engineer. Inst. Can.* **2:** 91–94.

CHAPTER 9

Snow as a defining character of the alpine/subalpine fauna

KEN GREEN and WILLIAM S. OSBORNE

INTRODUCTION

THE relatively low elevations of the Australian Alps mean that it is the presence of snow rather than the effects of altitude which determines the composition of much of the fauna (Green and Osborne 1994). Formozov (1964) in his classic work on the interaction between snow and fauna, defined three types of animals with respect to snow: chionophiles (animals with characteristic adaptations as a result of snow cover playing a large part in their evolution), chioneuphores (species that can withstand winters with considerable snow) and chionophobes. For this latter group "snow cover...is the most important element of environmental resistance and the struggle against this particular element is almost beyond (their) ability. Such species do not inhabit snowy regions" (Formozov 1964). An examination of the Australian fauna reveals many chionophobes, fewer chioneuphores and an unknown number of chionophiles. In this chapter we examine elements of the fauna occurring in the Australian mountains, both above and below the winter snowline, with respect to their fitness for survival in a snow-covered landscape.

Snow plays an important role in protecting some animals from winter cold and at the same time limits opportunities for other animals. The way in which the snow determines the composition of the fauna is dependent largely on: 1) the ability of each species to withstand or avoid cold; 2) its ability to complete a full life-cycle, or that part which is vulnerable to cold, during the snow-free period; 3) its means of locomotion and how appropriate this is in the snow; and 4) its means of gathering food and the availability of that food.

Depth, density and duration of snow have important influences on animal life. Depth and density are important factors in insulation for animals beneath the snow (Chapter 8), and influence the ease with which animals can both travel over the snow and, if needs be, tunnel through it to reach their burrow or food. Duration of the snowpack affects the time that food is covered, the time that animals beneath the snow are insulated from the cold air of autumn, winter and early spring, and the time for which animals in and on the ground are exposed in the summer to complete their life cycles. Snow also acts to mediate changes in the environment so that late-lying snow cover modifies atmospheric temperature; while snow is present much of the early summer heat is taken up as latent heat of fusion of snow and ice and only after this does the air and soil temperature rise (Mani 1990). The snow cover therefore helps to stabilise niche conditions, particularly for soil animals (Mani 1990).

In the Snowy Mountains (which has the most complete data record for snow cover in Australia) the winter snow cover begins its unbroken presence usually in June, but sometimes not until July with the biggest build up in July and August, leading to peak snow depths in August or September (Chapter 3). The thaw usually begins in earnest in September or October with the ground becoming free of snow at Spencers Creek in late October or early November. This pattern is similar for Victoria but differs in Tasmania where winter snow cover is not assured. It is also quite common for a temporary cover of snow to lie at any time following a summer snowstorm. These periods of snow, both winter and summer, are important determinants of the composition of the fauna, and any changes to either the longevity of the snowpack or its depth is likely to change the ecological conditions under which the present fauna has become adapted.

MAMMALS

Within the latitudinal band of south-eastern mainland Australia which encompasses the Australian Alps, about half of the fifty or so species of mammals between the coast and the western slopes occur in areas of winter snow cover (Table 1). Most of these are also common at low altitudes with two exceptions; the Mountain Pygmy-possum (*Burramys parvus*) and the Broad-toothed Rat (*Mastacomys fuscus*). The Mountain Pygmy-possum is found only in the vicinity of, or above, the level of the winter snowline in New South Wales and Victoria. The Broad-toothed Rat does occur in moist habitat down to sea level but is most common above about 1000 m in New South Wales and Victoria whereas in Tasmania it is more common at lower altitudes (Green and Osborne 1994). At lower altitudes in New South Wales similar moist habitat is occupied by the Swamp Rat (*Rattus lutreolus*) (Newsome and Catling 1979). In Tasmania, the number of mammal species (31) is lower than on the mainland, but a higher proportion (over two-thirds) can be found in the mountains (Green and Osborne 1994). This is largely due to the maritime climate of Tasmania, with consequent shorter duration of the snowpack, greater frequency of winter rain to melt the snow, generally higher winter minimum temperatures and therefore greater density of the snow, thus making travel easier.

A comprehensive data base for mammals in south-eastern New South Wales is provided by the Atlas of NSW Wildlife. Of 46 species of terrestrial mammal with 50 or more NSW Atlas records between 35° south and the Victorian Border, and from 147°

TABLE 1

NSW Atlas records for terrestrial vertebrates recorded between 35° south and the Victorian Border, and from 147° east to the coast, for fauna in 500m altitudinal bands. (Data compiled by Cate Gillies)

Altitudinal band	Mammals	Birds	Reptiles	Amphibians
0–500	47	148	17	9
501–1000	45	143	17	10
1001–1500	42	128	15	7
>1500m	25	103	15	6

east to the coast, seven species have not been recorded above 1000m. The remaining 39 species either increase or decrease in the frequency of recording between 1000–1500m and altitudes greater than 1500m. The numbers of three species increased: Broad-toothed Rat, Dusky Antechinus (*Antechinus swainsonii*) and Bush Rat (*R. fuscipes*). Of the 33 that decreased, 11 were bats (9 insectivorous, 2 fruit bats), seven were possums largely dependent upon hollow trees, and one was the Koala (*Phascolarctos cinereus*) which might be excluded because of lack of suitable food trees. Additionally, the House Mouse (*Mus domesticus*), being a specialised granivore which can tolerate cold environments (Formozov 1964), might be excluded by lack of suitable food. Snow may be the major factor in the reduced numbers of three species of macropod, the Wombat (*Vombatus ursinus*), Echidna (*Tachyglossus aculeatus*), Platypus (*Ornithorhynchus anatinus*), feral Cat (*Felis cattus*) and Spotted-tailed Quoll (*Dasyurus maculatus*); also of animals that need access to ground-based food year-round such as bandicoots, Rabbit (*Oryctolagus cuniculus*) and Pig (*Sus scrofa*). The reduction in numbers of feral Dogs (*Canis familiaris*) (which probably reached their peak under grazing) might be a result of the loss of larger prey species, but additionally, dogs are not well suited to snow (Formozov 1964) and have been replaced by the Red Fox (*Vulpes vulpes*). The Brown Antechinus (*A. stuartii*) might be excluded because of competition with the Dusky Antechinus, a species better adapted to the snow (Dickman *et al.* 1983, Green 1988). Competition with other rats within its preferred habitat is thought to be the reason for the scarcity of the Broad-toothed Rat at lower altitudes (Happold 1995), conversely, Broad-toothed Rats replace Swamp Rats at higher altitudes. The Eastern Pygmy-possum (*Cercartetus nanus*) is present in the snow country of Tasmania where seasonal conditions are quite different to the mainland and a possible competitor, the Mountain Pygmy Possum, does not occur. The reduced number of records of Foxes above 1500m might be due to lack of reporting because we know (Bubela 1995) that they are in equivalent densities above 1500m and in tableland farmland. Two other well-known species stand out as under-represented in the Atlas records, possibly due to under-reporting: Water Rat (*Hydromys chrysogaster*) and Mountain Pygmy Possum. These and their known distribution have been incorporated into Table 1.

There are many reasons why there are fewer species of mammals with increasing altitude. These generally relate to the adaptive behaviour and morphology of the individual species. Mammals may be excluded from winter activity in areas of deep snow cover by cold, dampness, lack of food, depth of snow or difficulty in locomotion due to low snow density. Some mammals inhabit areas of higher elevation in summer, but because they are precluded from winter activity, may migrate to lower altitudes or spend winter in hibernation. Other mammals which are found at lower altitudes, but which are excluded from winter activities at higher altitudes, may not be able to migrate the distance needed to reach subalpine and alpine habitats in summer. Territorial mammals, for example, would be unlikely to leave established territories to move to higher altitudes.

Mammals might also be excluded from the snow-covered higher altitudes because of their lesser ability to compete for resources, particularly in snow. Such competition may well have existed between two introduced mammals, the Fox and Dingo *(Canis lupus)*, leading to the general exclusion of the latter from areas above the level of the winter snowline (Green and Osborne 1994). Competition between the Dusky Antechinus and the Brown Antechinus is thought to have led to the lower numbers, and almost exclusion, of the Brown Antechinus at higher altitudes above the snowline (Dickman *et al.* 1983). In both cases the effects of competition are thought to be strengthened by the presence of snow in the winter months.

Mammals permanently inhabiting areas with a deep and persistent snow cover can be placed into two groups; firstly those that hibernate during the period that snow covers the ground, and secondly, those mammals that remain active throughout the year. Mammals that hibernate are represented by the bats, Mountain Pygmy-possum and Echidna. The mammals that remain active within the snow-covered area during winter include those species whose locomotion and food gathering is not generally affected by the presence of snow (for example arboreal species such as the Ringtail Possum *Pseudocheirus peregrinus*), and species (generally ground-dwelling) whose existence depends upon their ability to overcome the problems of locomotion and food gathering presented by the snow cover.

Cold tolerance and body size

To remain in snow-covered environments, an active mammal must either be small enough to take advantage of the subnivean space for insulation and foraging throughout the winter (see Chapter 8) or large enough to be able to bear sufficient insulation to keep itself warm above the snow. As mammals decrease in size they become less able to carry thick fur; once below a certain size, however, a further reduction in size may be advantageous. With the two species of Antechinus both the body weights and the head-body length decrease with an increase in altitude. The decrease is greater in the larger of the two species (the Dusky Antechinus) and this leads to a convergence in body size. This convergence also occurs between the sexes which are normally of a different size. The net result is that the two species and the two sexes are smaller and closer in size at higher altitude (Dickman *et al.* 1983). A smaller mammal is better able to exploit the restricted subnivean space and to reduce its total food intake (Mezhzerin 1964).

The size of a mammal is an important factor in deciding whether movement should be on the snow surface or in the subnivean space. In fact it has been suggested by Coulianos and Johnels (1962) that the size of the smaller mammals in snow-covered environments must have been influenced by the size of the subnivean space, and that there is a maximum size above which a species cannot inhabit this restrictive environment. If a mammal is large enough to live above the snow it must have some means of reaching its food despite the snow cover, and it must be able to travel efficiently over the snow.

For small mammals, the limits to subnivean movement are provided by the physical dimensions and continuity of the subnivean space together with their tunnelling ability; this is dealt with more fully in Chapter 8. Small mammals prefer areas in which an extensive subnivean space is built up so that most of their mid-winter movement is below the snow. Movement over the snow by small mammals is generally confined to the time when the snow cover is building up and to the period of the thaw, when areas of open ground are exposed and the distances across patches of snow are reduced. Mammals may return to the snow surface at this time to feed on wind-blown insects immobilised on the snow surface.

Movement over the snow

At a certain depth and density, snow cover becomes an important factor in resisting the movement of mammals attempting to travel over its surface. Ease of movement by mammals over the snow is influenced by their shape and ground pressure. A human sinking 20 cm into the snow is inconvenienced and will become tired rapidly (Formozov 1964) while a Wombat sinking the same distance is severely impeded in its progress because of its shorter legs. The presence of furrows caused by Wombats dragging their body through the snow is evidence of how poorly they are adapted to soft snow (Green and Osborne 1994). However, on even moderately compacted snow they will travel for many kilometres. The physical effort of movement through deep and soft snow may have been a contributing factor to the high winter mortality of Wombats in the mainland mountains in the deep snow of 1981.

To visualise the difficulties of moving through snow it is instructive to compare a human on foot with one on cross-country skis. A person on foot with 80 kg on 170 square centimetres of support surface has a surface pressure of 470 $gm.cm^{-2}$ and such a person may sink 20–25 cm into the snow. On cross country skis, however, with a supporting surface of 1200 square centimetres the ground pressure is only 67 $gm.cm^{-2}$ and the skier will sink only a centimetre or two.

Kangaroos and wallabies normally have a pentapedal gait when moving slowly with the fore feet and tail alternating with the hind feet in bearing the weight. In each case the ground pressure is about 230 $gm.cm^{-2}$. A kangaroo in the bipedal gait (hopping) would, however, be more severely impeded than in the pentapedal gait. Although the weight does not change, on a two-metre hop the animal hits the snow with about double the force and sinks into the snow to about the same depth as a human on foot.

The difficulty in moving their hind legs independently also makes kangaroos and wallabies vulnerable to deep snow and when trapped in such snow they may be unable to extricate themselves. In Tasmania, where snow is neither as deep nor as soft as on the mainland, Bennett's Wallabies (*Macropus rufogriseus*) are more commonly seen above the treeline than are kangaroos on the mainland (Green and Osborne 1994). In deep snow they have been observed compacting the snow with small hops before attempting a larger leap and then, on landing, repeating the process. In the past few years there have been more frequent sightings of Swamp Wallabies (*Wallabia bicolor*) in the snow on the mainland, particularly on north-facing slopes where the snow surface may be hardened at night by freeze-thaw cycles and where shallow snow makes food more available (K. Green pers. obs.).

Of the introduced mammals in the snow-country, the Brown Hare (*Lepus capensis*) is well suited to travelling on snow with a ground pressure of 22 to 24 $gm.cm^{-2}$ but may still sink deeply in very soft snow (Formozov 1964). In these conditions it may follow trails (Pruitt 1970), although these conditions are not common in Australia. With a ground pressure of 40 to 42 $gm.cm^{-2}$ Foxes (Formozov 1964) are more commonly seen using trails in soft snow, particularly those which are highly compacted by cross country skiers or oversnow vehicles (K. Green pers. obs.).

Species poorly fitted for travel on deep, soft snow include those with small, narrow hooves. Sheep (*Ovus aries*) for example are poorly adapted to areas subjected to deep winter snow and high mortality usually occurs if a flock gets caught in heavy falls with prolonged snow cover. Pigs, which occur in the Arctic are well adapted to cold but are not so well adapted to deep snow, except for hard spring snow and are not commonly seen in areas of the Australian Alps where snow exceeds about 30 cm in depth. Among the carnivores, the Cats are not adapted to movement on deep soft snow and the Dog is not as well adapted as is the Fox (Formozov 1964).

Feeding opportunities in snow cover

Large herbivores that may be seen in the summer in the mainland mountains appear to be poorly adapted to obtaining sufficient food in the winter. Horses, for example, have difficulty in getting food from under deep snow, and periodic decimations of wild horses in the snow country in the past are a testament to this. Kangaroos also lack any great ability to feed under heavy snow cover although, Swamp Wallabies are able to remain on the verges of the mainland snow country, particularly in years of shallow snow such as 1997, because they are able to feed on exposed shrubs. Wombats, even though they are unsuited for travel in deep, soft snow cover, manage to survive because their prodigious digging ability enables them to reach food deep beneath the snow in all but exceptionally deep snow cover as in 1981 on the mainland when many perished (Green and Osborne 1994). Heavy snowfalls occurring late in the winter may be more damaging than those occurring in the early winter when animals are fitter and better fed (Formozov 1964). At the end of the season, the snow becomes more compacted and may have a thick crust, requiring a great deal of effort from an animal to break through it to reach a food source.

Foxes are well adapted to snow-covered mountains because of their ability to withstand the climatic conditions, to move easily over the snow and to hunt enough food over the winter to carry them through to the spring (although with some loss of condition). Cats are uncommon away from ski resorts because of their poor ability as hunters in the snow. In the wild they may occasionally be seen in summer and even in spring on snow-covered ground, but generally they are poorly adapted to hunting in the snow. By contrast, Foxes hunt by a combination of sound and scent and have the ability to dig through the snow to reach their prey, or as observed by Norment (1989) they can pounce through a shallow snow cover to catch small mammals unawares. Cats hunt either by lying in wait, or by stalking, using both hearing and vision which become useless when the prey is hidden beneath a layer of snow. Their feet with narrow, strongly curved claws are not suited to digging in snow to obtain buried prey (Formozov 1964).

Rabbits are uncommon above the level of the winter snowline on mainland mountains but may exist in isolated pockets especially where exotic grass species have been sown or where patches of ground are free of winter snow (such as near ski resorts or along roads). Bryant (1971) stated that the reasons rabbits had not spread into the Guthega catchment area of the Snowy Mountains were the combination of low palatability and low nutrient status of snowgrass, the presence of winter snow and low temperatures. Further to the north-east, Snowy Plains (1370 m) is regarded as climatically marginal for rabbits (Dunsmore 1974). Even there, rabbits survive through periods when the area is snow-covered due to the presence of Wombats which, through their digging, keep open access to ground vegetation (D. Wimbush pers. comm.). Snow usually lies on Snowy Plains for a few days each year but a heavy snowfall in 1964, which covered the plain for a number of weeks, resulted in a 90% mortality of rabbits, and little breeding the following spring (Dunsmore 1974). The comparative lack of snow on Tasmanian mountains allows higher numbers to occur.

In the mountains, the autumnal onset of cold conditions reduces the number of flying insects earlier than at lower altitudes, resulting in an earlier disappearance of insectivorous bats. With the coming of the first of the permanent winter snow cover, even cold hardy insects such as grasshoppers (Green 1983), become unobtainable. The reduced availability of insects at high altitudes would be expected to lead to competition between ecologically similar insectivores which occur in the same habitat. It has been suggested that mammals have less dietary overlap at alpine altitudes and that, therefore, they may not compete as intensely for food (Hoffman 1974). This is probably true for some of the smaller alpine mammals. The Bush Rat is a generalist fungivore, the Broad-toothed Rat is herbivorous and the Mountain Pygmy-possum eats a variety of seeds and fruits as well as insects. The two species of Antechinus, however, are almost entirely insectivorous.

The distribution of the two species of Antechinus is of interest in the mainland snow-country. Both species are quite common up to about 1600 m, but above this altitude the Brown Antechinus becomes less common while the Dusky Antechinus remains relatively abundant (Green and Osborne 1981). The question of competition between the two species is complicated but it does illustrate some interesting features of the ecology of mammals in snow. To understand competition between these two small mammals requires an examination not only of their food but also their use of habitat, their body size and their time of breeding.

At lower altitudes Dusky Antechinuses mate about four weeks earlier than Brown Antechinuses (Dickman *et al.* 1983). With increasing altitude, mating is delayed in both species, but this is more pronounced in the Dusky Antechinus. The difference in mating time at low altitudes reduces competition for food because times of high food demand such as lactation and weaning are staggered. With a longer period of high diversity and abundance of food at lower altitudes this is possible. Above 1500m, however, the presence of snow through to spring reduces food abundance and diversity and also confines the peak in abundance to a short period in early summer. For this reason, both species mate synchronously, even breeding under the snow, so that the time of peak food demand coincides with the peak in food availability. Above 1500m, mating must be early enough to allow access to the peak in food supply, yet late enough for a late snow thaw not to affect food availability greatly. The convergence in mating times, however, is likely to result in increased competition for food (Dickman *et al.* 1983).

At lower altitudes the Dusky and Brown Antechinuses specialise to a certain extent in the prey that they take. Brown Antechinuses feed above ground on arboreal prey and on a wider variety of prey types than do Dusky Antechinuses which are more terrestrial in their habits. At about 1800m, however, both Antechinuses take the same kind of prey and the size of the prey is about the same (Green 1989). At higher altitudes, therefore, dietary separation by the two species appears to be difficult because, with decreasing tree cover, more of the prey tend to be terrestrial.

The changes in both morphology (both species get smaller at higher altitude) and breeding time of the two species occur in a linear fashion from sea level until about 1500m (Dickman *et al.* 1983). Above this, the rate of convergence in the time of mating and body sizes is much less marked, and Dusky Antechinuses become more abundant than Brown Antechinuses (Fig. 1). An explanation for this convergence seems to lie in the suite of characters that Formozov (1964) called the nival factors, or those factors appertaining to snow. It may be suggested that the decreased numbers of the Brown Antechinus are due to differences in the physiological adaptations to low environmental temperature possessed by that species. However, because the animal dwells under the snow in winter (a somewhat sheltered environment), the effects of cold should not differ between snow-covered areas where it is still common to 1600m and the treeline, where it is virtually absent, (Green 1988) so cold *per se* does not appear to be the answer.

Above the level of the winter snowline, both species are largely restricted for a third to half of the year to foraging in the subnivean space, presumably in the litter and soil where some species of invertebrate are still active (Chapter 8). Because of this spatial restriction, there is a greater chance of interspecific encounters. The potential

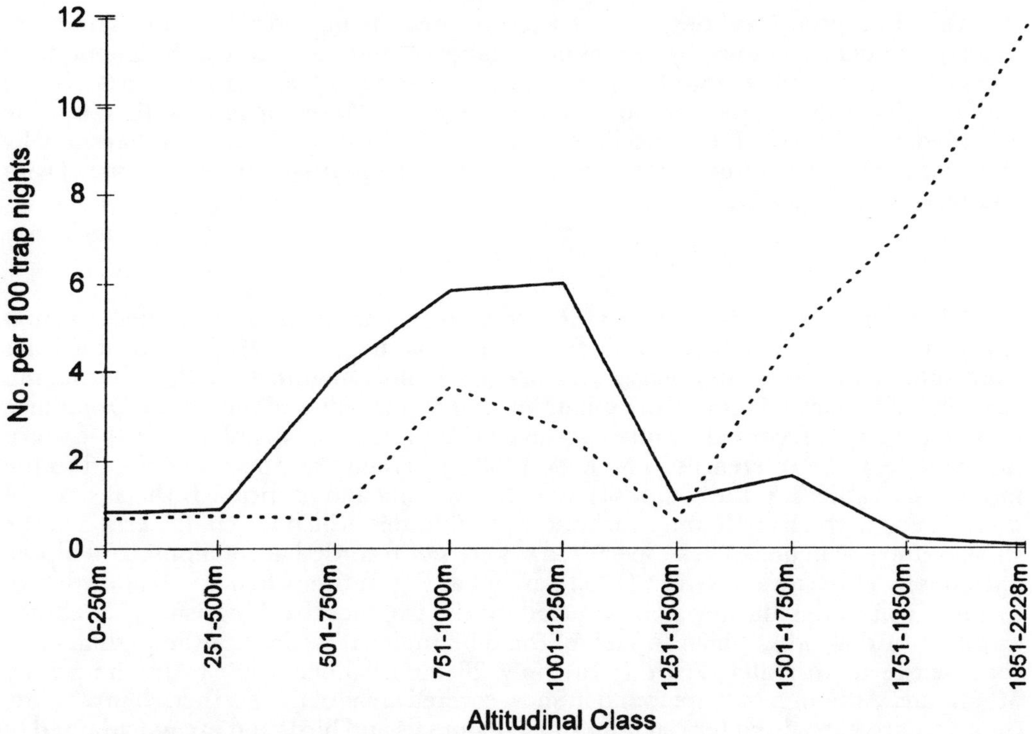

Fig. 1. The relative numbers of *Antechinus swainsonii* and *Antechinus stuartii* with increasing elevation, based on a total of 67 382 trapnights from sea level to the summit of Mt. Kosciuszko.

for competition at lower altitudes is minimised because the Brown Antechinus can forage in trees (Hall and Lee 1982), but in the mountains a ceiling of snow restricts upward movement and the potential for competition is thus enhanced. The presence of snow, because it reduces the diversity of available foraging niches and confines the two species to foraging on the ground, may have led to competition and hence exclusion of the Brown Antechinus (Dickman *et al.* 1983).

There is, however, an alternative explanation which also has snow as its ultimate cause. At altitudes below the snowline, the Brown Antechinus does not forage around the communal nest, but moves some distance from it before foraging (Lazenby-Cohen 1986). By contrast, at 1600 m, the Brown Antechinus forages around and between a variety of nests to increase its foraging range (Green and Crowley 1989). As the food supply is further reduced with increasing altitude (Green 1997) the nests must be further apart or must be used by fewer animals. Because animals are restricted in their movements beneath the snow, the distance between nests cannot always be increased and therefore their localities are restricted to areas of high productivity. The more solitary Dusky Antechinus, by not joining a large communal nest, can forage under a cover of snow and move into less productive areas such as grassland in winter (Green and Crowley 1989). The communally nesting Brown Antechinus may therefore be excluded, not because of interspecific competition, but because, by nesting communally and being restricted in movement by the snow, it is essentially limited to areas with relatively high food availability.

BIRDS

There have been few studies of the avifauna occurring in areas subject to snow cover in Australia. The best documented region is the Snowy Mountains which has been subject to two comprehensive studies (Gall and Longmore 1978, Osborne and Green 1992). The avifauna of the subalpine and alpine zones of the Snowy Mountains is similar to that recorded in other similar environments in mainland south-eastern Australia (e.g. see references in McEvey 1962). There are no species confined to the mountains, rather it is the composition of the avifauna and particularly the absence of many species, which are found nearby at lower altitudes, which is a characteristic of the snow-country avifauna. These species that were not recorded at subalpine and alpine altitudes by Osborne and Green (1992), except as vagrants or when accidentally blown to higher altitudes during storms include owls, frogmouths, kingfishers, lyrebirds, warblers, and woodswallows. In fact, of the 64 families recorded for the highlands of south-eastern Australia (Frith 1976) only 26 occur above 1500 m in the Snowy Mountains. Although owls are found in snow-covered areas of the northern hemisphere, their feet are specialised for catching small mammals and birds and are not adapted to an environment where most small mammals are hidden beneath the snow (Formozov 1964). Their existence in this environment is only possible because, with the build up of carbon-dioxide beneath the snow (see Chapter 8), voles are forced to construct ventilator shafts and are thus exposed to the owls when they come to the surface (see Pruitt 1960).

The number of bird species decreases markedly with increasing altitude (Table 1). This is reflected in both the total number of species that are known to use the subalpine zone (61) and alpine zone (36), and in the numbers of species and individuals recorded during counts made between altitudes of 1300 m and 2000 m in the Thredbo Valley (Osborne and Green 1992). Only Little Ravens (*Corvus mellori*) and Richard's Pipit (*Anthus novaeseelandiae*), species generally associated with treeless habitats, are more abundant in the alpine zone than the subalpine zone. Five species have been recorded breeding in the alpine zone compared to 23 species in the subalpine zone (Osborne and Green 1992).

The reduced numbers of birds observed in the alpine zone, when compared to the subalpine zone, is likely to be a result of the reduced structural diversity of the vegetation and hence fewer feeding positions. However, the reduced numbers of species in the subalpine zone (> 1500m) when compared to nearby montane habitats (1000–1500m) is not so easily explained. The following birds are found at montane altitudes throughout the year (Lamm and Wilson 1966) but generally avoid the subalpine and alpine zones (Osborne and Green 1992): Wonga Pigeon (*Leucosarcia melanoleuca*), Southern Boobook (*Ninox novaeseelandiae*), Laughing Kookaburra (*Dacelo novaeguineae*), Australian Owlet-nightjar (*Aegotheles cristatus*), Superb Lyrebird (*Menura novaehollandiae*), Eastern Yellow Robin (*Eopsaltria australis*), Crested Shrike-tit (*Falcunculus frontatus*), Spotted Quail-thrush (*Cinclosoma punctatum*), Eastern Whipbird (*Psophodes olivaceus*), Superb Fairy-wren (*Malurus cyaneus*), Pilotbird (*Pycnoptilus floccosus*), Red-browed Treecreeper (*Climacteris erythrops*), and Satin Bowerbird (*Ptilonorhynchus violaceus*). Although other features of the subalpine woodlands, such as the absence of tall, rough-barked trees, may be important, the presence of an extensive cover of winter snow is probably the main factor restricting the foraging opportunities available to these birds (Osborne and Green 1992). These species are comparatively sedentary, selecting sites that meet their requirements throughout the year, and therefore would be displaced during periods when snow cover persisted. Formozov (1964) also recognised this aspect, noting that the effect of snow on the distribution of bird species was greatest on those that obtain their food from on or near the ground, and which were not capable of flying far. Taking the pheasant as an example of a sedentary terrestrial form, Formozov (1964) examined its distribution which nowhere in Russia extends into areas where deep snow occurs. The pheasant has much in common with the lyrebird and the factors that Formozov (1964) considered as causing exclusion of the pheasant from snow-covered areas could equally apply to lyre birds; both are sedentary, with heavy flight, both have an unwillingness to take to the air, both run well, spend most of their time on the ground and collect all of their food on or close to the ground, they fly into trees to roost but cannot obtain food from foraging along branches.

The observations by Osborne and Green (1992) of a reduction in numbers of bird species with increasing altitude are supported by the records obtained from the NSW Wildlife Atlas. Of 148 species of terrestrial birds with 50 or more atlas records between 35°S latitude and the Victorian border and 147° longitude to the coast (Table 1), 20 species were not recorded above 1000m altitude. Of the remaining 128 species, 30 increased and 82 decreased by more than 0.5% of all records between 1000–1500m and altitudes above 1500m. It is likely that greater reporting of some conspicuous species within popular areas of Kosciuszko National Park accounted for at least 14 of the species that increased, for example: Australian Kestrel (*Falco cenchroides*), Black-shouldered Kite (*Elanus notatus*), Common Starling (*Sturnus vulgaris*), Pied Currawong (*Strepera graculina*), Australian Magpie (*Gymnorhina tibicen*), Galah (*Cacatua roseicapilla*), Great Cormorant (*Phalacrocorax carbo*), and Australasian Grebe (*Tachybaptus novaehollandiae*). Of the 82 species that decreased with altitude, at least 30 species may be excluded from high altitude areas simply because suitable habitat is not available. Some (mainly honeyeaters), are altitudinal migrants that depend on seasonal flowering of shrubs and eucalypt blossom during spring and early summer (Osborne and Green 1992) and therefore are likely to be observed less frequently. Others are known to be sedentary and are therefore excluded from year-round residency because of snow. A

TABLE 2

Bird species that are year-round residents of montane forests up to the altitude of the permanent winter snow-line which most likely are excluded from higher altitude woodlands by the presence of a permanent winter snow cover. Data interpreted from NSW Wildlife Atlas records and Osborne and Green (1992).

Species
Australian Owlet-nightjar *Aegotheles cristatus*
Spotted Quail Thrush *Cinclosoma punctatum*
White-winged Chough *Corcorax melanorhamphos*
Laughing Kookaburra *Dacelo novaeguineae*
Superb Lyrebird *Menura novaehollandiae,*
Southern Boobook *Ninox novaeseelandiae*
Tawny Frogmouth *Podargus strigoides*
White's Thrush *Zoothera dauma*
Wonga Pigeon *Leucosarcia melanoleuca*
Eastern Yellow Robin *Eopsaltria australis*
Eastern Whipbird *Psophodes olivaceus*
Pilotbird *Pycnoptilus floccosus*
Satin Bowerbird *Ptilonorhynchus violaceus*

number of summer migrants to the region appear to be restricted by a combination of higher altitudes and forest structure, and do not move far from the shelter of moist montane forests: these include Cicada Bird (*Coracina tenuirostris*), Black-faced Monarch (*Monarcha melanopsis*), Satin Flycatcher (*Myiagra cyanoleuca*), Golden Whistler (*Pachycephala pectoralis*), and Rufous Fantail (*Rhipidura rufifrons*). Other birds are known to move into the subalpine zone each summer (Osborne and Green 1992), but the Atlas records do not reflect this movement by Fan-tailed Cuckoos (*Cuculus pyrrhophanus*), Striated Thornbill (*Acanthiza lineata*), White-faced Heron (*Ardea novaehollandiae*) and Black-faced Cuckoo Shrike (*Coracina novaehollandiae*).

In concluding this discussion, and based on our own observations (Osborne and Green 1992; unpublished data) and the records of the Wildlife Atlas, we suggest that nine year-round residents of adjacent montane forests and woodlands (Table 2) are excluded from subalpine habitats year-round because of the influence of snow on winter food availability for these species.

Above 1500 m in the Snowy Mountains, seasonal migration is very apparent; out of the 62 species that have been recorded in summer, only 12 species are resident throughout the year (Osborne and Green 1992). A similar number of wintering species (11) was recorded in the subalpine zone at Falls Creek in Victoria by Marsland (1977). There is some evidence that birds, which are not major latitudinal migrants (see Recher

& Holmes 1985) but which still move away from the high-country during the colder months, may move to nearby tableland locations. Recher *et al.* (1983) noted that flocks of Crimson Rosellas (*Platycercus elegans*) and Gang Gang Cockatoos (*Callocephalon fimbriatum*) moved into their study area near Bombala in late autumn and winter. Further north, in the Canberra region, numbers of Gang Gang Cockatoos, Richard's Pipits, Flame Robins (*Petroica phoenicea*), Golden Whistlers (*Pachycephala pectoralis*), White-eared Honeyeaters (*Lichenostomus leucotis*), Crescent Honeyeaters (*Phylidonyris pyrrhoptera*), Eastern Spinebills (*Acanthorhynchus tenuirostris*), Spotted Pardalotes (*Pardalotus punctatus*), Striated Pardalotes (*P. striatus*) and Silvereyes (*Zosterops lateralis*) are much higher in winter (Taylor 1983), suggesting a movement of these species into the region from higher altitudes.

Influence of snow on movement

For most birds, snow does not present a great impediment to movement because they can fly above it. One exception of course is the Emu (*Dromaius novaehollandiae*) which is an occasional visitor to higher altitude areas but is rarely observed there in winter. Birds which feed mainly at ground level, however, may have their food supply cut off by a covering of snow.

For some species, the timing of the annual migration to the mountains appears to be determined in part by the spring snow thaw when areas free of snow became available for feeding and nesting. This is particularly noticeable with Richard's Pipit which moves into the alpine zone as soon as areas of ground become exposed on ridges, often before the ground has been exposed in the lower subalpine zone (Osborne and Green 1992). This early arrival may be an insurance against a poor season and Hendricks and Norment (1992) recorded a 79% mortality of nestlings of American Pipits (*Anthus rubescens*) in the alpine zone of the Beartooth Mountains in Wyoming following a midsummer snowstorm. The advantage of early arrival would be that they could lay a second clutch if the first was lost.

The influence of cold

Cold itself does not appear to be a major factor in causing the migration of birds from snow-covered areas. It has been found in the mountains of Tasmania (Ratkowsky and Ratkowsky 1987) that birds provided with additional food are able to remain throughout the winter in areas normally abandoned. Hence, several bird species recorded by Green and Osborne (1994) as autumnal migrants from the mountains or as being uncommon throughout the winter, may be seen regularly around areas inhabited by humans (where presumably more food is available). Thus, in winter, it is not uncommon to see more wintering birds in and near the ski resorts than can be seen further from human habitation. This in particular applies to scavenging birds such as Ravens, pest species such as Starlings taking advantage of both food and shelter, and birds such as Crimson Rosellas that are attracted to lodges where they are fed. Cold can, however, be a problem for small birds, and species such as White-browed

Scrubwrens (*Sericornis frontalis*) and Brown Thornbills (*Acanthiza pusilla*) may roost beneath the snow at night (Schulz 1991) in order to escape from the low temperatures on cold, clear nights. Resistance to cold takes time to develop and following a midsummer snowstorm all American Pipit nestlings older than 11 days (the age of endothermy) survived 24 hours buried in 30 cm of snow whereas among nestlings less than 10 days old there was an 89% mortality (Hendricks and Norment 1992). After the age of endothermy, however, continued exposure to a snow-covered landscape led to a 73% mortality through starvation in this ground-foraging species..

Birds such as the Silvereye are able to drop their body temperature by about 5°C overnight, thus saving on up to 50% of resting metabolic costs (Maddocks and Geiser 1996). This may help them over short periods of food deprivation - such as are imposed upon birds by the extremes of mountain weather. However, Silvereyes, which spend a considerable amount of the summer in the mountains searching through understorey shrubs feeding on insects attracted to the flowers, migrate away from the mountains for winter (Green & Osborne 1994). Maddocks and Geiser (1996) concluded that successful occupation by Silvereyes of areas with harsh weather conditions may be partially due to their use of nocturnal hypothermia. However, despite this ability to adapt to short periods of inaccessibility of food, the combination of repeated and more prolonged food shortage during winter, means that the birds have to undertake a seasonal migration.

Where snow has its greatest impact is in preventing access to invertebrate prey at ground level. That snow-cover has a major influence on bird populations is supported by studies elsewhere (e.g. Child 1978, Smith and Andersen 1985, Heil *et al.* 1988), however, there have been no comparable studies undertaken in Australia. One small study on Mt. Wellington in Tasmania recorded a reduction in the number of birds on a transect from 135 to 62 after snow had fallen, followed by an increase to 121 after the snow had melted (Thomas 1987). The birds most affected were whistlers, robins, honeyeaters and pardalotes; birds which also retreat from the Snowy Mountains in winter.

Influence of snow on foraging opportunities

Whilst cold undoubtedly reduces food availability and places a major physiological demand on birds, Osborne and Green (1992) concluded that it is likely that winter snow-fall, through its depressing influence on the extent and availability of winter foraging niches, reduces bird populations below that of nearby montane habitats. The total numbers of birds detected during monthly counts in the subalpine zone by Osborne and Green (1992) showed marked seasonality, declining from November to June and increasing in the following spring before declining again. Bird species that overwinter are either generalist feeders such as the Little Raven, or birds such as Brown Thornbills and Grey Shrike-thrushes (*Colluricincla harmonica*) that take a significant proportion of their food from tree trunks and branches. Virtually no ground-feeding birds remain above the snowline during winter, but a few species, such as

White-throated Treecreepers (*Climacteris leucophaea*) and Grey Shrike-thrushes that feed on trees or in bushes manage to overwinter each year. Other resident species such as Brown Thornbills which do not migrate altitudinally to any great extent, are very mobile in winter, moving in small flocks, presumably between areas of local food abundance. Larger birds such as Little Ravens and Australian Magpies (*Gymnorhina tibicen*) prey on invertebrates caught on the snow surface and on human refuse. Crimson Rosellas and Gang Gang Cockatoos feed on the mature fruit and galls of snowgums and Yellow-tailed Black Cockatoos (*Calyptorhynchus funereus*) extract wood-boring larvae from tree trunks and branches, and thus are able to use the subalpine woodlands to some extent throughout the year. During winter, when snow cover is extensive, birds such as White-browed Scrubwrens, which feed on the ground and in the shrub layer, remain in the snow-country but retreat slightly to lower altitude areas where shrubs still project through the snow (Osborne and Green 1992). Most terrestrial and arboreal bird species have feet totally unsuited for digging in the snow so that the presence of snow cover itself is enough to separate many bird species from their food supply. In winter, the main foraging substrates identified by Osborne and Green (1992) as being used by birds were rocks, tree trunks and tree branches, whereas in summer the main foraging positions were tree foliage, shrubs, ground and air, with early spring arrivals such as Flame Robins, Richard's Pipits and Little Ravens also feeding on snow and meltback areas.

Snow also determines the availability of summer food and indirectly affects breeding (Smith and Anderson 1985, Osborne and Green 1992). In early spring in the Snowy Mountains, the numbers of birds increases rapidly, coinciding with the presence of insects on the snow surface and the early spring flowering of the shrub Royal Grevillea (*Grevillea victoriae*) (Osborne and Green 1992). The snow surface becomes a profitable foraging substrate in spring, when insects metamorphosing at lower altitudes are blown up into the mountains and become immobilised on the snow (Edwards 1973). This insect fall-out on the snow in early spring provides an abundant food source for early migrants such as Flame Robins and Little Ravens. Utilisation of such resources by birds has also been observed elsewhere (Mani 1962, Pattie and Verbeek 1966). The earliest Crescent Honeyeaters arrive when snow melt exposes the flowers of the Royal Grevillea, even when snow cover beneath the shrubs is extensive. An example of the importance of the duration of the snowpack was shown in the reduced snow cover in 1982 (Chapter 3), flowering by these shrubs commenced earlier, resulting in the recording of more honeyeaters than in 1983 when flowering was later (Osborne and Green 1992). Smith and Anderson (1985) in a study in the USA found that at 2500m the breeding behaviour of the passerine *Junco hyemali* is affected by yearly variations in snowmelt and the date at which meadows become free of snow. They found that in years of early snowmelt the breeding season is extended and more birds appear to produce two clutches; whereas in years of late snowmelt the birds delay breeding, usually breed only once, and produce larger clutches. This response is believed to relate to insect availability.

REPTILES

The number of reptile species in the Australian Alps decreases markedly with increasing altitude (Green and Osborne 1994). This is not , however, very obvious from the results of the Wildlife Atlas because of the general lack of reporting of reptile species (Table 1). In the mainland mountains, only three families of reptiles are represented above the snowline; the dragons (Agamidae), with one species, the skinks (Scincidae), represented by ten species, and the front-fanged snakes (Elapidae), with two species (Green and Osborne 1994). This contrasts with about 25 species of reptiles which occur in the nearby montane zone from 1000 to 1500 m. In Tasmania, only two families of reptiles are represented in alpine areas which have persistent winter snow. These are the skinks with four species and the snakes with one species (Green and Osborne 1994).

Several species of reptiles are confined to high-mountain areas. On the mainland these are the High-plains Skink (*Pseudemoia cryodroma*), the Alpine Water Skink (*Eulamprus kosciuskoi*), the Alpine Oak Skink (*Cyclodomorhus praealtus*) and two undescribed species related to White's Skink (*Egernia whitii*); the Snowy Mountains Rock Skink (*Egernia* species 1) and the Tan-backed Rock Skink (*Egernia* species 2). In Tasmania, there are two high-altitude endemics; the Northern Snow Skink (*Niveoscincus greeni*), and the Southern Snow Skink (*N. microlepidotum*).

All of the reptiles found in the snow-country of Australia are heliotherms (sun-basking species). The major advantage of heliothermy is that at higher altitudes they are able to raise their body temperatures above air and substrate temperatures which in alpine areas may be low, even on sunny days. Viviparity is the major form of reproduction at high altitudes because live-bearing species can maintain their developing embryos at optimal temperatures by basking in the sun. By contrast, the eggs of oviparous species are susceptible to cold which can retard development or cause the death of the embryos (Spellerberg 1972a). Species found in colder environments generally have lower critical minimum temperatures (the lowest temperature at which a reptile can undertake limited movement and thus avoid predation or freezing) (Spellerberg 1972b, Rawlinson 1975). In fact, the Mountain Log Skink (*Pseudemoia entrecasteauxii*) has the lowest recorded critical minimum temperature for any Australian skink (Beesley 1980) and is the only lizard species occurring above 2000 m in Australia (Green and Osborne 1994).

A number of species of reptiles reach their upper altitudinal limit at about the lower limit of the winter snowline; these include the Tiger Snake (*Notechis scutatus*), Delicate Skink (*Lampropholis delicata*), Penny Lizard (*L. guichenoti*), Weasil Skink (*Nannoscincus maccoyi*), Litter Skink (*Niveoscincus coventryi*) and Blotched Bluetongue (*Tiliqua nigrolutea*). However, this feature is more likely to result from the general decline in the number of reptile species with increasing altitude, a feature of the thermoregulatory requirements of the group (Green and Osborne 1994). This feature, together with the degree of endemism, and other adaptations to a mountain existence, are all likely to be a result of selection for animals with adaptations for cold rather than for the seasonal presence of snow. There is little that is known or can be inferred about the importance of snow in the life of reptiles in the Australian Alps. Snow is important in terms of its insulating effect on the ground, thus protecting hibernation sites (Chapter 10), and the duration of snow lie at different altitudes affects the period available for activity each season. This may be particularly important in the alpine zone where reptiles may have to avoid hibernating in areas subject to late -lying snow patches. The winter hibernation sites of reptiles in the Australian Alps are poorly known, although the few observations available of winter hibernation deep within rotting logs, in burrows or under thick mats of vegetation and soil would indicate that reptiles are well protected from cold and contact with ice. It is noteworthy that above about 2000 m altitude in the Snowy Mountains there is only one small species of skink, the Mountain Log Skink (Green and Osborne 1994). Larger-bodied lizards such as *Egernia* spp., *Eulamprus kosciuskoi* and *E. tympanum* occur below 2000 m, where a combination of earlier snow-melt and slightly higher temperatures may favour their survival. Thus, we suggest that snow is likely to be a factor in the survival of reptiles at the very highest altitudes in Australia. In most other situations (i.e. where the length of the snow-free season is not restrictive) low temperatures in combination with habitat features are likely to be more important determining factors in the composition of the herpetofauna, rather than the presence of snow. Further research on the behavioural and physiological adaptations of reptiles to high altitudes in Australia should provide useful insight into the evolution of cold-tolerance and the effect of snow on reptiles in this area.

AMPHIBIANS

As with reptiles, there are few records of frogs in the NSW Wildlife Atlas (Table 1). Eight species of frogs are found above the winter snowline on the Australian mainland with one further species in Tasmania (Green and Osborne 1994). Of these, only two species, the Common Eastern Froglet (*Crinia signifera*) and the Alpine Tree Frog (*Litoria verreauxii alpina*), are found (or rather have been found) above the tree-line on the mainland. In the last 15 years, the Alpine Tree Frog has disappeared from most of its former range in the subalpine and alpine zone (Hunter *et al.* 1997) giving rise to concern that it may have been extirpated by increased levels of UV-B radiation (Broomhall 1997; see Chapter 13). One species, the Baw Baw Frog (*Philoria frosti*) is found only on the Baw Baw plateau in Victoria, the Victorian Froglet (*Geocrinia victoriana*) is found on the Bogong High Plains and Mt. Baw Baw in Victoria, and two species of *Pseudophryne*, the Southern and Northern Corroboree Frogs (*P. corroboree* and *P. pengilleyi* respectively) are restricted to the high montane and subalpine regions of the Snowy Mountains and do not occur in Victoria. Several species occur at montane altitudes but apparently do not survive in areas subject to lasting snow-cover. These include the Tasmanian Tree Frog (*Litoria burrowsii*) which occurs at altitudes up to about 1000 m in Tasmania, the Brown Tree Frog (*Litoria ewingii*) which occurs on the mainland and in Tasmania, and the Plains Brown Tree Frog (*L. paraewingi*) which is widespread on the plains of north-eastern Victoria but also occurs on the Mt. Buffalo Plateau. The Southern

Toadlet (*P. dendyi*), a lowland species, occurs up to the margins of the snow country to about 1600m altitude where it appears to be replaced by the two species of Corroboree Frogs.

Whilst empirical observations are lacking, there is little doubt that snow cover is important in the ecology of alpine frogs. The slow spring and summer melt of the snowpack, particularly in snowdrifts, provides a sequential series of ephemeral pools for breeding by species such as the Common Eastern Froglet (D. Hunter pers. comm.). It is also likely that snow melt may help to maintain pools containing tadpoles of species with more prolonged periods of larval development, such as the two species of Corroboree Frog. The actual extent of any dependency on snow melt water should be a priority for research given concerns about the effects of global warming. Both species have obligatory over-wintering tadpoles or eggs. It is not known what effect lack of winter snow cover has on these tadpoles, other than to note that in pools subject to ice cover the tadpoles still appear to be mobile beneath the ice. In the absence of an insulating snow-cover, however, deeper icing up of the shallow breeding pools may put the tadpoles at risk of freezing (K. Green unpubl. data).

Although there has been little work on the cold tolerance of Australian frogs, it is known that the Northern Corroboree Frog is slightly more cold tolerant than the Southern Corroboree Frog (Zentelis 1992). One possible reason for this is that, since the Southern Corroboree Frog is generally protected under the snow through the winter, it may not need to be as cold tolerant as the Northern Corroboree Frog, which occurs in a region subject to only a few days or weeks of snow cover each winter.

INVERTEBRATES

Green (1987) recorded 175 families of insects and 979 species for the area above the winter snowline in the Snowy Mountains. The invertebrate fauna is less well known in Victoria with species of 75 families recorded from the "alpine area", an area wider than that dealt with here (Land Conservation Council 1983). No comparable record is available for the Tasmanian snow-country. Generally, very few orders of insect have successfully colonised the alpine zone of high mountains worldwide, with the dominant high altitude orders being the Plecoptera (stoneflies), Coleoptera (beetles), Lepidoptera (moths and butterflies), Diptera (flies), and Collembola (springtails) (Mani 1968). The pattern is similar in the considerably lower Australian mountains except that the Orthoptera (grasshoppers) and Hymenoptera (particularly ants and wasps) are more important than they are in the higher mountains overseas.

Alpine animals can be divided into four groups: eualpine, tychoalpine and xenoalpine (Hesse 1924) and regular summer alpines (Green and Osborne 1994). Eualpine species are those that are exclusively alpine; tychoalpine species extend from lower altitudes into the alpine zone and are able to complete their life cycles there; xenoalpine species consist of those that find there way to the alpine zone but are unable to become established; and regular summer alpines are those such as the Bogong

Moth (*Agrotis infusa*) which regularly use the zone at an ecologically favourable time. With so little known of the alpine invertebrates particularly those (the insects) which have great mobility due to their powers of flight, it is difficult to give a breakdown by species according to the groups that they fall into, or for that matter how many species are affected by snow cover. Snow cover is one of the most important factors determining the distribution and abundance of insects in higher latitudes of the northern hemisphere (Formozov 1964). Its main impacts on the invertebrate fauna relate to its importance for winter insulation and in setting the duration of the summer growing season.

Three possible overwintering strategies are exhibited by alpine insects. They may migrate to lower altitudes, they may overwinter as eggs or pupae or they may overwinter in an otherwise mobile stage of their life cycle. Insects that remain in the snow-country may avoid death from freezing in three ways: by supercooling, by tolerating freezing or by sheltering in the soil or leaf litter (Chapter 10). Insects may be totally dependent upon the winter cover of snow and at similar ambient temperatures high winter mortality of insect eggs occur in seasons of relatively shallow snow cover (Riegert 1967). Winter cold, however, is not the only factor affecting the invertebrates. The decreasing length and predicability of the growing season with increasing altitude is also important (Dearn 1977). The major pressure upon an alpine insect during the growing season is the number of day degrees (the number of days during which the temperature is above the threshold of development multiplied by the number of degrees by which that threshold is exceeded). The alpine growing season starts later than at lower altitudes, because the snow cover must melt before the insect can recommence development, and it is further reduced by adverse weather, including snowfall, during the alpine summer. Phytophagous insects are dependent upon the level of primary productivity and a key component of vegetative biomass production is the length of the growing season (inversely related to snow cover) (Körner 1994). Snow cover therefore determines the growing season both for invertebrates and for their food, and many invertebrates would be excluded from snow-covered landscapes because of their lack of adaptation to such short growing seasons.

Despite the relatively low altitudes of the Australian Alps, insects in this region still display many alpine adaptations. One adaptation of the grasshoppers to the short growing season of high mountains and high latitudes of North America, Europe and Asia is a reduction in the number of instars from five or six to only four (Alexander and Hilliard 1964). This strategy appears to be absent in the Australian mountains, however, the average adult body size of the grasshopper *Kosciuscola usitatus* (which is common within the alpine zone) becomes smaller with increasing altitude (Dearn 1977). Both smaller size and fewer instars may be regarded as means of abbreviating the life cycle if the time to reach maturity is also reduced (Green 1983). Alternatively, development during the growing season may be accelerated by the maintenance of a high metabolic rate. During cold periods within the growing season, the ability to increase metabolic rate, and therefore development, is advantageous to a grasshopper and may enable it to complete growth and reproduction in an abnormally short growing season. The Alpine Thermocolour Grasshopper (*K. tristis*) is able to do this by increased heat due to changing to darker colouration at low temperature and increased reflectivity by becoming paler as temperature increases (Key and Day 1954). Three of six resident mainland alpine species in the Snowy Mountains have evolved an extended life cycle, that is, they accumulate sufficient day-degrees by extending their life cycle through several seasons (Green 1983). Species in which more than one year is needed to complete a single generation are apparently favoured at high altitude (Mani 1968). Insects with the ability to tolerate cold and to enter diapause in almost all instars are not restricted to development in one season. A shorter than average growing season which may decimate annual species, is not so significant for individuals with an extended life-cycle as they can simply enter diapause and await a more favourable season. Development, therefore, need not be completed in a set calendar time thus circumventing the problem of a short growing season by treating the available developmental days as physiological rather than calendar time. Five of six alpine grasshoppers from the Snowy Mountains are therefore known to have specialised life cycles to deal with the short growing season imposed in the main by the winter snow cover.

Other species may be dependent upon the winter snow cover for another reason. Fourteen species of stoneflies occur above or just slightly below the treeline, seven from the Snowy Mountains, eight from the Victorian Alps and three from Tasmania (Campbell *et al.* 1986). Stoneflies have a long nymphal life spent in the water and together with mayflies prefer cool, clear waters. The existence of these mountain species may, therefore, be tied to the lower water temperatures found in streams from snowmelt. Invertebrates generally are adapted no more than they need to be for the environment they inhabit. Because of this, large deviations from the norm such as shallow winter snow and low temperatures, or deep and long-lasting snowcover reducing the growing season may have large impacts on the fauna and therefore help to determine the composition of that fauna. The importance of snowfall in the growing season is not to be taken lightly in this regard and Green (1997) documented an increase in the numbers of invertebrates at low subalpine altitudes after the thaw with a subsequent decrease after a heavy October snowfall. Ehrlich *et al.* (1972) also documented the local extinction of a butterfly population after a summer snowfall in the subalpine environment in Colorado.

CONCLUSION: KEY ROLES OF SNOW IN THE LIFE OF ANIMALS IN THE AUSTRALIAN MOUNTAINS AND POSSIBLE CONSEQUENCES OF GLOBAL WARMING

- A slowly-melting snowpack provides water for animal activity into the summer, particularly in drought years when there is no other source. This is important both for providing pools for frogs and providing cold-water stream flow. Altitudinal zonation of insects in streams may be related to water temperature and this could vary with duration of snow cover and whether the water arrives as snow or rain.

- Snow provides protection against winter cold and a barrier against predators. This not only applies to active forms but also to species quiescent in the soil which are dependent on not having the ice front reach them in winter. The alternative of penetrating to great depth in the soil may be difficult and energetically expensive (Danks 1978). Snow also provides a moderating influence in niche conditions in the summer, particularly in the soil.
- Snow provides a medium upon which many animal species cannot operate easily and which further reduces the ability of some species to overwinter by cutting off their food supply.
- Snow provides a catchment for high energy food (mainly insects) for insectivorous animals at the beginning of the breeding season, and for specialist snow foraging and parasitic species.
- Currently animals are adapted no more than they need to be. The absence of snow may, therefore, paradoxically select for cold-adapted species. These may be lower altitude species. For example *P. pengilleyi* appears to be better adapted to the cold than *P. corroboree* even though it occurs at a lower altitude. Other examples of species that are well adapted to cold which are currently excluded from the higher altitudes because of their reduced ability to survive in snow are pigs and cats.
- The altitudinal impoverishment of the fauna and the lack of niche overlap means there is room for generalist invaders should conditions ameliorate.
- The wombat may be an animal on the brink of survival at the edge of its ecological range (it is rare at higher altitudes in the snow) and is disadvantaged in years of heavy snow. Such a species may be favoured by less snow.
- Phytophagous animals (mainly insects in the alpine zone) are dependent upon the amount of primary productivity. This is quite low in alpine areas with only deserts being lower because of the reduced period of the year available for growth, low temperatures and occasional drought stress (Billings 1979). This may increase with global warming.

REFERENCES

Alexander, G. and Hilliard, J. R., 1964. Life History of *Aeropedellus clavatus* (Orthoptera: Acrididae) in the alpine tundra of Colorado. *Ann. Ent. Soc. Amer.* **57:** 310–317.

Beesley, P., 1980. "Some aspects of the thermal biology of *Leiolopisma entrecasteauxii* from a cold temperate environment." Unpublished Honours thesis, Australian National University, Canberra.

Billings, W. D., 1979. High Mountain Ecosystems: Evolution, structure, operation and maintenance. Pp. 97–125 *in* "High Altitude Geoecology" ed by P. J. Webber. Westview Press, Boulder, Colorado.

Broomhall, S. D., 1997. "Comparative effects of ultraviolet-B (UV-B) radiation on two sympatric species of Australian anurans, *Litoria verreauxii alpina* and *Crinia signifera*". Unpublished Honours thesis, University of Canberra.

Bubela, T. M., 1995. "Social effects of sterilising free-ranging vixens (*Vulpes vulpes* L.) in subalpine Australia." Unpublished Ph.D. thesis, University of Sydney.

Bryant, W. G., 1971. Deterioration of vegetation and erosion in the Guthega catchment area, Snowy Mountains, New South Wales. *Soil Con. J.* **27:** 62–81.

Campbell, I. C., McKaige, M. E. and Lake, P. S., 1986. The fauna of Australian high mountain streams, zoogeography and evolution. Pp. 83–104 in "Flora and Fauna of Alpine Australasia" ed by B. A. Barlow. CSIRO, Melbourne.

Child, P., 1978. Fluctuations in birdlife in a subalpine basin. *Notornis* **25**: 71–83.

Coulianos, C. C. and Johnels, A. G., 1962. Note on the subnivean environment of small mammals. *Arkiv For Zool.* **15**: 363–370.

Danks, H. V., 1978. Modes of seasonal adaptation in the insects, 1. Winter survival. *Can. Ent.* **110**: 1167–1205.

Dearn, J. M., 1977. Variable life history characteristics along an altitudinal gradient in three species of Australian grasshopper. *Oecologia* **28**: 67–85.

Dickman, C. R., Green, K., Carron, P. L., Happold, D. C. D. and Osborne, W. S., 1983. Coexistence, convergence and competition among *Antechinus* (Marsupialia) in the Australian high country. *Proc. Ecol. Soc. Aust.* **12**: 79–99.

Dunsmore, J. D., 1974. The rabbit in subalpine south-eastern Australia. 1. Population structure and productivity. *Aust. Wildl. Res.* **1**:17–26.

Ehrlich, P. R., Breedlove, D. E., Brussard, P. F. and Sharp, M.A., 1972. Weather and the "regulation" of subalpine populations. *Ecology* **53**: 243–247.

Formozov, A. N., 1964. "Snow Cover as an Integral Factor of the Environment and its Importance in the Ecology of Mammals and Birds (English translation)." Boreal Institute for Northern Studies, University of Alberta, Edmonton.

Frith, H. J., (ed.) 1976. "Birds in the Australian High Country." Reed, Sydney.

Gall, B. C. and Longmore, N. W., 1978. Avifauna of the Thredbo valley, Kosciusko National Park. *Emu* **78**: 189–196.

Green, K., 1983. Alternative strategies in the life cycles of alpine grasshoppers (Orthoptera: Acridoidea). *Proc. Ecol. Soc. Aust.* **12**: 125–133.

Green, K., 1987. "An annotated list of the fauna occurring above the winter snowline in Kosciusko National Park." Unpublished report to Kosciusko National Park.

Green, K., 1988. "A study of *Antechinus swainsonii* and *Antechinus stuartii* and their prey in the Snowy Mountains". Unpublished Ph.D. thesis. Australian National University, Canberra.

Green, K., 1989. Altitudinal and seasonal differences in the diets of *Antechinus swainsonii* and *A. stuartii* (Marsupialia: Dasyuridae) in relation to the availability of prey in the Snowy Mountains. *Aust. Wildl. Res.* **16**: 581–592.

Green, K., 1997. Inter-annual, seasonal and altitudinal differences in invertebrate activity in the Snowy Mountains. *Vict. Nat.* **114**: 222–229.

Green, K. and Crowley, H., 1989. Energetics and behaviour of active subnivean insectivores *Antechinus swainsonii* and *A. stuartii* (Marsupialia: Dasyuridae) in the Snowy Mountains. *Aust. Wildl. Res.* **16**: 509–516.

Green, K. and Osborne, W. S., 1981. The diet of foxes, *Vulpes vulpes* (L.) in relation to abundance of prey above the winter snowline in New South Wales. *Aust. Wildl. Res.* **8**: 349–360.

Green, K. and Osborne, W. S., 1994. "Wildlife of the Australian Snow-Country." Reed, Sydney.

Hall, S. and Lee, A. K., 1982. Habitat use by two species of *Antechinus* and *Rattus fuscipes* in tall open forest in southern Victoria. Pp. 209–220 *in* "Carnivorous Marsupials" ed by M. Archer. Royal Zoological Society of New South Wales, Sydney.

Happold, D. C. D., 1995. Broad-toothed Rat. Pp. 562–564 *in* "The Mammals of Australia" ed by R. Strahan. Reed, Sydney.

Heil, S. J., Verner, J. and Balda, R. P., 1988. Weather and bird populations in true fir forests of Sierra Nevada, California. *Condor* **90**: 561–574.

Hendricks, P. and Norment, C. J., 1992. Effects of a severe snowstorm on subalpine and alpine populations of nesting American pipits. *J. Field Ornith.* **63**: 331–338.

Hesse, R., 1924. Tiergeographie auf Okologischer Grundlage Jena , Gustav Fischer.

Hoffman, R. S., 1974. Terrestrial vertebrates. Pp. 475–568 *in* "Arctic and Alpine Environments" ed by J. D. Ives and R. G. Barry. Methuen, London.

Hunter, D., Osborne, W. S. and Smith, M. J., 1997. Distribution and abundance of the Alpine Tree Frog (*Litoria verreauxii alpina*) in the Australian Alps National Parks: Report on the first season's survey (1996–97). Report to the Australian Alps National Parks Liaison Committee, Canberra.

Körner, C., 1994. Impact of atmospheric changes on high mountain vegetation. Pp. 155–166 *in* "Mountain Environments in Changing Climates" ed by M. Beniston. Routledge London.

Lamm, D. W. and Wilson, S. J., 1966. Seasonal fluctuations of birds in the Brindabella Range, Australian Capital Territory. *Emu* **65:** 183–207.

Land Conservation Council, 1983. "Alpine Region review." Land Conservation Council, Melbourne.

Lazenby-Cohen, K., 1986. "Social behaviour and home range of *Antechinus stuartii* at Monga State Forest, New South Wales." Unpublished Honours thesis, Australian National University, Canberra.

McEvey, A., 1962. Zoology of the High Plains: part II birds and mammals. *Proc. Roy. Soc. Vict.* **75:** 315–317.

Maddocks, T. A. and Geiser, F., 1996. Nocturnal hypothermia in silvereyes (*Zosterops lateralis*). Pp. 109–114 *in* "Adaptations to the Cold: Tenth International Hibernation Symposium" ed by F. Geiser, A. J. Hulbert, and S. C. Nicol. University of New England Press, Armidale.

Mani, M. S., 1962. "Introduction to High Altitude Entomology". Methuen, London.

Mani, M. S., 1968. "Ecology and Biogeography of High Altitude Insects." Dr. W. Junk, The Hague.

Mani, M. S., 1990. "Fundamentals of High Altitude Biology (2nd edition)" Aspect Publications, London.

Marsland, A., 1977. Birds in winter in the Kiewa valley, Victoria, above 1530 metres. *Emu* **77:** 33.

Mezhzerin, V. A., 1964. Dehnel's phenomenon and its possible explanation. (In Russian with English and Polish summaries) *Acta theriol.* **8:** 95–114.

Newsome, A. E. and Catling, P. C., 1979. Habitat preference of mammals inhabiting heathlands of warm temperate coastal, montane and alpine regions of southeastern Australia. Pp. 301–316 *in* "Heathlands and Related Shrublands of the World." ed by : R.L. Specht. Elsevier, Amsterdam.

Norment, C. J., 1989. "In the North of Our Lives." Down East Books, Camden, Maine.

Osborne, W. S. and Green, K., 1992. Seasonal changes in composition, abundance and foraging behaviour of birds in the Snowy Mountains. *Emu* **92:** 93–105.

Pattie, D. L. and Verbeek, N. A. M., 1966. Alpine birds of the Beartooth Mountains. *Condor* **68:** 167–76.

Pruitt, W. O., 1960. Animals in the snow. *Sci. Amer.* **202:** 61–68.

Pruitt, W. O., 1970. Some ecological aspects of snow. Pp 83–99 *in* "Proceedings of the 1960 Helsinki Symposium on Ecology of the Subarctic Regions." UNESCO, Paris.

Ratkowsky, A. V. and Ratkowsky, D. A., 1987. Mountain birds of Tasmania. *Tas. Bird. Rep.* **16:** 21–25.

Rawlinson, P. A., 1975. Two new lizard species from the genus *Leiolopisma* (Scincidae: Lygosominae) in southeastern Australia and Tasmania. *Mem. Nat. Mus. Vict.* **36:** 1–16.

Recher, H. F. and Holmes, R. T., 1985. Foraging ecology and seasonal patterns of abundance in a forest avifauna. Pp. 79–96 *in* "Birds of Eucalypt Forests and Woodlands: Ecology, Conservation and Management " ed by A. Keast, H. F. Recher, H. Ford and D. Saunders. Surrey Beatty, Sydney.

Recher, H. F., Holmes, R. T., Schultz, M., Shields, J. and Kavanagh, R., 1985. Foraging patterns of breeding birds in eucalypt forest and woodland of southeastern Australia. *Aust. J. Ecol.* **10:** 399–419.

Riegert, P. W., 1967. Association of subzero temperatures, snow cover, and winter mortality of grasshopper eggs in Saskatchewan. *Can. Ent.* **99:** 1000–1003.

Schulz, M., 1991. Observations on birds in Snow Gum woodland and alpine heathland during mid winter in south-eastern Australia. *Aust. Bird Watch.* **14:** 95–102.

Smith, K. G. and Andersen, D. C., 1985. Snowpack and variation in reproductive ecology of a montane ground-nesting passerine, *Junco hyemalis*. *Ornis Scandinavia* **16:** 8–13.

Spellerberg, I. F., 1972a. Thermal ecology of allopatric lizards (*Sphenomorphus*) in Southeast Australia. II. Physiological aspects of thermoregulation. *Oecologia* **9**: 385–398.

Spellerberg, I. F., 1972b. Temperature tolerances of Southeast Australian reptiles examined in relation to reptile thermoregulatory behaviour and distribution. *Oecologia* **9**: 23–46.

Taylor, I., 1983. Annual bird report: 1 July 1981 to 30 June 1982. *Canb. Bird Notes* **8**: 5–52.

Thomas, D. G., 1987. The effect of hard weather on bird abundance. *Tas. Bird Rep.* **16**:17–20.

Zentelis, R. A., 1992. "An investigation of the specific status of the northern and southern form of the corroboree frog, *Pseudophryne corroboree* Moore, (Anura: Myobatrachidae). Unpublished Honours thesis, Australian National University, Canberra.

CHAPTER 10

Snow as a factor in animal hibernation and dormancy

MICHELLE WALTER and LINDA BROOME

INTRODUCTION

ANIMALS in Australian alpine and subalpine environments have a number of alternatives when faced with the prospect of low ambient temperatures, reduced food availability and a cover of snow during winter. Because the geographical extent of the area of winter snow cover is small and at high elevation, many mobile species which use the habitats above the snowline during summer are able to migrate to lower elevations before the onset of low temperatures in autumn. These include most bird species and bats. Those species which remain, including a high proportion of endemics (Chapter 9) are faced with low ambient temperatures, seasonal reduction in food availability and the physical prospect of living with varying degrees of snow cover which may last for periods of a few weeks at elevations of 1200–1400 m up to 4 months or more at elevations from 1800–2200m on the mainland (Whetton *et al.* 1996). The alternatives for these species are to tolerate the extremes of temperature and winter desiccation experienced above the snow; to take advantage of the relatively mild temperatures in the subnivean space, remain active and extract what food is available (if they are small enough — Chapter 8); or to reduce activity and energy demands to a minimum by seeking a sheltered refuge, and reducing metabolism until temperatures rise and food becomes available in spring. The focus of this chapter is the third strategy, that of hibernation.

In this chapter our aim is to evaluate how important snow cover is to fauna which hibernate in the Australian snow country. In order to do this it is necessary to have a good grasp of the concept of hibernation and its alternatives, and how overwinter survival and the physiology of hibernation are influenced by environmental temperatures. Therefore in this review we explain the physiology of hibernation in endothermic and ectothermic fauna species in some detail and identify those species which do hibernate, before returning to the question of the importance of snow cover. Given the predictions of climate change associated with global warming (Galloway 1988, Whetton *et al.* 1996), we also attempt to predict how this may affect Australian snow country hibernators. Where there is insufficient information to enable informed predictions we outline possibilities for further research.

HIBERNATION

The term hibernation is used extensively to refer to winter dormancy in animals from mammals to insects. Hibernation is derived from the Latin for winter quarters, so the general term is applicable to all animals which seek a local refuge (a hibernaculum) in which to winter (Ultsch 1989). By this definition, hibernation is a suitable term to describe winter dormancy in ectotherms (Gregory 1982, Ultsch 1989, Pinder *et al.* 1992, Leather *et al.* 1993). Hibernation in endotherms is quite different to that in ectotherms because it is inextricably related to the control of body temperature (Lyman 1982b). Nevertheless, the term hibernation continues to be used for both endotherms and ectotherms in most of the literature. In this review we consider both endotherms and ectotherms and use hibernation to refer to winter dormancy in both groups.

Hibernation in Endotherms

Endotherms (Gr. *endon*, within + *therme*, heat) have a body temperature determined by heat derived from the animal's own oxidative metabolism (Hickman *et al.* 1990). Large endotherms (> 10 kg) generally maintain a high, steady body temperature throughout their adult life, relying on the insulative properties of fat and fur, or behaviour to ameliorate ambient temperatures (homeotherms). However, as body size is reduced, metabolic rates and body surface area increase disproportionately and the costs of maintaining high body temperature during times of low ambient temperature or food shortage become prohibitively high. It becomes advantageous for endotherms in the smaller size range (<10 kg) to reduce body temperatures (T_b) and save energy by undergoing periods of torpor on a daily, seasonal or opportunistic basis (heterotherms). The benefits of torpor for small mammals are particularly high because the minimum maintenance metabolic rate (basal metabolic rate) is greater per unit body weight in smaller than larger mammals (Feist and White 1989) and, because metabolic rate during torpor is independent of body weight, the relative reduction in metabolic rate increases with decreasing size (Geiser and Ruf 1995). Nedergaard and Cannon (1990) also showed that daily torpor becomes uneconomic in animals larger than about 10 kg because the cost of rewarming exceeds the energy savings accrued from the period at low temperature.

Torpor in heterothermic mammals and birds can be separated physiologically into two categories, daily heterothermy and hibernation. During daily heterothermy, body temperatures (T_b's) are typically reduced for less than 24 hours from about 35°C to between 10–28°C, depending on the species. In many species this can occur in response to short periods of inclement weather or low temperatures at any time during the active period. Hibernation involves prolonged periods of torpor lasting several days to weeks at T_b's of less than 10°C during winter (Hudson 1973, Dawson 1989, Wang 1989, Geiser 1994). Hibernation enhances an endotherm's chances of winter survival by reducing energy consumption and eliminating dependence on a continuous food supply. Some hibernators can show brief torpor bouts at other times of the year, similar to daily heterothermy. However, these are short bouts of hibernation and differ physiologically from daily torpor. Strictly, daily heterotherms are unable to undergo prolonged torpor (F. Geiser, pers. comm.).

Which endotherms hibernate?

Hibernation has been observed in all three major mammal groups: monotremes, marsupials and eutherians (Lyman 1982a) and is physiologically similar across these groups (Bartholomew and Hudson 1962, Geiser 1987, 1993, Grigg *et al.* 1989). The only monotreme known to hibernate is the Short-beaked Echidna (*Tachyglossus aculeatus*) (Grigg *et al.* 1989). Of the marsupials, hibernation has been observed in all of the Australian pygmy-possums (Burramyidae), the feathertail glider (Acrobatidae) and a South American marsupial *Dromiciops australis* (Microbiotheriidae) (Geiser 1994). Hibernation in eutherians has been studied more extensively, and has been observed in bats (Chiroptera), hedgehogs and tenrecs (Insectivora), badgers and bears (Carnivora) and many rodents (Rodentia) inhabiting the temperate zone in the northern hemisphere, including the squirrel-like rodents (Sciuromorpha) (ground squirrels, chipmunks, prairie dogs and marmots) and the mouse-like rodents (Myomorpha) (hamsters, dormice, pocket mice and jumping mice) (Lyman 1982a, Geiser and Ruf 1995). Hibernation has also been documented in one bird species from the family Caprimulgiformes, the North American Poor-will (*Phalaenoptilus nuttalii*)(Geiser and Ruf 1995).

In the Australian snow country hibernation has been well documented in only two species, the marsupial Mountain Pygmy-possum (*Burramys parvus*) and the Short-beaked Echidna. Echidnas occur throughout the sub-alpine zone and are occasionally encountered above the tree-line, and hibernate during the winter in both snow-covered and snow-free environments (Grigg *et al.* 1992, Nicol and Anderson 1996). Several species of bats which occur in the area also probably hibernate, as does the Eastern Pygmy-possum (*Cercartetus nanus*) (Geiser 1993), which however, does not appear to extend its distribution above the lower sub-alpine areas on mainland mountains (Green and Osborne 1994). The Brown Antechinus (*Antechinus stuartii*), which occurs in wooded habitats in the subalpine zone, although not undergoing hibernation is known to enter daily torpor (Geiser 1988). It is likely that the Dusky Antechinus (*A. swainsonii*), which occurs in high numbers right up to the alpine zone, may enter daily torpor similar to *A. stuartii* under emergency situations (F. Geiser, pers. comm.).

The endangered Mountain Pygmy-possum is of particular interest because it is a classic seasonal hibernator which is endemic to alpine and subalpine environments (above 1400m) in mainland Australia. Its normal distribution does not extend below the winter snowline (Mansergh and Broome 1994). The primary habitat is boulderfields and the associated heathland vegetation which has a restricted and disjunct distribution (Broome and Mansergh 1989). This habitat covers a total area of approximately 8 km^2 in Kosciuszko National Park in New South Wales (Caughley 1986) and less than 3 km^2 in Victoria (Mansergh pers. comm.). The total breeding population is estimated to be about 2600 individuals (Mansergh and Broome 1994).

The physiology of hibernation in endotherms and the influence of environmental temperatures

Throughout the hibernation season endotherms undergo a sequence of torpor bouts interspersed with short, usually periodic, normothermic arousal periods. During torpor, body temperatures become readjusted to new, lower levels, and heart rate, metabolic rate and other physiological functions show corresponding reductions. However, throughout the season, spontaneous or induced arousal to normal levels is possible at all times (Hoffman 1964, Willis 1982). The reason for the arousal periods, which are energetically very costly is largely unknown, although possibly related to elimination of metabolic wastes, restoring metabolite imbalances, resetting the "internal clock" to maintain circadian rhythms, or some kind of continuation of normal periodicity (Fisher 1964, French 1985, Willis 1982, Geiser and Kenagy 1988).

Geiser and Ruf (1995) evaluated physiological variables of torpor in 104 species of mammals and birds. They showed that during hibernation, metabolic rate reduction averaged 5.1% of basal (normothermic) metabolic rate (BMR) and maximum torpor bout duration averaged 355.3 h. This was distinct from daily heterothermy, where metabolic rate reduction averaged 29.5% of BMR and torpor bouts averaged 11.2 h. They also demonstrated that the two groups were clearly distinguished on the basis of minimum metabolic rates, measured as rate of oxygen consumption (VO_2) (0.037 vs. 0.535 ml O_2 g^{-1} h^{-1}). The mean minimum body temperature (T_b), the threshold body temperature that is metabolically defended during torpor (Heller and Hammel 1972), was also lower in hibernators than daily heterotherms (average 5.8°C *vs.* 17.4°C).

Energy savings during hibernation can be substantial. Richardson's Ground Squirrels (*Spermophilus richardsonii*), for example, make energy savings of 88% for the hibernation season as a whole, and up to 96% in a month (Wang 1978). The total energy expenditure for an individual during the hibernation season is dependent on the frequency and length of activity periods, and the body temperature at which it hibernates (Geiser and Kenagy 1988, Geiser and Broome 1991). Minimum energy expenditure, and maximum torpor bout duration is achieved when the minimum T_b is at, or slightly above, the ambient temperature experienced during hibernation. At this point, metabolic rate and weight loss is lowest and arousal periods are infrequent. Exposure to temperatures above or below the minimum T_b results in shorter bouts of torpor and increased energetic demands (Geiser and Kenagy 1988, Geiser and Broome 1993) e.g. Figure 1. The minimum T_b is species-specific and increases with factors which include animal body mass and environmental temperatures experienced in the wild (Geiser and Broome 1993). Minimum T_b's for the Mountain Pygmy-possum were found to be around 2°C (Geiser and Broome 1991), Figure 1.

Hibernators respond to a drop in ambient temperature below the minimum T_b either by arousing from torpor, or increasing their metabolic rate to defend T_b above that of the surroundings (Geiser and Kenagy 1988, Geiser and Broome 1993). When ambient temperatures are high, energy expenditure is increased because body temperature and metabolic rates are higher, and energetically costly arousals become more frequent (Geiser and Broome 1993).

Arousal from torpor, and the following activity period, consume the most energy during the hibernation season (Wang 1978). Geiser and Broome (1991), for example, calculated that the metabolism of constantly torpid adult Mountain Pygmy-possums would have used up only 2.9 g of body fat in males and 2.3 g in females during 6 months of torpor, while one normothermic period of a single day would require about 1.56 g fat in males and 1.85 g fat in females, which is more than half the total energy used by torpid animals during approximately 155 days of hibernation. Clearly, to minimise energy expenditure the frequency of arousals must be minimised. To achieve this and most effectively conserve energy during the hibernation season an endotherm must select a hibernaculum where the ambient temperature remains stable and close to the minimum T_b threshold.

Despite the substantial energy savings achieved by hibernators, the hibernation season is a time of high mortality (e.g. Broome 1992). To maximize energy storage it is important that hibernators take advantage of the typically short summers by breeding early to allow for growth of young and pre-hibernation fattening before the next hibernation season. Therefore, there are advantages in terminating the hibernation season as early as possible. Sexual selection pressures for early emergence also apply to males of many species because their testes regress during hibernation. Early termination allows them to undergo spermatogenesis before females arouse which optimises their chance of mating females as soon as the females emerge (French 1988).

The factors which determine the onset and termination of hibernation are species-dependent. Retreat to hibernacula and commencement of hibernation may be cued by sufficient weight gain, diminishing food supply, low ambient temperatures, shortened daylengths or a combination of these factors (Lyman 1982a,b, Mrosovsky 1971). The onset and termination (hence duration) of the hibernation season in relatively large northern hemisphere hibernators such as sciurid rodents is predominantly governed by an endogenous circannual rhythm (Pengelly and Fisher 1957, 1963, Mrosovsky 1980). Entry into hibernation is therefore fairly synchronous and is cued by factors which change seasonally (changing daylength or ambient temperatures). Emergence in spring can often occur before snowmelt and fresh food becomes available. This is possible due to their relatively large body size which enables them to survive prolonged normothermic periods on stored fat. Adult male Belding Squirrels (*Spermophilus beldingi*) (475 grams), for example, often spontaneously terminate hibernation before food is

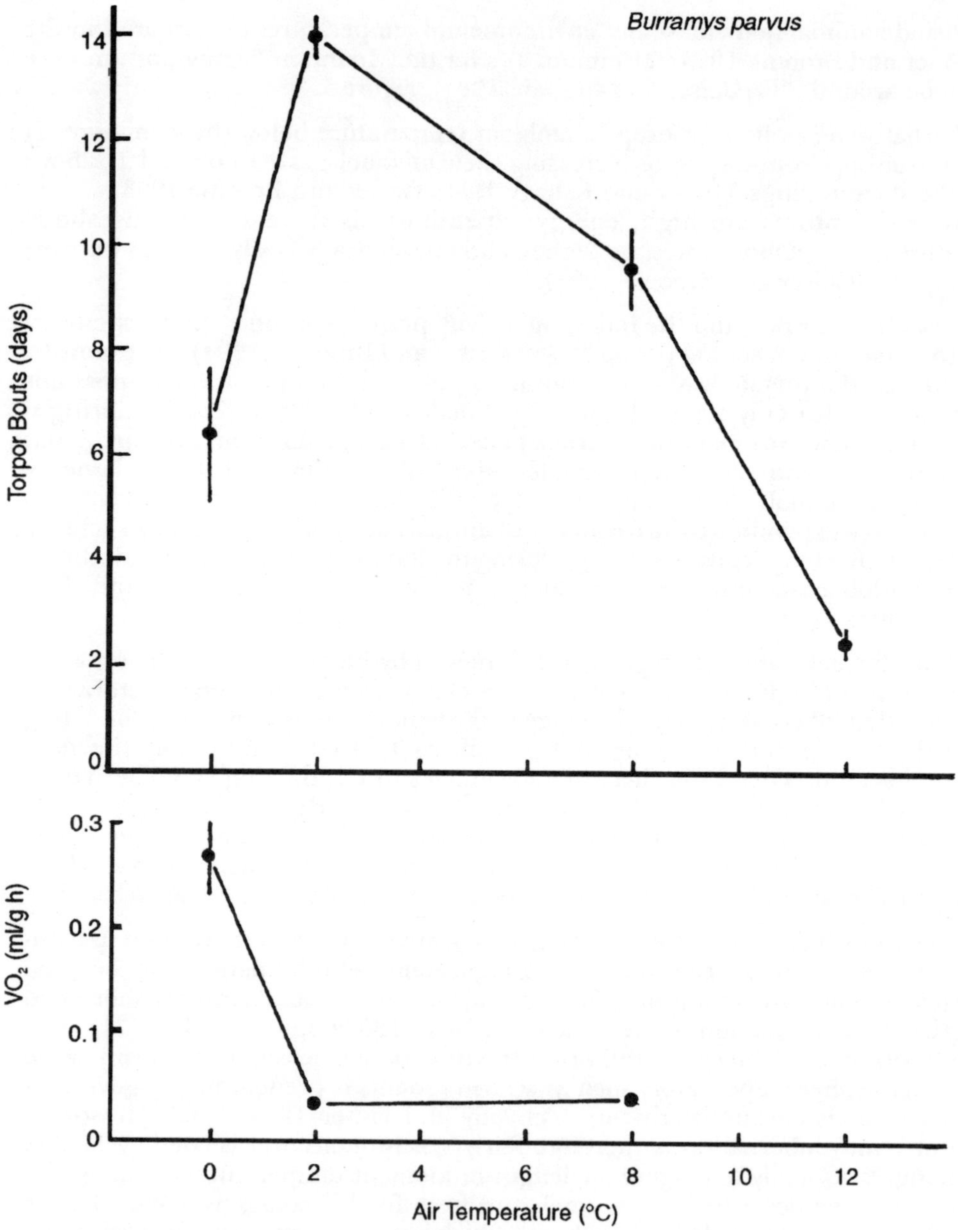

Fig. 1. The duration of torpor bouts (upper) and the metabolic rate (measured as oxygen consumption, VO_2) of hibernating Mountain Pygmy-possums. Values are means ± 1 standard error for five experimental animals. (After Geiser and Broome 1993).

available in spring, while females (380 grams) and yearlings (240 grams) do not (French 1988).

Small seasonal hibernators such as Mountain Pygmy-possums are constrained in their timing of emergence from hibernation because of their dependence on the immediate availability of a sufficient and reliable food supply and emergence therefore appears to be more opportunistic (French 1990). For Mountain Pygmy-possums the end of hibernation coincides with snowmelt and the settling of their main summer food source, the migratory Bogong Moth (*Agrotis infusa*) (Mansergh *et al.* 1990, Smith and Broome 1992) in the crevices of the boulderfields which provide their summer aestivation sites (Common 1954, Körtner and Geiser 1997). However, males still achieve reproductive advantages by their choice of hibernacula which, at least in years of early snowmelt, permit earlier emergence and earlier spring weight gain than for females (Broome 1992, Walter 1996, Körtner and Geiser 1997).

The onset of hibernation in Mountain Pygmy-possums also appears to be fairly opportunistic and independent of seasonal changes in daylength or ambient temperatures (Körtner and Geiser 1995). Entry varies considerably among individuals and years (Mansergh and Scotts 1990, Broome 1992, Körtner and Geiser 1997) and may depend on the fatty acid composition or abundance of food, and body mass (Geiser and Kenagy 1987, Geiser *et al.* 1992, Körtner and Geiser 1997).

Like most hibernators, hibernation in Mountain Pygmy-possums and Echidnas starts well before permanent snow cover is established. Individual Mountain Pygmy-possums undergo pre-hibernation fattening and enter hibernation from late February (adults) to May (juveniles). Pre-hibernation weight gain is rapid with adults doubling their body mass from 35–40g in summer to as much as 80 g prior to hibernation. Juveniles fatten later in the season and do not reach adult weight until their second summer (Mansergh and Scotts 1990, Broome 1992). Adult Mountain Pygmy-possums commence hibernation earlier and hibernate longer (about 7 months) than the smaller juveniles (about 5 months) (Geiser and Broome 1991). As is typical of hibernators, torpor bouts tend to be shorter and body temperatures higher at the start of the hibernation season (Geiser and Broome 1991, Broome and Geiser 1995), which reflects the higher temperatures in hibernacula (Walter 1996, Körtner and Geiser 1997), and are longest in the middle of the season (Geiser and Broome 1991, Walter 1996, Körtner and Geiser 1997) when a deep snow cover is established, and hibernacula temperatures have stabilised around 2–3°C (Walter 1996, Körtner and Geiser 1997). Under laboratory conditions it was established that Mountain Pygmy-possums hibernate most efficiently at an environmental temperature of about 2°C (Geiser and Broome 1993). Additionally, adult females usually hibernated at lower minimum body temperatures (2°C) than males (2.7°C) and torpor bouts for females were longer than for males (Geiser and Broome 1991). Similar trends were apparent in the wild, and reflect the differences in choice of hibernacula (Geiser and Broome 1991, Broome and Geiser 1995, Körtner and Geiser 1995, 1997, Walter 1996). Some juveniles never fattened in the laboratory and did not hibernate at all (Geiser and Broome 1991) and it is possible that not all individuals in the wild undergo seasonal hibernation, or that this could vary from year to year. Green and Osborne (1981), for example, recorded the presence of Mountain Pygmy-possum hairs in fox droppings in winter.

Hibernation in Echidnas is facultative, and varies from a classical hibernation pattern of prolonged torpor bouts of several weeks interspersed by short, periodic arousals in animals studied in Kosciuszko National Park and Tasmania (Grigg *et al.* 1989, 1992, Nicol and Andersen 1996), to short periods of seasonal torpor or reduced foraging activity during winter in warmer climates (Nicol and Andersen 1996). They also undergo daily heterothermy, with a pronounced cycling of body temperature from around 32°C during the daily active period to 29°C or lower overnight (Grigg *et al.* 1992). All 12 individuals studied in Kosciuszko National Park hibernated, with the duration of the hibernation season tending to be longer above 1300 m than below 1000m (average of 4.4 compared to 3.1 months). Minimum periods of hibernation ranged from 1.6 to 6.6 months, centred on autumn and winter. Body temperatures of hibernating Echidnas fell close to ambient, with the lowest T_b falling at 3.7°C at an ambient temperature of 0.8°C (Grigg *et al.* 1992). Although minimum T_bs have not been clearly defined, Nichol and Anderson (1993) found in a laboratory study that heat production increased when T_bs fell below about 5°C. Entry into hibernation occurred between March and May in the Kosciuszko study, as well as in central Tasmania (Nicol and Andersen 1996). Emergence was more variable, ranging from late June in sexually mature individuals, in time for the normal July-August breeding season, to October in juveniles (Beard *et al.* 1992, Nicol and Andersen 1996). Sexual selection pressures for early emergence appeared to operate in Echidnas as they do in Ground Squirrels and Mountain Pygmy-possums. Beard *et al.* (1992) found that large, sexually mature male Echidnas commonly aroused earlier than smaller younger individuals. Nicol and Andersen (1996) suggest that hibernation in Echidnas is not so much an escape from the cold as a response to a winter reduction in food energy. Only in areas of prolonged heavy snow cover is this seen as an actual food shortage.

Hibernation in ectotherms

Ectotherms (Gr. *ektos*, outside + *therme*, heat) have variable body temperatures derived from heat acquired from the environment (Hickman *et al.* 1990). In seasonally cold environments, they show pronounced seasonal activity patterns. In winter, most of them become inactive for all or part of the season.

A satisfactory definition of ectotherm hibernation is difficult to find. The term hibernation for example is used in contemporary literature in reference to amphibians and reptiles which over-winter in a hibernaculum even if they remain active (e.g. van Gelder *et al.* 1986), while it is often replaced with terms such as diapause for insects (Leather *et al.* 1993). Gregory (1982), Pinder *et al.*(1992) and Leather *et al.* (1993) all discuss terminology in more detail. This chapter will consider hibernating ectotherms to be those which retreat into insulated shelters and remain dormant for all or most of the winter. The majority of freshwater fish do not become torpid in winter in the same way as frogs and reptiles, readily responding to stimulation (Ultsch 1989) and therefore are not classified as hibernators.

Hibernation in ectotherms poses similar advantages and disadvantages to hibernation in endotherms. Hibernation is an advantage because it allows survival in a harsh environment by reducing metabolic rates and facilitating survival for long periods without the animal feeding (Gregory 1982, Pinder *et al.* 1992, Leather *et al.* 1993). Concealed hibernacula also offer protection from many winter predators such as birds. There are, however, disadvantages, ectothermic hibernators are typically immobilised and may have to endure less favourable conditions such as dessication, waterlogging or freezing (Leather *et al.* 1993).

Adaptations to the cold in ectotherms

Ectothermic animals can cope with very low environmental temperatures in a number of ways; they can: 1) avoid subzero temperatures by choosing a warm hibernaculum, 2) endure subzero temperatures but remain in a supercooled state, or 3) endure subzero temperatures with a capacity for freeze tolerance (Storey 1985).

Some ectotherms can cool to temperatures below 0°C and remain unfrozen in a supercooled state (Leather *et al.* 1993). Supercooling is the maintenance of the body fluids in the liquid phase below its normal freezing point temperature (Block 1991). Many insects have remarkable supercooling capabilities. This is primarily because the capacity for supercooling is inversely related to volume (Lee 1991). Small arthropods such as collembola, mites, and ants, as well as insect eggs, supercool to very low temperatures; supercooling points have been recorded at less than –40°C in such species (Lee 1991). Herpetofauna (reptiles and amphibians) have more limited supercooling capabilities (Storey 1985). Painted Turtle hatchlings (*Chrymys picta*) have been observed to supercool to as low as –8°C (Costanzo *et al.* 1995). The lowest supercooling points for other reptiles are above this (Lowe *et al.* 1971). Frogs have poorer supercooling capabilities, the lowest supercooling point observed being –2°C to –3°C (Schmid 1982, Storey 1985).

Animals in a supercooled state are at risk of inoculative freezing which occurs when an animal's body fluids (such as at mucous membranes) come into contact with environmental ice (Lee 1991, Costanzo *et al.* 1995). Frogs are particularly susceptible to inoculative freezing because their water-permeable skin is easily inoculated by environmental ice (hence their poor supercooling capacity) (Pinder *et al.* 1992). Inoculation by environmental ice is an issue for all supercooled organisms and is dependent on microenvironmental factors, including substrate composition, water content and moisture tension (Costanzo *et al.* 1995). Insects avoid inoculative freezing by constructing protective cocoons, or by selecting suitable hibernacula (Danks 1978).

Freeze tolerance is a phenomenon whereby organisms can tolerate extracellular freezing (Storey and Storey 1988). Intracellular ice formation however is lethal (Storey 1985). There is a limit to freeze tolerance which is a function of temperature and time.

Mountain Spotted Grasshoppers (*Monistria concinna*) from the Snowy Mountains showed no adverse reaction when exposed in the laboratory to temperatures between 0°C and –4°C for 20 days, but no individuals were able to withstand temperatures of –12°C to –14°C for two days (Green 1983). Natural freeze tolerance is well known in several invertebrate groups and a few vertebrate animals (Storey 1985, Storey and Storey 1988, Lee 1991, Pinder *et al.* 1992). Freeze tolerant terrestrial insects are commonly members of the Coleoptera, Diptera, Hymenoptera and Lepidoptera (Storey and Storey 1988). Survival of insects to –25 to –30°C is common, and examples of freeze tolerance can be found at all life stages (Storey and Storey 1988). The vertebrates known to exhibit freeze tolerance capabilities include four species of frog (Schmid 1982, Storey 1985), one salamander species (Berman *et al.* 1984) and two species of turtle (Costanzo *et al.* 1995). Limited freeze tolerance has been observed in Gartersnakes (*Thamnophis radix*) (Bailey 1949), and the European Common Lizard (*Lacerta vivipara*) (Grenot *et al.* 1996). All of the vertebrate species known to tolerate the natural freezing of body water during winter occur in the northern hemisphere, there are no data for Australia.

Supercooling and freeze tolerance capabilities are enhanced by antifreeze agents (cryoprotectants) (Gregory 1982, Lee 1991, Storey and Storey 1991). Glycerol is by far the most common cryoprotectant but there are many others including sorbitol (Storey and Storey 1991). Mountain Spotted Grasshoppers from the Snowy Mountains synthesize sorbitol when acclimated to freezing conditions but not if kept in warmer conditions (Green 1983). For freeze tolerant ectotherms, cryoprotectants act to regulate cell volume during extracellular ice formation to prevent the freeze concentration of cells from reaching lethal levels (Storey and Storey 1991). All cryoprotectants depress the freezing point and supercooling point. Supercooling points of insects have been extended to –40°C with cryoprotectants, but without them they could only tolerate temperatures as low as –12°C (Storey and Storey 1991). Factors other than cryoprotectants can also increase supercooling ability and include dehydration and pre-hibernation fasting (Lee 1991, Leather *et al.* 1993, Costanzo *et al.* 1995). Although supercooling and freeze tolerance allow animals to survive subzero body temperatures, they still have a limit to the minimum temperatures they can tolerate. Cold tolerance is not developed very much above the level which permits survival (Danks 1978).

Despite these examples of supercooling and freeze tolerance, the major strategy used by ectotherms in adapting to the cold is avoiding subzero temperatures by the choice of an insulated hibernaculum. The majority of resident invertebrate species move to more sheltered habitats for the winter (Danks 1991), as do most herpetofauna (reptiles and amphibians) (Storey 1985). Invertebrates which live in vegetation typically enter microsites protected by plant material and a cover of snow, while some litter inhabiting species burrow into the soil to depths of usually less than 15 cm (Danks 1978). Species which over-wintering in plant parts, including under bark, usually enter the deepest available fissures and crevices, and aquatic species from shallow water move to more stable deeper areas for the winter (Danks 1991).

The choice of a hibernation site can be critical to survival for all hibernators, even those which can tolerate subzero temperatures (Gregory 1982, Storey 1985, Pinder *et al.*

1992, Leather *et al.* 1993). A suitable hibernaculum can mediate the effects of low temperatures, rapidly changing temperatures, the chilling effect of wind, desiccation, predation and lack of oxygen in aquatic environments (Leather *et al.* 1993). Ectothermic hibernators will have the most successful hibernation season if they are in a hibernaculum where temperatures are low (to minimise metabolic rate), stable and remain above lethal temperatures. A good hibernaculum will also be influenced by emergence cues at the end of the hibernation season. Hibernacula may be terrestrial or aquatic. Terrestrial hibernacula include burrows and sites under logs, bark, leaf litter, rocks, vegetation and soil (Holmquist 1931, Danks 1991, Pinder *et al.* 1992, Leather *et al.* 1993).

Which herpetofauna hibernate?

There are numerous herpetofaunal species which over-winter in seasonally snow-covered environments. From the Class Reptilia, both snakes and lizards are present in the Australian snow country and hibernate for the entire winter (Green and Osborne 1994), while in the northern hemisphere, these groups are joined by turtles (Costanzo *et al.* 1995). In the Class Amphibia, salamanders, newts and anurans (frogs and toads) are found at high latitudes in the northern hemisphere (Pinder *et al.* 1992). In Australia, there are no salamanders or newts, and the 9 frogs which occur above the snowline are from the families Hylidae and Myobatrachidae (Tyler *et al.* 1993, Green and Osborne 1994).

Most of the Australian snow country frogs are not proficient burrowers. The exception being the Eastern Banjo Frog (*Lymnodynastes dumerilii*) which occur up to elevations of about 1600m, just above the lower limit of the winter snow line (Green and Osborne 1994). Terrestrial amphibians capable of burrowing form the majority of species which hibernate in cold northern hemisphere environments (Pinder *et al.* 1992). Those species burrowing below the frost-line are not freeze tolerant and are thought to be dependent on avoiding freezing.

The Alpine Tree Frog (*Litoria verreauxii alpina*), the Common Eastern Froglet (*Crinia signifera*) and the Tasmanian Froglet (*C. tasmaniensis*) all occur in the alpine zone of the Australian snow country. Neither they nor the subalpine species are adept burrowers (Green and Osborne 1994), however, we do not know if they are freeze tolerant. A number of northern hemisphere terrestrial frogs not adept at burrowing are freeze tolerant including *Rana sylvatica*, *Hyla versicolor*, *H. crucifer* and *Pseudacris triseriata* (Pinder *et al.* 1992). All of these frogs hibernate on land remaining at the soil surface but well-covered by leaf litter, logs, stones or tree roots (Storey 1985). Most frogs occurring in the Australian snow country are thought to use similar hibernacula (Green and Osborne 1994).

There are 18 species of reptiles which regularly occupy the alpine or sub-alpine zones of mainland Australia and/or alpine environments of Tasmania. Fifteen of these are skinks, two are snakes and the last an agamid lizard, the Mountain Heath Dragon (*Tympanacryptis diemensis*) (Green and Osborne 1994). Like the frogs of the Australian

Snow Country, they are poorly represented by species which burrow into the soil. Three possible exceptions are the Tan-back Rock Skink (*Egernia* sp.), the Snowy Mountains Rock Skink (*Egernia* sp.) and the Alpine Water Skink (*Eulamprus kosciuskoi*). The two species of rock skinks have been observed sheltering in burrow networks beneath partially embedded or buried boulders, and the Alpine Water Skink is thought to occupy burrows made by freshwater crayfish (Green and Osborne 1994) although it is not known if they over-winter in these burrows (W. Osborne pers. comm.). Precise over-wintering sites used by reptiles are poorly-known, however, reptiles probably hibernate under rocks, logs, vegetation and leaf litter. The micro-climate of such sites are poorly-known as are the adaptations to the cold possessed by these reptiles to survive the winters.

In the Australian snow country, there are no known aquatic reptile hibernators, and the only known amphibians to over-winter in water (ponds) are the tadpoles and/or eggs of the Southern Corroboree Frog (*Pseudophryne corroboree*), the Northern Corroboree Frog (*P. pengilleyi*), the Southern Toadlet (*Pseudophryne dendyi*), the Victorian Froglet (*Geocrinia victoriana*) and the Eastern Banjo Frog (*Limnodynastes dumerilli*) (Green and Osborne 1994).

Which invertebrates hibernate?

Invertebrates have an array of responses to winter conditions. Some species remain active throughout most of the season, others cease activity altogether and many adopt a mixed strategy depending on weather conditions (Leather *et al.* 1993). Most of the insects in the Australian snow country, like other species occupying regions where winters are long and often severe, either migrate to lower altitudes in winter (e.g. the migratory Bogong Moth, Common 1954) or over-winter in an immobile stage of their life cycle (such as pupa, eggs or lavae). The remainder over-winter in an otherwise mobile stage of their lifecycle. Of the second group, some remain active beneath the snow (Green 1982, 1983, 1997, Green and Osborne 1994).

THE IMPORTANCE OF SNOW TO HIBERNATORS

The depth and duration of snow cover can vary greatly due to influences ranging from global effects and annual variation in climate to very small scale differences in microsite. In any one year, large scale factors such as elevation, continentality and latitude are important (Danks 1991). Within a region, some sites naturally receive more snow due to topographic influences and weather patterns (Formozov 1964). Of particular influence on snow accumulation in the Snowy Mountains and Victorian Alps are elevation and exposure to wind. After the snow has fallen, aspect, slope and tree cover will affect insolation and therefore snowmelt (Formozov 1964). Microsite differences occur due to local variation in topography and aspect. A hollow for example may accumulate snow whereas an exposed rise will lose snow quickly. Snow cover can also vary greatly between years (Chapter 3).

A blanket of winter snow restricts the movement of large animals over its surface (Chapter 9), it also creates a subnivean space in which small animals such as *Antechinus* can remain active all winter (Chapter 8). For the inactive hibernators, the blanket of snow plays a role as an insulator and determines the length of the hibernation season.

Snow as an insulator of hibernacula

Snow is a poor conductor of heat and hence a good insulator (Formozov 1964). Snow density varies widely, as does the thermal conductivity which in compacted snow is 3–4 times greater than that of light, new snow (Geiger 1959). Once a complete cover of dry, new snow reaches about 20cm deep, air temperatures at the ground surface become reasonably stable because it is insulated from external or above-snow air temperature fluctuations. When these depths are reached, the insulative properties of snow typically moderate temperatures in all hibernacula which are covered, including those above the ground surface, such as under rocks, rotting logs and on vegetation (Holmquist 1931, Leather *et al.* 1993) and those beneath the soil (Mail 1930, Mani 1962, Barash 1973, Turnock *et al.* 1983, Breitenbach *et al.* 1984).

Temperatures were measured by Turnock *et al.* (1983) at 5cm depth in soil in Canada at several plots. At two plots, snow was allowed to accumulate, and two adjacent plots were kept snow-free. In the snow-covered plots, temperatures generally remained above –5°C throughout the winter while air temperatures remained below –10°C for extended periods. Soil temperatures in snow-free plots were lower and more variable. At the 5cm depth in the snow-free plot, the lowest mean daily temperature dropped to –19.2°C. Turnock *et al.* (1983) found that survival of Bertha Armyworm (*Mamestra configurata*) pupae, which over-winter at depths of 2–15cm in the soil, was above 90 percent in snow covered areas, while in snow-free plots it was less than one percent. It can be concluded from this study that, despite these pupae having supercooling points that range from –18 to –22°C, snow played a critical role in determining overwinter survival of their populations.

Another species known to hibernate in the soil at high northern latitudes is the Painted Turtle (*Chrymys picta*) which often hibernate as hatchlings. Breitenbach *et al.* (1984) found that soil temperatures in snow-covered sites at the depth of their nests were significantly higher than at sites without snow. Over a five-year study, they found that no hatchlings died during a winter of deep snow cover, whereas the highest mortality rate occurred during the two years with days where minimum temperatures were at or below 0°C and there was no snow cover. The soil temperatures in the absence of snow fell to or below the lower supercooling limit of the hatchlings, so again it appears that the insulative properties of snow are a key factor in the annual success of the species.

The insulative properties of snow in association with arthropod hibernacula were clearly illustrated in a study by Holmquist (1931). Continuous temperature records were kept over winter in leaf litter and under bark at several points (e.g. under a log, on top and in areas of different states of decay). It was found that in the open air the mean daily temperatures of both the log and the leaves were only partly buffered from the mean air temperatures, and followed a similar pattern of fluctuation. With about 10 cm of snowcover, however, the hibernacula temperatures became very stable at around zero, even though the air temperatures were well below zero. The lowest air temperature was –16.8°C; the corresponding record for the temperature of leaves was only –1.5°C, and under the bark of the log –3.3°C. Snow cover was not permanent over the hibernacula. When the snow melted, hibernacula temperatures fluctuated with air temperatures but returned to constant zero temperatures as the snow again covered the hibernacula. Snow benefits arthropods in hibernacula by maintaining very stable temperatures close to zero, preventing them from dropping too low. Both of these factors should increase the chance of survival of the resident arthropods in snow covered locations (Holmquist 1931).

The insulative properties of snow also appear to play an important role in determining the location of endotherm hibernacula. Hibernation burrows of Yellow-bellied Marmots, (*Marmota flaviventris*), for example, were situated where snow accumulation was greatest. Highest mortality over winter occurred where snow accumulation was least and snow melt earliest (Svendsen 1974). Similarly, over-winter mortality of the Olympic Marmot (*M. olympus*) correlated inversely with snow cover. Mortality was lowest during a winter of heavy snow cover and high when the snow cover was sparse and rainfall was high (Barash 1973).

Hibernating animals can also receive insulation from soil and may not always require snow cover for overwinter survival. The degree of insulation depends on depth of hibernacula. Temperatures close to the surface of the soil closely follow air temperatures in the absence of snow. However, with increasing depth of soil, temperatures become more stable (Mail 1930, Bailey 1949, Costanzo *et al.* 1995). An obvious strategy for hibernators to avoid lethal temperatures would be to burrow deeper into the soil. The Common Toad (*Bufo bufo*), an adept burrower does this (van Gelder *et al.* 1986) as do some insects (Leather *et al.* 1993). Common Toads have been observed performing a considerable number of vertical movements in the soil profile throughout the winter, always remaining at temperatures above 0°C. Downward movements mostly coincided with a decreasing temperature gradient from the surface as temperatures dropped, whereas when the summer soil warming reached the depth of the toad, it moved upwards (van Gelder *et al.* 1986). Similarly, Common European Lizards (*Lacerta vivipara*) in highland populations hibernate in much deeper burrows than lowland populations (Grenot *et al.* 1996). Hibernating Columbian Ground Squirrels (*Spermophilus columbianus*) appear to have a reduced chance of survival in shallower and less well constructed hibernacula because this leads to higher metabolic costs (Young 1990). There are however, trade offs for hibernating deeper in the soil. Energetic costs of burrowing are increased with burrow depth, and the ability to detect external cues, such as light changes and improved spring conditions becomes increasingly difficult (Danks 1978). Additionally, in areas where the soil depth is shallow, it may not be possible to excavate a deep burrow.

In Australia, Mountain Pygmy-possums hibernate below the ground surface and begin to hibernate in well buffered hibernacula well before the accumulation of snow cover. However, snow cover appears to be essential to buffer hibernacula from extreme winter temperatures and to prevent soil temperatures from falling below the thermal threshold of hibernating animals as winter progresses (Körtner and Geiser 1996, 1997, Walter 1996). In many hibernacula, more than 20–50 cm of snow cover may be necessary for the insulative properties of snow to be effective. A large proportion of the Mountain Pygmy-possum population, especially females, hibernate under boulders within boulderfields. Other individuals hibernate some distance from the main boulderfields, under isolated emergent boulders or sometimes under the soil and vegetation surface where there are no boulders evident, but presumably in connection with crevasses associated with buried boulders. (Mountain Pygmy-possums are not capable of digging burrows and no hibernacula have ever been excavated). In some boulderfields the boulders are very large, some greater than 2 metres in diameter. A deep snow cover (at least a metre or more) is required to fill in all of the gaps and prevent cold air drainage between the boulders.

Temperatures in hibernacula have been monitored in conjunction with the monitoring of body temperature of hibernating Mountain Pygmy-possums with temperature-sensitive radio-collars. Ambient temperatures within hibernacula (T_{hib}) were recorded when animals either slipped their collars or died (Körtner and Geiser 1996, 1997, Walter 1996). Although ambient temperatures were buffered to a large degree in hibernacula, it was not until a complete snow cover was formed that extreme temperature fluctuations were eliminated and a stable T_{hib} plateau of 1.5–3°C (depending on site) was reached (Körtner and Geiser 1996, 1997, Walter 1996). The buffering effect of snow on a well chosen hibernaculum is illustrated in Figure 2. Ambient temperatures showed extreme fluctuations on the soil surface under heath vegetation but were buffered to a degree within the boulderfields, and this effect increased with depth. Temperature fluctuations were considerably less extreme and the average minimum temperature was higher in the hibernaculum. However, the elimination of temperature fluctuations did not occur until snow covered the hibernaculum and this did not occur under the heathland vegetation or in the boulderfields until snow depth was greater than 1 m (Fig. 2). Broome (1992) also found that surface temperatures did not become stable until snow depth was greater than 80–100 cm on Mt. Blue Cow, and it is likely in uneven terrain that an unbroken snow cover is as important as snow depth in maintaining temperature (Chapter 8).

The effect of snow cover was also illustrated by hibernating Mountain Pygmy-possums e.g. Figure 3. Body temperatures during torpor did not stabilise and reach a minimum plateau until a complete snow cover was formed. This point was shown to vary between years for the same individual depending on when snow accumulated (Körtner and Geiser 1997).

Snow depth also appeared to determine the quality of Mountain Pygmy-possum hibernacula. Those individuals which hibernated in areas of heavy snow accumulation, at high elevation and/or on south-easterly aspects e.g. on the south east aspect near the peak of Mt. Blue Cow (1980 m), experienced short arousal periods (less than one

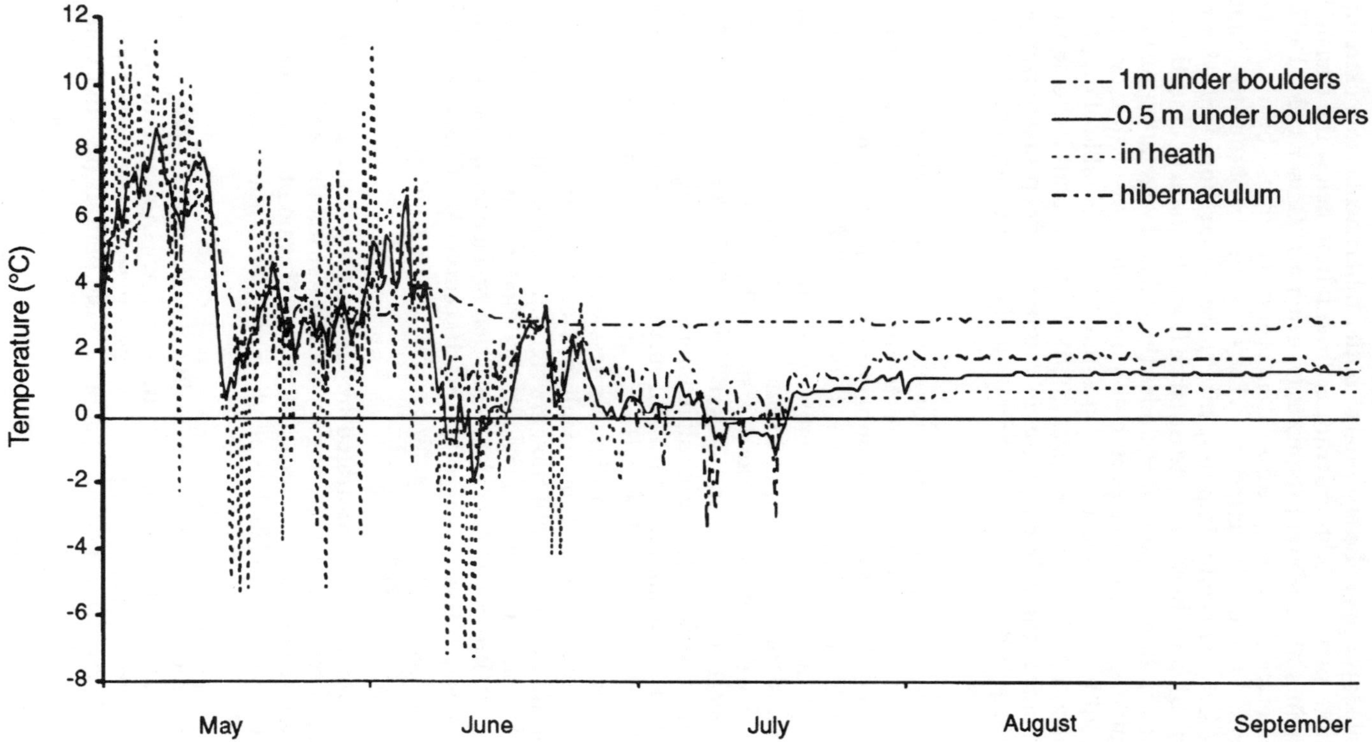

Fig. 2. Average daily temperature measured in Mountain Pygmy-possum habitat at the Paralyser, Kosciusko National Park, autumn to spring 1996. Environmental temperatures were measured from temperature probes leading to a data logger. The hibernaculum temperature was from a temperature-sensitive radio collar (Walter 1996).

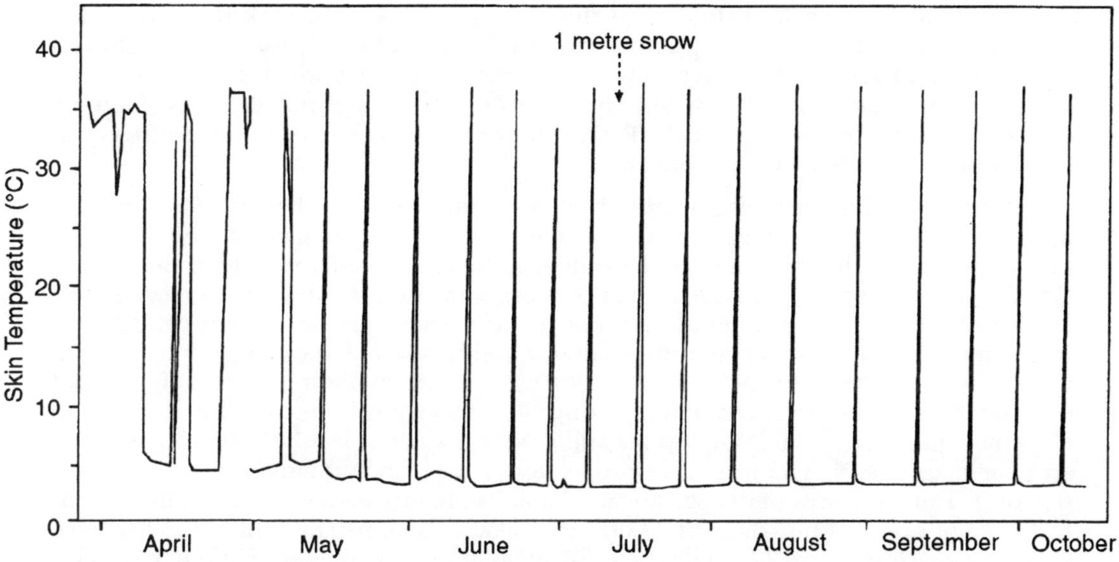

Fig 3. The pattern of hibernation in an adult female Mountain Pygmy-possum. Temperature was measured half hourly as skin temperature under the throat from a temperature-sensitive radio collar (Walter 1996).

day) without leaving the hibernacula (Broome 1992, Broome and Geiser 1995). Individuals at lower elevations on south westerly aspects changed sites up to 12 times and often left the site and remained active for several days (Broome 1992, Körtner and Geiser 1997, Walter 1996). The differences in quality of hibernacula were reflected in length of survival of radio-tracked individuals. Those individuals who survived the winter showed little energy demanding activity and were seldom observed to change hibernacula (Broome 1992, Walter 1996, Körtner and Geiser 1997).

Differential site quality of hibernacula was also reflected in the survival rates of male compared to female Mountain Pygmy-possums. Although there is some overlap between males and females, males tend to hibernate in more marginal habitat at lower altitudes or on north-westerly aspects where boulder fields are shallower, snow cover is less reliable and melts earlier (Mansergh and Broome 1994, Walter 1996, Körtner and Geiser 1997). Females hibernate in locations where the boulderfields are deep and winter snow readily accumulates and is longer-lasting. That males are compromised by their choice of lower quality wintering sites is indicated by higher mortality rates over winter and shorter life expectancies than females and the resultant female biased sex ratios at some breeding sites (Mansergh and Scotts 1990, Broome 1992). Arnold (1990) also showed that low quality of hibernacula was linked to increased rates of weight loss and mortality in Alpine Marmots (*Marmota marmota*).

Hibernacula used by Echidnas in the Kosciuszko National Park study included cavities in the bases of tree stumps, hollow logs, wombat and rabbit warrens, and cavities under soil and rocks. These were always pre-existing cavities which provided secure shelter, rather than burrows dug entirely by the animal, such as nursery sites (Augee *et al.* 1992). No details were given of the sites above the snowline, or of temperatures in hibernacula or assessments of quality.

Aquatic habitats are often utilised by over-wintering ectotherms. The primary advantage of over-wintering in aquatic environments is avoidance of freezing. The temperature in a shallow pond can be similar to that at considerable depth in the soil (Danks 1978). However, without a cover of snow, some ponds (such as those once used by Corroboree Frogs at higher altitudes at the southern extreme of their range) may freeze throughout the water column making winter survival of animals problematic (K. Green pers. comm.). In fact, winterkill of ectotherms including turtles, frogs and invertebrates due to the complete freezing of the water column, has been reported elsewhere, particularly in drought years and/or severe winters (Danks 1978, Christiansen and Bickham 1989). The insulative properties of snow should ameliorate such pond freezing. Temperatures in deeper aquatic habitats are less severe and variable than at the edge of ponds. Animals on the bottom of glacial lakes usually will not experience temperatures less than 4°C. While the substrate of ponds are favoured, open water and aquatic vegetation are used by very few over-wintering invertebrates (Danks 1991).

Aquatic hibernators are susceptible to anoxia and complete freezing of the water column. Snow can influence both of these threats. A complete snow cover of 15–25 cm over a lake or pond, as commonly occurs on Blue Lake in Kosciuszko National Park, may reduce light penetration so that photosynthesis is virtually arrested, then the water becomes anoxic which can lead to winterkill of hibernators (Ultsch 1989), and is especially injurious to fish (Danks 1978). The impact is reduced if the water body is covered with transparent ice or not completely covered with snow, however, due to the influence of wind depositing snow in the cirques this is uncommon in Australia. Snow may, however, be advantageous to aquatic hibernators. Fewer problems are encountered by animals occupying running water (Pinder *et al.* 1992).

Duration of the snow season and time of snowmelt

It appears that, as far as is known, all hibernators in the Australian snow country are opportunistic with regard to the onset and end of hibernation, except perhaps Echidnas (Grigg *et al.* 1992). At the end of winter, the melting of snow and increased insolation cues the beginning of the active season for many species. Körtner and Geiser (1997) suggest that decreases in ambient temperature (T_a) as a result of rainfall and snowmelt may be the cue for final emergence of Mountain Pygmy-possums from hibernation. Insects continue to hibernate as long as the snow cover lasts and terminate only when the winter snow has melted away from the vicinity of the hibernaculum allowing the insolation to penetrate the ground (Mani 1990). Terrestrial herpetofauna become active because of the increased ambient temperatures. The snow melt leads to the commencement of breeding, and the rush to develop young to a winter resting stage before the next bout of cold weather in the following winter.

The timing and duration of the snow season and the amount of snow cover present appears to affect survival and breeding success in at least some species. The timing of snowmelt and therefore the onset of breeding in the Mountain Pygmy-possum can vary by up to 6 weeks from year to year (Fig. 4). Females carry their one litter of four pouch young for approximately one month, after which young are left in the nest from about mid December to mid January (depending on when breeding commenced). Weight gain in autumn is rapid, however, the juveniles do not reach adult size until the end of their second summer (Mansergh 1984, Mansergh and Scotts 1990, Broome 1992). Increased survival of both adults and juveniles over the following winter appears to be related to early breeding (Broome unpubl. data), probably due to the time and energy constraints of nurturing the young and gaining sufficient weight for the ensuring hibernation season.

Insect species emerge from hibernation earliest in spring at sites where snow melts first (Danks 1978). Green (1997) found invertebrates at low elevations in the Australian snow country emerged earlier than those at high elevations because of earlier snowmelt. However, invertebrates at lower elevations subsequently suffered a decline after a late fall of about 30 cm of snow, while those still beneath the snow at higher elevations were not affected.

Coping without snow — a diminishing niche?

Many hibernators including mammals, skinks, frogs and invertebrates are endemic to the snow country (Green and Osborne 1994). Snow is obviously an important aspect of their ecology. But some, if not all of them can survive in captivity in snow-free environments. Why then do they not occur at lower elevations in the wild? It is quite possible that their adaptations to the snow country gives them a competitive advantage over more generalist species, but this advantage is lost at lower elevations in the absence of snow. In areas where there is no prolonged period of snow cover Mountain Pygmy-possums, for example, may be out-competed by species such as Bush Rats (*Rattus fuscipes*) or Dusky Antechinus, species which do not hibernate. The snow may also exclude potential predators e.g. Kookaburras which prey on herpetofauna. The continued presence of snow is almost certainly vital in maintaining a niche for these endemic species.

Global warming, due to the enhancement of the greenhouse effect, is the greatest apparent long-term threat to the alpine and subalpine environments (Whetton *et al.* 1996). Predicted decreases in the duration and amount of snowcover are likely to have a significant effect on populations of alpine hibernators. This will likely result both from diminishing availability and quality of hibernacula, due to a decrease in the insulative benefits of snow, and exposure to severe temperature fluctuations particularly early in the season, and from invasion of generalist species from lower elevations. Although these predictions are speculative for most species due to lack of data, available evidence suggests that the current Mountain Pygmy-possum population is limited by

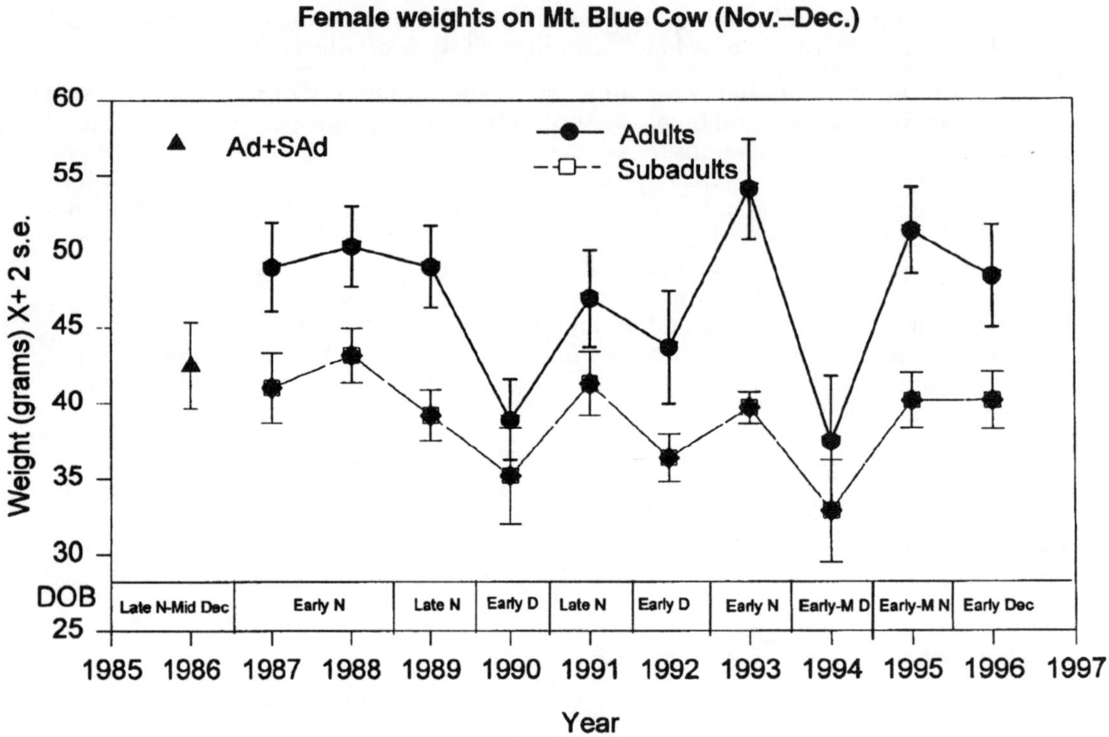

Fig. 4. The weights of female Mountain Pygmy-possums trapped on Mt. Blue Cow, Kosciusko National Park, during the first 2 weeks of December, 1986–1996, and the approximated dates of birth (DOB) of their young (Broome, upubl. data).

the availability of good quality hibernacula and by the limit of the snowline. The same hibernacula are used by the same or different individuals in successive years (Broome 1992, Körtner and Geiser 1997). Individuals which occupy more marginal sites may not survive in years of poor snow cover because of the energetic costs associated with frequent arousals and thermoregulating at sub-optimal ambient temperatures (Geiser and Broome 1991, 1993). Fossil records indicate that Mountain Pygmy-possums had a much wider distribution during the last glaciation. It is reasonable to assume that the current populations of Mountain Pygmy-possums, which occur in isolated and disjunct mountainous areas above the winter snowline, are relict populations which have shrunk as Australia's climate has slowly warmed (Caughley 1986, Mansergh and Broome 1994). This further suggests that snow is a critical element of Mountain Pygmy-possum habitat which restricted their distribution in the past and will continue to do so in the future.

Although early snowmelt appears to be beneficial for some species, increases in winter rainfall which are predicted to accompany rising temperatures (Whetton *et al.*

1996) are likely to outweigh these benefits. Mountain Pygmy-possum T_{hib}'s, for example, were generally steady after snowcover was established except when heavy rainstorms promoted snowmelt and sudden decreases in T_a (Körtner and Geiser 1996). Years of heavy rainfall may reduce winter survival, as found for example in Olympic Marmots (Brash 1973), due to frequent snowmelts and accompanying low T_as. The effect would be exacerbated in Mountain Pygmy-possums if these arousal periods occurred when food sources, e.g. Bogong Moths, were scarce.

Given the tight energy budget of hibernating Mountain Pygmy-possums (Geiser and Broome 1991), the strong influence of ambient temperature on energy expenditure (Geiser and Broome 1993) and the observation that their distribution is so strictly limited by the presence of snow, it appears that permanent snow cover for most of the winter months is crucial for the survival of Mountain Pygmy-possum populations in the Australian Alps (Körtner and Geiser 1997). Without snow cover, especially if replaced by heavy winter rainfall, the Mountain Pygmy-possum is likely to become extinct. Under an analysis of the current bioclimate (Bennett *et al.* 1991) predicted that a suitable bioclimate for the Mountain Pygmy-possum will disappear with a 1°C rise in mean temperature.

Our understanding of the over-winter ecology of the remaining Australian snow country hibernators is poor. Even for Mountain Pygmy-possums, data on the variation in over-winter survival rates and factors affecting over-winter survival are scant. However, given the importance of snow as an insulator and the observation that many alpine species are endemic to the alpine and subalpine environments, we predict that most endemic hibernators will experience a contraction in range under current global warming scenarios. More generalist, mobile species such as Echidnas, may move from subalpine to alpine areas as the snow cover recedes. Since no species of reptiles or amphibians are confined to the alpine zone, it is possible that subalpine and alpine species will contract in their range but survive in available habitat in the present alpine zone. However, this may not be possible for frogs if this subjects them to lethal levels of UV-B radiation (W. Osborne, pers. comm.) The only unaffected hibernators may be those which hibernate in areas where snow does not accumulate, such as below the winter snow line, high up in trees, or in windswept locations.

FUTURE RESEARCH

The lack of definitive information about the overwintering strategies and survivorship of most Australian alpine and subalpine fauna species, limits predictions concerning the effects of changes in snow cover or duration. An understanding of over-winter ecology is particular important for species endemic to the Australian snow country, because global warming may threaten their long term survival. The factors limiting their distribution to the snow country need to be identified. They could, for example, be dependent on the insulative properties of snow, the surge of water experienced following snow melt, or something more indirect such as endemic food supplies, lack of predators, or competition from more generalist species from lower elevations (see Chapter 9).

The physiology of hibernation in the only endemic endothermic hibernator, the Mountain Pygmy-possum, is relatively well understood both from laboratory studies and in the field. Most of the research on survival at low temperature of ectotherms however, has been laboratory-based examining biochemical and physiological mechanisms at the species level. The implications of ecological factors such as variable or decreasing snow cover on winterkill by freezing in natural populations still needs to be investigated (Leather *et al.* 1993). Field research is needed to determine how adaptations such as hibernation actually benefit free-living animals. Environmental factors which influence the success of hibernators also need to be identified. Snow is likely to be a critical element for hibernators in snow-covered environments because it has such outstanding effects on fauna. There is evidence to suggest that the insulative effects of snow can be critical for hibernators during winter, however there are very few studies which deal with this topic specifically, and none of them are from the Australian snow country. Factors such as annual variability of snow fall, for example, may have a large impact on the composition of communities. Current research acknowledges the role of snow in animal hibernation, primarily as an insulator but also as a factor controlling timing of emergence from hibernation. However, very few studies attempt to determine optimal snow conditions for hibernation, or the long term implications of successive years of poor snow cover on hibernators, such as those predicted under global warming scenarios.

For all species there is still a lack of knowledge on the variation in over-winter survival rates and the factors which affect over-winter survival, and how this affects the composition of the subalpine and alpine animal communities. Due to the high degree of topographical, microsite and climatic variability in the Australian alps some of this information can be gained from comparative studies across a range of study sites. Predictions can then be made about future scenarios. However, monitoring of populations at selected study sites is needed to determine long-term trends. Monitoring should initially focus on those species endemic to the Australian snow country, in conjunction with their potential competitors and predators. Three alpine/subalpine hibernating species are currently being monitored in mainland Australia. Populations of Mountain Pygmy-possums, the Southern Corroboree Frog and the Baw Baw Frog, have been monitored over the last 10–15 years (Mansergh and Broome 1994, Osborne 1996, K. Green, pers. comm.). However, little work has been conducted on invertebrates, and for many taxa even the degree of endemism is not fully known. Once these species have been identified, population sizes at a number of sites which reflect the variation in environmental conditions over the species' range should be determined, then monitored annually with the aim of determining over-winter survival rates. Over-winter survival rates can then be related to annual environmental factors such as snow depth, length of snow season, timing of snow melt and the severity of the winter. Another component of this research would involve the identification of hibernacula, measurement of temperatures in hibernacula, and determination of environmental factors affecting the hibernacula such as aspect, elevation, and the duration and depth of snow cover. Such information is elementary in determining the potential influence of global warming on Australian snow country fauna, and in aiding future decisions on their management.

CONCLUSION

Although snow cover varies greatly from year to year in the Australian Alps and is not a prerequisite for hibernation, it appears to be important for all classes of hibernators. Information on the hibernation strategies and tolerance to low ambient temperatures of most Australian Alpine species is scant or lacking. Retreat to a hibernaculum which is insulated against extreme temperature fluctuations is usual. However, some of the Australian grasshoppers are tolerant of freezing and it is probable that other ectotherms are as well (K. Green, pers. comm.). A common trend across all hibernators is an adaptation to the cold which reflects the environment they live in. That is, optimal hibernation temperatures reflect long term environmental temperatures typically experienced at the site. In general, species cannot tolerate winter temperatures below those typically experienced in their hibernacula. Winter temperatures in the Australian Alps are mild compared to many northern hemisphere alpine areas, and soils do not generally freeze for extended periods. Early in winter, soil temperatures remain relatively high due to retained heat from summer insolation. As long as snow cover forms before extreme overnight low temperatures are experienced in late July and August, soil temperatures remain above freezing. Hibernacula which are located in the litter or soil therefore experience very stable, low temperatures. These conditions appear to be ideal for both endothermic and ectothermic hibernators. When temperatures in hibernacula drop below minimum T_b's, mammalian hibernators consume more energy, while temperature may become lethal for immobilised ectotherms.

The effects of long-term increases in average winter temperatures, or a substantial contraction of the snow season (Whetton *et al.* 1996) are most likely to impact on hibernating animals. Severe chilling in winter hibernacula, resulting from loss of snow cover and low overnight minimum temperatures or extended periods of freezing potential (e.g. Breitenbach *et al.* 1984), are likely to be lethal to most species. Given the already small, disjunct populations of many of the alpine and subalpine fauna species, it is probable that extinction rates will accelerate dramatically under a scenario of diminishing snow cover.

REFERENCES

Augee, M. L., Beard, L. A., Grigg, G. C. and Raison, J. K., 1992. Home range of echidnas in the Snowy Mountains. Pp. 225–231 *in* "Platypus and Echidnas" ed by M. L. Augee. Royal Zoological Society of New South Wales, Sydney.

Arnold, W., 1990. The evolution of marmot sociality: Why disperse late? *Behav. Ecol. Sociobiol.* **27:** 229–237.

Bailey, R. M., 1949. Temperature toleration of gartersnakes in hibernation. *Ecology* **30:** 238–242.

Barash, D. P., 1973. The social biology of the Olympic Marmot. *Animal Behav. Mono.* **6:** 171–245.

Bartholomew, G. A. and Hudson, J. W., 1962. Hibernation, estivation, temperature regulation, evaporative waterloss, and heart rate of the pigmy possum, *Cercartetus nanus*. *Physiol. Zool.* **35:** 94–107.

Beard, L. A., Grigg, G. C. and Augee, M. L., 1992. Reproduction by echidnas in a cold climate. Pp. 93–105 *in* "Platypus and Echidnas" ed by M. L. Augee. The Royal Zoological Society of New South Wales, Mosman.

Bennett, S., Brereton, R., Mansergh, I., Berwick, S., Sandiford, K. and Wellington, C., 1991. The Potential Effect of the Enhanced Greenhouse Climate Change on Selected Victorian Fauna. Technical Report Series no. 123. Arthur Rylah Institute, Victoria.

Berman, D. I., Leirikh, A. N. and Mikhailova, E. I., 1984. Winter hibernation of the Siberian salamander, *Hynobius keyserlingi*. *J. Evol. Biochem. Physiol.* **3:** 323–327.

Block, W., 1991. To freeze or not to freeze? Invertebrate survival of subzero temperatures. *Funct. Ecol.* **5:** 284–290.

Breitenbach, G. L., Congdon, J. D. and van Loben Sels, R. C., 1984. Winter temperatures of *Chrysemys picta* nests in Michigan: effects on hatchling survival. *Herpetology* **40:** 76–81.

Broome, L. S., 1992. Ecology and Management of the Mountain Pygmy-Possum *Burramys parvus* (Marsupialia: Burramyidae) in Kosciusko National Park: Impact of the Mt. Blue Cow Ski Resort. A Report to the NSW National Parks and Wildlife Service, NSW.

Broome, L. and Geiser, F., 1995. Hibernation in free-living mountain pygmy-possums, *Burramys parvus* (Marsupialia:Burramydae). *Aust. J. Zool.* **43:** 373–379.

Broome, L. S. and Mansergh, I. M., 1989. The mountain pygmy possum *Burramys parvus* (Broom) an alpine endemic. Pp. 241–264 *in* "The Scientific Significance of the Australian Alps" ed by R. B. Good. Australian Academy of Science, Canberra.

Caughley, J., 1986. Distribution and abundance of the mountain pygmy-possum, *Burramys parvus* Broom, in Kosciusko National Park. *Aust. Wildl. Res.* **13:** 507–516.

Christiansen, J. L. and Bickham, J. W., 1989. Possible historic effects of pond drying and winterkill on the behaviour of *Kinosternon flavescens* and *Chrysemys picta*. *J. Herpetol.* **23:** 91–94.

Common, I. F. B., 1954. A study of the ecology of the adult Bogong moth, *Agrotis infusa* (Boisd.) (Lepidoptera: Noctuidae), with special reference to its behaviour during migration and aestivation. *Aust. J. Zool.* **2:** 223–263.

Costanzo, J. P., Iverson, J. B., Wright, M. F. and Lee, R. E., 1995. Cold hardiness and overwintering strategies of hatchlings in an assemblage of northern turtles. *Ecology* **76:** 1772–1785.

Danks, H. V., 1978. Modes of seasonal adaptation in the insects I. Winter Survival. *Can. Ent.* **110:** 1167–1205.

Danks, H. V., 1991. Winter habitats and ecological adaptations for winter survival. Pp. 231–259 *in* "Insects at Low Temperature" ed by R. E. Lee and D. L. Denlinger. Chapman & Hall, New York.

Dawson, T. J., 1989. Responses to Cold of Monotremes and Marsupials. Pp. 255–288 *in* "Advances in Comparative Physiology" ed by L. C. H. Wang. Springer-Verlag, Berlin.

Feist, D. D. and White, R. G., 1989. Terrestrial mammals in cold. Pp. 328–360 *in* "Advances in Comparative and Environmental Physiology" ed by L. C. H. Wang. Springer-Verlag, Berlin.

Fisher, K. C., 1964. On the mechanisms of periodic arousal in the hibernating ground squirrel. *Ann acad Sci, Fenn, Ser A* **71:** 143–156.

Formosov, A. N., 1964. "Snow Cover as an Integral Factor of the Environment and its Importance in the Ecology of Mammals and Birds." Boreal Institute for Northern Studies, University of Alberta, Edmonton.

French, A. R., 1985. Allometries of the durations of torpid and euthermic intervals during mammalian hibernation: A test of the theory of metabolic control of the timing of changes in body temperature. *J. Comp. Pysiol B* **156:** 13–19.

French, A. R., 1988. The patterns of mammalian hibernation. *Am. Sci.* **76:** 569–575.

French, A. R., 1990. Age-class differences in the pattern of hibernation in yellow-bellied marmots, *Marmota flaviventris*. *Oecologia* **82:** 93–96.

Galloway, R. W., 1988. The potential impact of climate changes on Australian ski fields. Pp. 428–437 *in* "Greenhouse: Planning for Climate Change" ed by G. I. Pearman. CSIRO/Brill, Melbourne.

Geiger, R., 1959. "The Climate Near the Ground." Harvard University Press, Cambridge.

Geiser, F., 1988. Daily torpor and thermoregulation in Antechinus (Marsupialia): influence of body mass, season, development, reproduction, and sex. *Oecologia* **77:** 395–399.

Geiser, F., 1993. Hibernation in the eastern pygmy possum, *Cercartetus nanus*. *Aust. J. Zool.* **41:** 67–75.

Geiser, F., 1994. Hibernation and daily torpor in marsupials: a review. *Aust. J. Zool.* **42:** 1–16.

Geiser, F. and Broome, L., 1991. Hibernation in the mountain pygmy possum *Burramys parvus* (Marsupialia). *J. Zool., Lond.* **223:** 593–602.

Geiser, F. and Broome, L., 1993. The effect of temperature on the pattern of torpor in a marsupial hibernator. *J. Comp. Physiol.* **A163:** 133–137.

Geiser, F. and Kenagy, G. J., 1987. Polyunsaturated lipid diet lengthened torpor and reduces body temperature in a hibernator. *Am. J. Physiol.* **252:** R897–R901.

Geiser, F. and Kenagy, G. J., 1988. Torpor duration in relation to temperature and metabolism in hibernating ground squirrels. *Physiol. Zool.* **61:** 442–449.

Geiser, F. Stahl, B., and Learmonth, R. P., 1992. The effect of dietary fatty acids on the pattern of torpor in a marsupial. *Physiol. Zool.* **65:** 1236–1245.

Geiser, F. and Ruf, T., 1995. Hibernation versus daily torpor in mammals and birds: physiological variables and classification of torpor patterns. *Physiol. Zool.* **68:** 935–966.

Green, K., 1982. Notes on winter-active invertebrates beneath the snow. *Vic. Nat.* **99:** 144–146.

Green, K., 1983. Alternative strategies in the life cycles of alpine grasshoppers (Orthoptera: Acridoidae). *Proc. Ecol. Soc. Aust.* **12:** 125–133.

Green, K., 1997. Inter-annual, seasonal and altitudinal differences in invertebrate activity in the Snowy Mountains. *Vic. Nat.* **114:** 222–229.

Green, K. and Osborne, W. S., 1981. The diet of foxes, *Vulpes vulpes* (L.) in relation to abundance of prey above the winter snowline in New South Wales. *Aust. Wildl. Res.* **8:** 349–360.

Green, K. and Osborne, W., 1994. "Wildlife of the Australian Snow-Country." Reed Books, Sydney.

Gregory, P. T., 1982. Reptilian hibernation. Pp. 153–154 *in* "Biology of the Reptilia" ed by C. Gans and F. H. Pough. Academic Press, London.

Grenot, C. J., Garcin, L. and Tsere-Pages, H., 1996. Cold-hardiness and behaviour of the European Common Lizard from French populations in winter. Pp. 115–122 *in* "Adaptations to the Cold Tenth International Hibernation Symposium" ed by F. Geiser, A. J. Hulbert and S. C. Nicol. University of New England Press, Armidale.

Grigg, G. C., Beard, L. A. and Augee, M. L., 1989. Hibernation in a monotreme, the echidna (*Tachyglossus aculeatus*). *Comp. Biochem. Physiol.* **92A:** 609–612.

Grigg, G. C., Augee, M. L. and Beard, L. A., 1992. Thermal relations of free-living echidnas during activity and in hibernation in a cold climate. Pp. 160–173 *in* "Platypus and Echidnas" ed by M. L. Augee. Royal Zoological Society of New South Wales, Sydney.

Heller, H. C. and Hammel, H. T., 1972. CNS control of body temperature during hibernation. *Comp. Biochem. Physiol.* **41A:** 349–359.

Hickman, C. P., Roberts, L. S. and Hickman, F. M., 1990. Biology of Animals. Times Mirror / Mosby College, St. Louis.

Hoffman, R. A., 1964. Terrestrial animals in cold: hibernators. Pp. 379–403 *in* "Handbook of Physiology. Section 4: Adaptation to the Environment" ed by D. B. Dill, E. F. Adolf and C. G. Wilber. American Physiological Society, Washington.

Holmquist, A. M., 1931. Studies in arthropod hibernation. III. Temperatures in forest hibernacula. *Ecology* **12:** 387–400.

Hudson, J. W., 1973. Torpidity in mammals. Pp. 98–166 *in* "Comparative Physiology of Thermoregulation" ed by G. C. Whittow. Academic Press, New York.

Körtner, G. and Geiser, F., 1995. Effect of Photoperiod and ambient temperature on activity patterns and body weight cycles of mountain pygmy-possums, *Burramys parvus* (Marsupialia). *J. Zool., Lond.* **235:** 311–322.

Körtner, G. and Geiser, F., 1996. Hibernation of mountain pygmy-possums (*Burramys parvus*) in the Australian alps. Pp. 31–38 *in* "Adaptations to the Cold. Tenth International Symposium" ed by F. Geiser, A. J. Hulbert and S. C. Nicol. University of New England Press, Armidale.

Körtner, G. and Geiser, F., 1997. Ecology of natural hibernation in the marsupial mountain pygmy-possum (*Burramys parvus*). *Oecologia* (In Press).

Leather, S. R., Walters, K. F. A. and Bale, J. S., 1993. "The Ecology of Insect Overwintering." Cambridge University Press, Cambridge.

Lee, R. E., 1991. Principles of insect low temperature tolerance. Pp. 17–46 *in* "Insects at Low Temperature" ed by R. E. Lee and D. L. Denlinger. Chapman and Hall, New York.

Lowe, C. H., Lardner, P. J. and Halpern, E. A., 1971. Supercooling in reptiles and other vertebrates. *Comp. Biochem. Physiol.* **39A:** 125–135.

Lyman, C. P., 1982a. Who is Who Among the Hibernators. Pp. 12–36 *in* "Hibernation and Torpor in Mammals and Birds" ed by C. P. Lyman, J. S. Willis, A. Malan and L. C. H. Wang. Academic Press, New York.

Lyman, C. P., 1982b. Why bother to hibernate? Pp. 1–11 *in* "Hibernation and Torpor in Mammals and Birds" ed by C. P. Lyman, J. S. Willis, A. Malan and L. C. H. Wang. Academic Press, New York.

Mail, G. A., 1930. Winter soil temperatures and their relation to subterranean insect survival. *J. Agric. Res.* **41:** 571–592.

Mani, M. S., 1962. "Introduction to High Altitude Entomology. Insect Life Above the Timber-line in the North-west Himalaya." Methuen & Co. Ltd, London.

Mani, M. S., 1990. "Fundamentals of High Altitude Biology". Aspect Publications, London.

Mansergh, I. M., 1984. Ecological studies and conservation of *Burramys parvus*. Pp. 545–552 *in* "Possums and Gliders" ed by A. Smith and I. Hume. Surrey Beatty and Sons Pty Ltd, Chipping Norton.

Mansergh, I. and Broome, L., 1994. "The Mountain Pygmy-Possum." New South Wales University Press, Kensington.

Mansergh, I., Baxter, B., Scotts, D., Brady, T. and Jolley, D., 1990. Diet of the mountain pygmy-possum, *Burramys parvus* (Marsupialia: Burramyidae) and other small mammals in the alpine environment at Mt. Higginbotham, Victoria. *Aust. Mammal.* **13:** 167–177.

Mrosovsky, N., 1971. "Hibernation and the Hypothalamus." Appleton Century Crofts: New York.

Mrosovsky, N., 1980. Circannual cycles in golden-mantled ground squirrels: experiments with food deprivation on effects of temperature on periodicity. *J. Comp. Physiol.* **136**: 355–360.

Nedergaard, J. and Cannon, B., 1990. "Mammalian Hibernation." *Philos. Trans. Roy. Soc. Lond. B* **326:** 669–686.

Nicol, S. and Anderson, N. A., 1996. Hibernation in the echidna: not an adaptation to cold? Pp. 7–12 *in* "Adaptations to the Cold. Tenth International Symposium" ed by F. Geiser, A. J. Hulbert and S. C. Nicol. University of New England Press, Armidale.

Pinder, A. W., Storey, K. B. and Ultsch, G. R., 1992. Estivation and Hibernation. Pp. 250–274 *in* "Environmental Physiology of the Amphibians" ed by M. E. Feder and W. W. Burggren. University of Chicago Press, Chicago.

Osborne, W. S. 1996. Recovery plan for the Southern Corroboree Frog *Pseudophryne corroboree*. New South Wales National Parks and Wildlife Service, Sydney.

Osborne, W. S., Zentelis, R. A., and Lau, M. 1996. Geographical variation in Corroboree frogs, *Pseudophryne corroboree* (Moore) (Anura: Myobatrachidae): A reappraisal supports recognition of *P. pengilleyi* Wells & Wellington. *Aust. J. Zool.* **44:** 569–587.

Pengelley, E. T. and Fisher, K. C. 1957. Onset and cessation of hibernation under constant temperature and light in the golden-mantled ground squirrel (*Citellus lateralis*). *Nature* **180:** 1371–1372.

Pengelley, E. T. and Fisher, K. C. 1963. The effect of temperature and photoperiod on the yearly hibernating behaviour of captive golden-mantled ground squirrels (*Citellus lateralis tescorum*). *Can. J. Zool.* **41:** 1103–1120.

Schmid, W. D., 1982. Survival of frogs in low temperature. *Science* **215:** 697–698.

Schmidt-Neilsen, K., 1975. "Animal Physiology." Cambridge University Press, Cambridge.

Smith, A. P. and Broome, L., 1992. The effects of season, sex and habitat on the diet of the mountain pygmy-possum (*Burramys parvus*). *Wildl. Res.* **19:** 755–768.

Spellerberg, I. F., 1972. Temperature tolerances of southeast Australian reptiles examined in relation to reptile thermoregulatory behaviour and distribution. *Oecologia* **9:** 23–46.

Storey, K. B., 1985. Freeze tolerance in terrestrial frogs. *Cryo-Lett* **6:** 115–134.

Storey, K. B. and Storey, J. M., 1988. Freeze tolerance in animals. *Physiol. Rev.* **68:** 27–84.

Storey, K. B. and Storey, J. M., 1991. Biochemistry of cryoprotectants. Pp. 64–93 *in* "Insects at Low Temperatures" ed by R. E. Lee and D. L. Denlinger. Chapman and Hall, New York.

Svendsen, G. E., 1974. Behavioral and environmental factors in the spatial distribution and population dynamics of a Yellow-bellied Marmot population. *Ecology* **55:** 760–771.

Turnock, W. J., Lamb, R. J. and Bodnaryk, R. P., 1983. Effects of cold stress during pupal diapause on the survival and development of *Mamestra configurata* (Lepidoptera: Noctuidae). *Oecologia* **56:** 185–192.

Tyler, M. J., Davies, M. and Watson, G. F., 1993. General description and definition of the class Amphibia. Pp. 1–9 *in* "Fauna of Australia" ed by C. J. Glasby, G. J. B. Ross and P. L. Beesley. Australian Government Publishing Service, Canberra.

Ultsch, G. R., 1989. Ecology and physiology of hibernation and overwintering among freshwater fishes, turtles and snakes. *Biol. Rev.* **64:** 435–516.

van Gelder, J. J., Olders, J. H. J., Bosch, J. W. G. and Starmans, P. W., 1986. Behaviour and body temperature of hibernating common toads *Bufo bufo*. *Holarctic Ecol.* **9:** 225–228.

Walter, A.-M. J., 1996. The Ecology of Hibernating Mountain Pygmy-possum *Burramys parvus* in Kosciusko National Park. Honours, University of Canberra.

Wang, L. C. H., 1978. Energetics and field aspects of mammalian torpor: the Richardson's ground squirrel. Pp. 109–146 *in* "Strategies in Cold: Natural Torpidity and Thermogenesis" ed by L. C. H. Wang and J. W. Hudson. Academic Press, New York.

Wang, L. C. H., 1989. Ecological, physiological and biochemical aspects of torpor in mammals and birds. Pp. 361–401 *in* "Advances in Comparative and Environmental Physiology" ed by L. C. H. Wang. Springer-Verlag, Berlin.

Whetton, P. H., Haylock, M. R. and Galloway, R., 1996. Climate change and snow cover duration in the Australian Alps. *Climatic Change* **32:** 447–479.

Willis, J. S., 1982. The mystery of the periodic arousal. p. 317 *in* "Hibernation and Torpor in Mammals and Birds" ed by C. P. Lyman, J. S. Willis, A. Malan and L. C. H. Wang. Academic Press, New York.

Young, P. J., 1990. Hibernation Patterns of free-ranging Columbian ground squirrels. *Oecologia* **83:** 504–511.

PART 3

Global Threats

Global warming

Ozone depletion

CHAPTER 11

Climate change impacts on the spatial extent of snow-cover in the Australian Alps

PETER WHETTON

INTRODUCTION

CHANGES in regional climate due to enhancement of the global greenhouse effect have the potential to affect the natural environment of the Australian Alps significantly as well as many mountain-based human activities. In particular, a major physical impact expected due to climate change is a reduction in natural snow-cover as a consequence of higher temperatures. This would affect skiing activities and storage of water for irrigation among other impacts.

Whetton *et al.* (1996) made an assessment of how climate change may affect the duration of natural snow-cover throughout the Alps using a version of the snow-cover model of Galloway (1988) and a set of climate-change scenarios for the region (CIG 1992). Under a worst-case climate change scenario, simulated duration of snow-cover across the Alps was substantially reduced by 2030 and approached zero, even at the highest sites by 2070. Under a best-case climate change scenario, the simulated impact of natural snow-cover was small relative to natural variability.

CSIRO's scenarios of future regional climate change are revised and updated as new climate modelling results and other relevant information come to hand. An updated set of scenarios was released in November 1996 (CIG 1996). The range of projected regional warming was substantially reduced in these scenarios compared to the previous scenarios used by Whetton *et al.* (1996) in assessing impacts on snow-cover.

This Chapter presents new simulations of future snow-cover in the Australian Alps based on the revised scenarios of CIG (1996). Thus this chapter represents an update to Whetton *et al.* (1996), taking into account revisions to projections of future climate change. In addition, the chapter also briefly considers impacts of climate change on the alpine environment.

THE ENHANCED GREENHOUSE EFFECT AND CLIMATE CHANGE

Current scientific understanding predicts with confidence some global warming in the decades to come. This is due to the well-understood warming effect of greenhouse gases in the earth's atmosphere and the high confidence that the concentrations of these gases will continue to grow in the future (see IPCC 1996). This growth is due to the expectation that anthropogenic emissions of greenhouse gases (mainly carbon dioxide) will not be dramatically reduced. A range of "scenarios" are produced, rather than a single prediction, because there are many uncertainties and alternative assumptions that still need to be allowed for when projecting regional climate change. While it is possible that greenhouse-gas concentrations may be stabilised in the future in response to international legislation, this scenario is not considered here.

The latest Intergovernmental Panel on Climate Change assessment (IPCC 1996) presented scenarios of global warming for 1990–2100. In these scenarios, global warming ranges from 0.4–0.8°C in 2030 and 0.7–2.1°C in 2070. This range takes into account the range of global climate model (GCM) estimates of the sensitivity of the global climate system (an equilibrium global average warming of 1.5–4.5°C for a doubling of the present CO_2 concentration) and the range of plausible future greenhouse gas emission scenarios (IS92a-e) as determined by IPCC (1992). The scenarios also include an estimate of the cooling effect of increasing sulfate aerosol concentration. The latest CSIRO climate change scenarios (CIG, 1996) are based on these global warming scenarios, and include estimates of how patterns of temperature and precipitation may change per degree of global warming in the Australian region. These estimates are taken from a range of GCMs, including a set of models which have a simple 50 m deep slab ocean ("slab ocean GCMs") and a set of models each of which are coupled to a full ocean model ("coupled GCMs"). GCMs are computer models which solve mathematical equations based on the laws of physics, leading to a representation of atmospheric and oceanic processes relevant to climate.

Full details of these latest CSIRO scenarios are given in CIG (1996). Here, as in Whetton *et al.* (1996), the best-case scenario for snow (least increase in temperature and greatest increase in precipitation in winter consistent with the scenarios for the study area) and the worst-case scenario (greatest warming, greatest decrease in precipitation in winter) are drawn from these scenarios. The climate scenarios relevant to the Australian Alps are given in Table 1.

Note that for simplicity, the scenarios in Table 1 were drawn from the scenarios in CIG (1996) based on coupled GCMs. Had the slab-model-based scenarios been included as well, the best-case and worst-case scenarios would differ negligibly.

TABLE 1

Climate change scenarios for the Australian alpine area based on CIG (1996).

	best-case 2030	worst-case 2030	best-case 2070	worst-case 2070
Temperature	+ 0.3°C	+ 1.3°C	+ 0.6°C	+ 3.4°C
Precipitation	0 %	– 8 %	0 %	– 20 %

The rate of warming in the worst-case scenario is substantially less than the corresponding case used in Whetton *et al.* (1996) (e.g. a warming of 3.4°C versus one of 5°C in 2070). This downward revision in the high case scenario stems mainly from changes in the IPCC global warming scenarios. This revision is due to a number of reasons, including the inclusion of the cooling effect of sulfate aerosol pollution, and model development.

Scenarios for changes in average temperature and average winter precipitation are based on climate modelling information of coarse horizontal resolution (about 300–500 km between grid points). At this resolution, the mountains of southeastern Australia (the Great Dividing Range) are not resolved by the model. However, some assessment of how much the scenarios could differ if the mountains had been represented can be made based on some high resolution climate modelling recently undertaken by CSIRO (Whetton *et al.* in press a). This study used a regional model of a horizontal resolution of 60 km which was forced at the model boundaries by the results from the coarse resolution global model of CSIRO. A horizontal resolution of 60 km allows the Great Dividing Range to be represented in the model (although the maximum elevation of the range in the model is still substantially less than in reality). The nested experiment indicated that enhanced greenhouse warming over the mountains does not differ substantially from the broad regional warming simulated by the GCM. The experiment did, however, indicate that more realistic topography led to small simulated winter precipitation increases (up to 5% in the best-case in 2070) along the south side of the Divide in Victoria which were not present in the GCM. This suggests that the best-case scenario given above may be a little pessimistic with regards to locations such as Mt. Baw Baw and Lake Mountain in Victoria.

The impact of the above scenarios on change in mean temperature and precipitation and thus on natural snow-cover in the Australian Alps will be considered in detail in this Chapter. However, it is noted that changes in climate may have significant impacts on the alpine environment in ways that do not directly involve snow-cover. For example, changes in temperature and rainfall can be expected to influence plant and animal distribution and (along with changes in wind speed and relative humidity) fire frequency (Busby 1988; Beer and Williams 1995). There is now a large body of evidence, based on the analysis of climate model output, that suggests that very heavy rainfall events may become more intense under enhanced greenhouse conditions (Gordon *et al.* 1992; Whetton *et al.* 1993; Fowler and Hennessy 1995). This is likely to increase the frequency or intensity of flash flooding, and, in the mountain environment, may increase erosion and the likelihood of landslide (see Breiling and Charazma 1994).

MODELLING SNOW-COVER IMPACTS

A range of studies both overseas (e.g. Martin *et al.* 1994, Bultot *et al.* 1994) and in Australia (Galloway 1988, Hewitt 1994, Whetton *et al.* 1996) have revealed the likelihood of marked reductions of natural snow-cover as a result of climate change. Furthermore, this sensitivity is due mainly to increases in temperature rather than accompanying changes in precipitation. A variety of approaches have been used in these studies ranging from simple snow models driven by mean precipitation and temperature (Galloway 1988, Whetton *et al.* 1996) to approaches that allow for GCM simulated changes in the frequency of snow-bearing circulation systems (Hewitt 1994).

Here the Whetton *et al.* (1996) version of the Galloway snow model (model B in Galloway (1988)) is used. The model is semi-empirical and simulates the duration of the snow-cover and the depth of snow through the season in water equivalent terms using mean monthly precipitation and temperature and standard deviation of temperature. Full details of the model can be found in Whetton *et al.* (1996) who also developed a gridded climate data set of high resolution for the Australian alpine region so that the model could be applied in a spatially distributed manner. The observed data used station observations interpolated to a 2.5 km grid using the three dimensional interpolation method of Hutchinson (1989) which allows for the effect of elevation on climate. A long-term mean version of the data set was produced which used all available records as well as a version which included the interannual variability of 1966–85. In Whetton *et al.* (1996), the model was driven by this data set unmodified, to simulate current snow conditions across the Alps, and then modified according to the CIG (1992) scenarios, to simulate possible future snow conditions. Here the same model and observed data are used but the CIG (1996) climate change scenarios are applied.

Results comparing simulated current snow conditions against observed snow conditions are not presented here. This was covered in Whetton *et al.* (1996), although a paucity of suitable observations made comparison difficult. The model was found to simulate reasonably well the observed altitudinal variation in average snow-cover duration in the region (an effect dominated by changes in temperature), and to simulate observed interannual variations in snow-cover adequately (where large changes in precipitation are important). This suggests the relative sensitivities of the model to changes in temperature and precipitation are approximately correct.

However, Whetton *et al.* (1996) identified a tendency of the model to simulate durations that were too short at marginal, low elevation sites, where observed snow-cover can often be discontinuous through the winter. This means that short simulated durations (less than around 30 days) should be viewed as "marginal" and not interpreted literally. This model deficiency was attributed to the use of data resolved monthly as input to the model. A version of the model has been developed which can accept daily data (Schreider *et al.* in press) and, in some limited tests, significantly improved model performance at marginal sites has been achieved.

Other caveats associated with the approach used here that should be noted are:

- Changes in interannual variability and precipitation under enhanced greenhouse conditions are not taken into account. However analysis of changes in variability under enhanced greenhouse conditions recently undertaken for the region using nested model results (Whetton *et al.* in press b) indicates little significant change in this variability.
- The scenarios take into account our best current estimates of future climate change. However, these estimates will change in the future as the science develops and our ability to model future climate improves.

For selected sites, Whetton *et al.* (1996) also explored the sensitivity of simulated snow-cover duration to a broad range of changed climates (temperature changes of –2°C through to +4°C, and precipitation changes of –80% through to +80%). It was noted that in terms of their effect on simulated duration of snow-cover, a 0.5°C change in temperature was typically equivalent to a 40–50% change (of opposite sign) in precipitation. The scenarios of climate change used by Whetton *et al.* (1996), and those to be used here give precipitation changes of the order of 5% for each 0.5°C of warming. This means that that snow-cover duration simulated for the climate change situation is driven mainly by changes in temperature.

SIMULATED NATURAL SNOW-COVER IN THE ALPS UNDER CLIMATE CHANGE CONDITIONS

Figure 1 shows simulated snow-cover duration over the Australian Alpine area for current climate, and for the best and worst-case scenarios for 2030 and 2070. These figures may be compared with the corresponding diagrams in Whetton *et al.* (1996).

The areal extent of simulated snow-cover under best-case conditions in 2030 (Fig. 1b) is smaller than the equivalent for current climate conditions (Fig. 1a). For example the region with simulated snow-cover duration in excess of thirty days (which may be considered the region where snow is a significant component of the local environment) has an area of 4540 km² in 2030 compared to 5540 km² in the current climate case. This is a reduction of 18%. This area further declines to 3770 km² by 2070 (see Fig. 1c), which represents a 39% reduction in area compared to current climate.

The areal extent of snow-cover declines much more markedly under the worst-case scenario for 2030 and 2070 (Figs 1d and 1e). The area of simulated snow-cover duration in excess of thirty days in 2030 is 1870 km² (a 66% reduction from the current climate area) and in 2070 is 195 km² (96% reduction).

These simulated reductions in area may be compared to year to year variations in the past. Figure 2 shows the simulated snow-cover in two extreme years within the data period of 1966–85. The year 1973 (Fig. 2a) represents a year of very poor snow-cover,

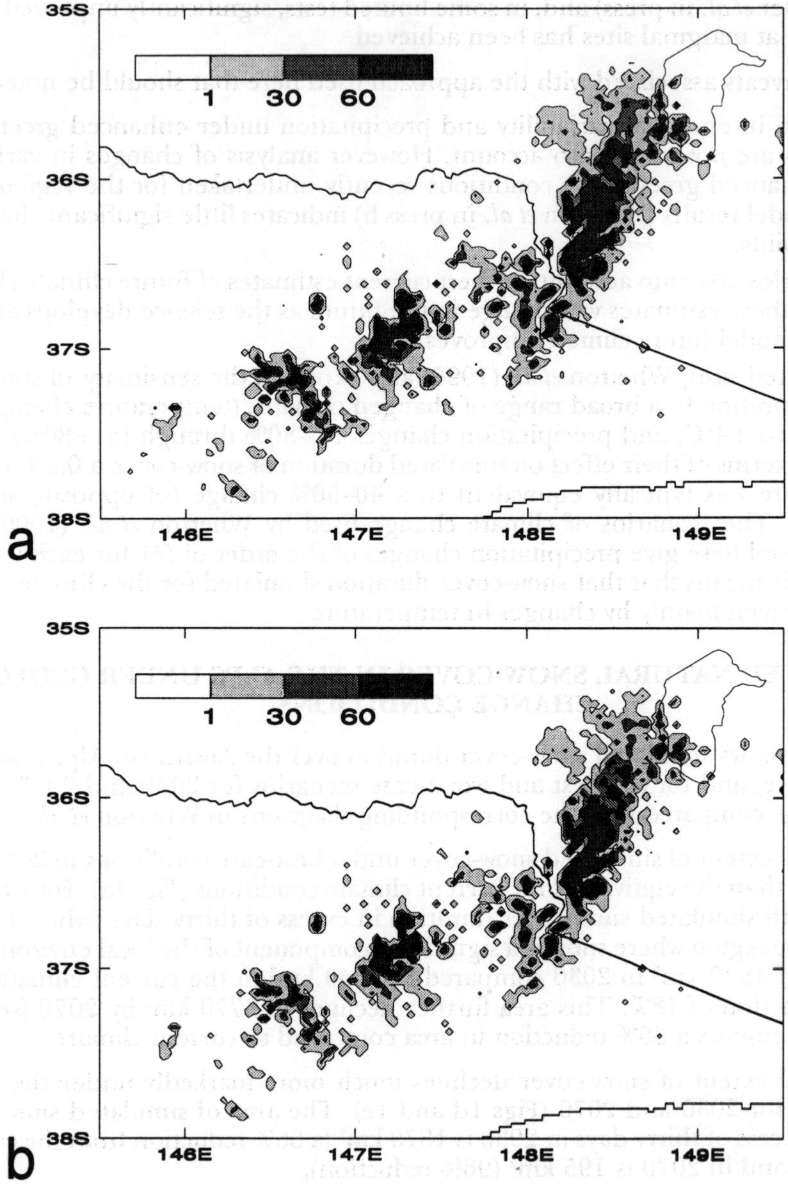

Fig. 1. **Simulated average snow-cover duration in days for (a) current climate, (b) best-case scenario for 2030. (Horizontal bar shows number of days).**

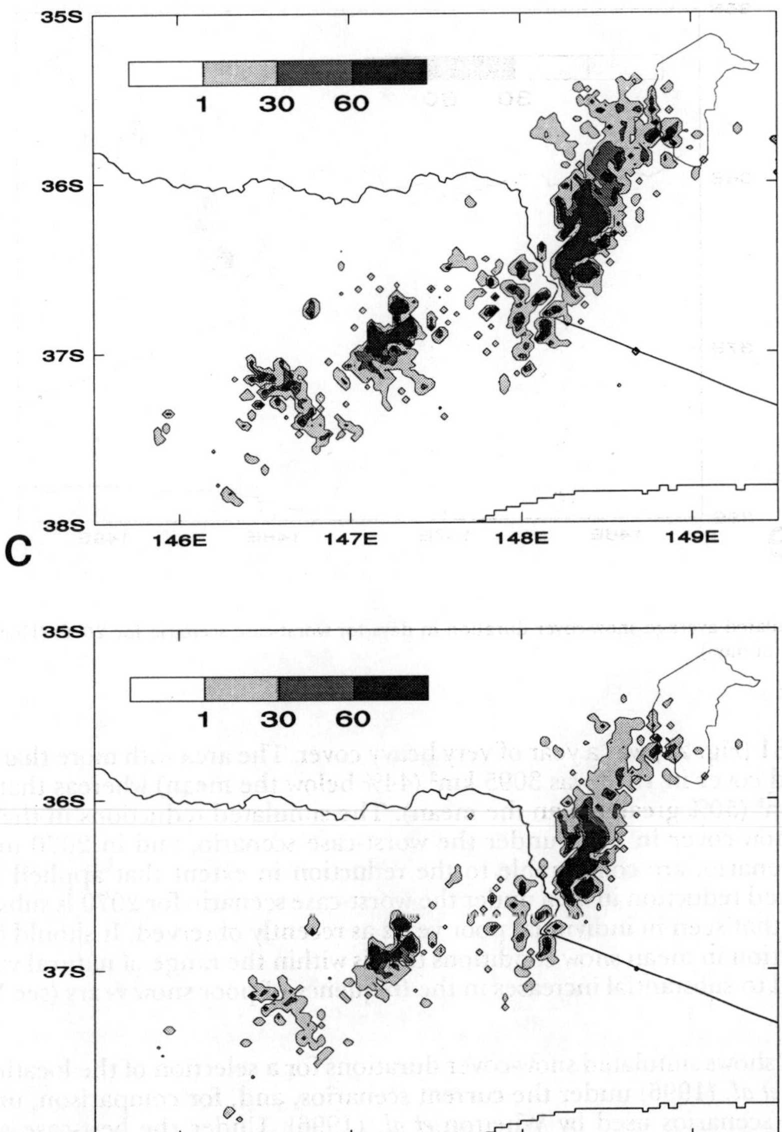

Fig. 1. **Simulated average snow-cover duration in days for (c) the best-case scenario for 2070, (d) worst-case scenario for 2030. (Horizontal bar shows number of days).**

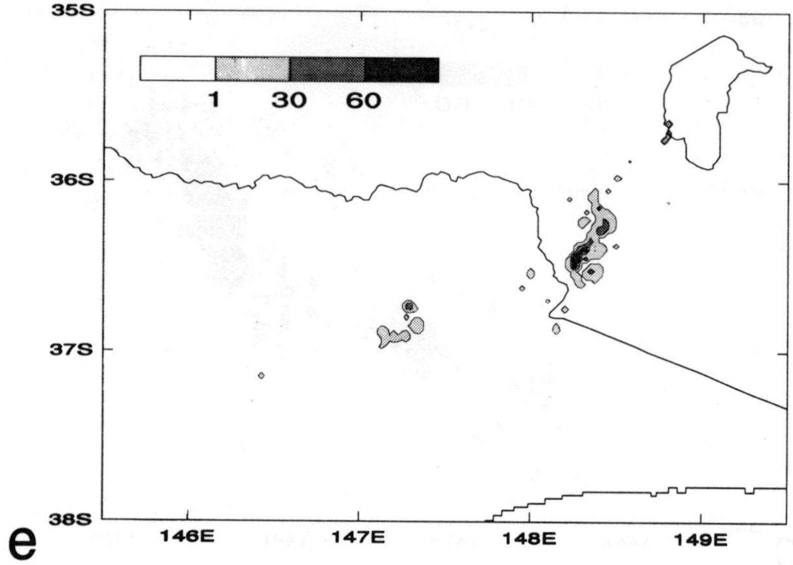

Fig. 1. (e) Simulated average snow-cover duration in days for worst-case scenario for 2070. (Horizontal bar shows number of days).

whereas 1981 (Fig. 2b) was a year of very heavy cover. The area with more than 30 days of simulated cover in 1973 was 3095 km² (44% below the mean) whereas that of 1981 was 8310 km² (50% greater than the mean). The simulated reductions in the average extent of snow-cover in 2030 under the worst-case scenario, and in 2070 under the best-case scenario, are comparable to the reduction in extent that applied in 1973. The simulated reduction in area under the worst-case scenario for 2070 is substantially larger than that seen in individual poor years as recently observed. It should be noted that a reduction in mean snow conditions that is within the range of natural variability can still lead to substantial increases in the frequency of poor snow years (see Whetton *et al.* 1996).

Table 2 shows simulated snow-cover durations for a selection of the locations used in Whetton *et al.* (1996) under the current scenarios, and, for comparison, under the CIG (1992) scenarios used by Whetton *et al.* (1996). Under the best-case scenario, simulated duration declines in 2030 by around ten days and in 2070 by around twenty days. The decreases are more important in proportional terms at the lower sites. The decreases are also very similar to those simulated using the best-case scenario from CIG (1992). Under the worst-case scenario in 2030, simulated duration is zero at around 1400 m and at higher sites is around 40–50 days shorter than current simulated

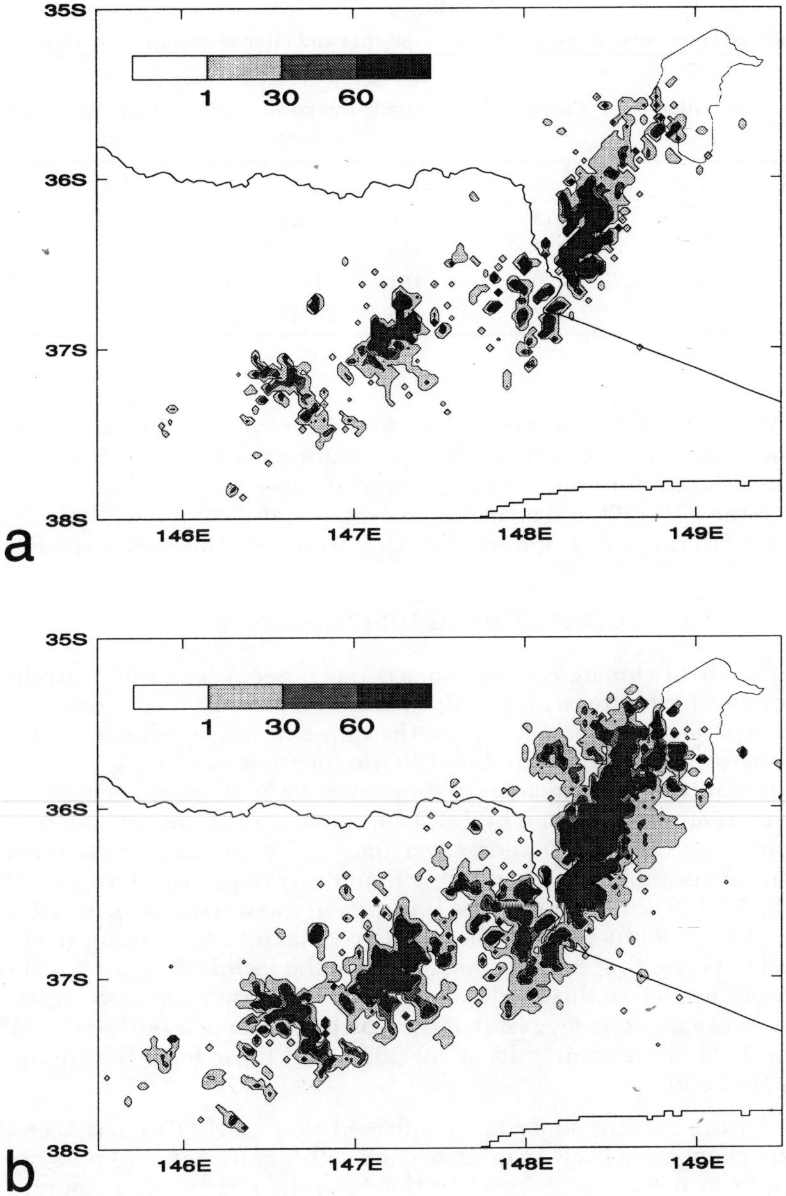

Fig. 2. Simulated snow-cover duration for two selected extreme years: (a) 1973, a year of poor snow-cover, and (b) 1981, a year of very good snow-cover. (Horizontal bar shows number of days).

TABLE 2

Simulated snow-cover durations for selected sites and climate change scenarios.

Location	Elevation (m)	Current	Simulated snow-cover duration (days) CIG (1992) scenarios 2030	2070	CIG (1996) scenarios 2030	2070
Lake Mountain	1400	29	0–16	0–7	0–17	0–12
Falls Creek	1643	113	36–101	0–88	64–100	0–92
Mt. Buller	1805	127	52–116	0–103	80–115	5–107
Mt. Bogong	1986	159	103–151	1–142	123–150	45–144
Mt. Kosciusko	2228	187	135–178	29–170	152–177	92–172

durations. In 2070, simulated durations are around zero up to 1800 m elevation and around 100 days shorter than current climate durations at higher elevations. These worst-case reductions in duration are not as large as those presented in Whetton *et al.* (1996) based on the CIG (1992) scenarios. For example, the 2030 worst-case durations are around twenty to thirty days longer than they were with the previous scenarios.

CONCLUDING DISCUSSION

This short study of climate change impacts on snow-cover in the Australian Alps updates the results of Whetton *et al.* (1996) to take into account recent revisions to the CSIRO climate change scenarios. Amongst the range of future climates indicated in the CSIRO scenario, the most favourable scenario for snow conditions (the "best-case scenario") leads to moderate reductions in snow-cover by 2030 (the area with simulated cover of more than 30 days declines by 18%) and more significant reductions by 2070 (39% reduction). The changes in snow-cover under a worst-case scenario are much more severe (the area simulated to have more than thirty days cover is reduced by 66% by 2030, and by 96% by 2070), but not as severe as in the worst-case scenario used by Whetton *et al.* (1996). Reductions in the predicted maximum warming between the CIG (1992) and CIG (1996) scenarios, due in part to the inclusion of the cooling effect of sulfate aerosols, has led to this moderation in the worst-case scenario impacts. Very broadly, the latest calculations suggest that the worst-case impacts outlined in Whetton *et al.* (1996) for 2030 are now more likely for 2050, and those for 2070 are now more likely to apply for 2090.

In summary, the current study has confirmed that CSIRO's latest scenarios of regional climate change are likely to be associated with significant, and possibly rather severe, reductions in natural snow-cover in the Australian Alps. Such changes would affect alpine ecosystems, hydrology and water resources. They would also have important implications for the skiing industry because an increased frequency of years of poor natural snow-cover may reduce skier numbers (Chapter 12).

With regards to skiing, further calculations using the Galloway model and the current scenarios suggest that, under the worst-case scenario, there would be insufficient natural snow for viable ski operations by 2030 at the lower Australian resorts (e.g. Mt. Baw Baw) and at all resorts under this scenario by 2070, but that higher resorts may be still be able to operate as late as 2070 under the best-case scenario (U. Koenig, pers. comm.). In addition Koenig (Chapter 12) has shown that ski operations may be able to adapt through the greater use of artificial snow making, although this in itself is affected by climate change. Increasing temperature is likely to reduce the frequency of nights cold enough for snow-making, and would melt artificial (as well as natural) snow more quickly. The latter would mean that a larger water supply would be required. The viability of snow making as an adaptation to climate change requires further research.

ACKNOWLEDGEMENTS

Thanks are due to Urs Koenig, Kevin Hennessy, Cher Page and Ramasamy Suppiah for their assistance in preparing the manuscript. The author remains grateful to Robert Galloway for making available the details of his snow model. The work contributes to the CSIRO Climate Change Research Program.

REFERENCES

Beer, T. and Williams, A., 1995. Estimating Australian forest fire danger under conditions of doubled carbon dioxide concentrations. *Climatic Change* **29**: 169–188.

Breiling, M. and Charamza, P., 1994. Localizing the threats of climate change in mountain environments. Pp. 341–365 *in* "Mountain Environments in Changing Climates" ed by M. Beniston. Routledge, London.

Bultot, F., Gellens, D., Schadler, B. and Spreafico, M., 1994. Effects of climate change on snow accumulation and melting in the Broye catchment (Switzerland). *Climatic Change* **28**: 339–364.

Busby, J. R., 1988. Potential impacts of climate change on Australia's flora and fauna. Pp. 387–398 *in* "Greenhouse: Planning for Climate Change" ed by G. I. Pearman. CSIRO, Melbourne.

CIG, 1992. "Climate Change Scenarios for the Australian Region." Climate Impact Group, CSIRO Division of Atmospheric Research.

CIG, 1996. "Climate Change Scenarios for the Australian Region." Climate Impact Group, CSIRO Division of Atmospheric Research, Melbourne.

Fowler, A. M. and Hennessy, K. J., 1995. Potential impacts of global warming on the frequency and magnitude of heavy precipitation. *Natural Hazards* **11**: 283–303.

Galloway, R. W., 1988. The potential impact of climate changes on Australian ski fields. Pp. 428–37 *in* "Greenhouse: Planning for Climate Change" ed by G. I. Pearman. CSIRO, Melbourne.

Gordon, H. B., Whetton, P. H, Pittock, A. B., Fowler, A. M. and Haylock, M. R., 1992. Simulated changes in daily rainfall intensity due to the enhanced greenhouse effect: Implications for extreme rainfall events. *Climate Dynamics* **8**: 83–102.

Hewitt, S. D., 1994. The impact of enhanced-greenhouse conditions on the Australian snowpack. *in* "Proceedings of Symposium on Snow and Climate, Geneva, 22–23 Sept. 1994." Department of Geography, University of Geneva.

Hutchinson, M. F., 1989. A new objective method for spatial interpolation of meteorological variables from irregular networks applied to the estimation of monthly mean solar radiation, temperature, precipitation and windrun. *CSIRO Division of Water Resources Tech. Memo.* **89/5**: 95–104.

IPCC, 1992. "Climate change 1992: the supplementary report to the Intergovernmental Panel on Climate Change scientific assessment" ed by J. T. Houghton, B. A. Callander and S. K. Varney. Working Group 1, Cambridge University Press, Cambridge.

IPCC, 1996. "Climate change 1995: Contribution of Working Group 1 to the Second Assessment Report of the Intergovernmental Panel on Climate Change" ed by J. T. Houghton, L. G. Meira Filho, B. A. Callander, N. Harris, A. Kattenberg, and S. K. Varney. Cambridge University Press, Cambridge.

Martin, E., Brun, E. and Durand, Y., 1994. Sensitivity of the French Alps snow-cover to the variation of climatic variables. *Annales Geophysicae* **12**: 469–477.

Schreider, S. Yu., Whetton, P. H., Jakeman, A. J. and Pittock, A. B., in press. Runoff modelling for snow-affected catchments in the Australian Alpine region, eastern Victoria. *J. Hydrol.*

Whetton, P. H., Haylock, M. R. and Galloway, R. W., 1996. Climate change and snow-cover duration in the Australian Alps. *Climatic Change* **32**: 447–479.

Whetton, P. H., Fowler, A. M., Haylock, M. R. and Pittock, A. B., 1993. Implications of climate change due to the enhanced greenhouse effect on floods and droughts in Australia. *Climatic Change* **25**: 289–317.

Whetton, P. H., Hennessy, K. J., Wu, X., McGregor, J. L., Katzfey, J. J. and Nguyen, K., in press a. "Fine Resolution Assessment of Enhanced Greenhouse Climate Change in Victoria: Part 2", Consultancy report by CSIRO for the Victorian Department of Natural Resources and Environment.

Whetton, P. H., Katzfey, J. J., Nguyen, K., McGregor, J. L., Page, C. M. Elliott, T. I. and Hennessy, K. J., in press, b. "Fine Resolution Climate Change Scenarios for New South Wales, Part 2: Climatic Variability", Consultancy report by CSIRO for the New South Wales Environment Protection Authority.

CHAPTER 12

Climate change and the Australian ski industry

URS KÖNIG:

INTRODUCTION

IN many alpine areas, the winter tourism industry is a major contributor to the regional economy, with many businesses heavily dependent on visitation by skiers. In Australia, for example, the ski industry contributes approximately A$410m per annum to the Victorian and New South Wales (NSW) state economies and creates around 12,000 full-time jobs for the winter season (Buckby *et al.* 1993, KPMG Management Consulting 1994). The commercial viability of winter tourism, however, depends on snow conditions, namely snow-depth and duration and extent of snow cover. In past poor snow seasons in Australia (e.g. 1982, 1988, 1993), lift companies, resorts and all other related commercial activities such as accomodation, restaurants, ski schools, ski shops and ski retailers faced considerable decline in their revenues (for a detailed discussion about the impacts of past poor snow seasons on the Australian winter tourism industry see König 1997).

If the assumptions of impacts of climate change due to enhanced greenhouse effect hold true, snow-cover in the Australian Alps is expected to diminish (e.g. Haylock *et al.* 1994, Whetton *et al.* 1996), and skiing is expected to concentrate on the resorts at higher altitude, while ski fields at lower altitude may lose their ski tourism (König 1997). Research into possible impacts of climate change on tourism, has concentrated on the impact of climate change on the snow-reliability of ski fields (e.g. Barringer 1990, Koch and Rudel, 1990, Wall 1993, Bultot *et al.* 1994). Very little has been written about possible response strategies of the tourism industry to climate change (e.g. Abegg 1996), and no study has investigated how tourists might substitute one activity or one location for another if climate change was to occur. This chapter aims at addressing this research gap, it presents the results of a skier survey undertaken in the NSW ski resorts and discusses possible response strategies of the Australian alpine tourism industry to climate change.

THE DEMAND SIDE: WHAT WILL SKIERS DO?

Methods

A skier survey was undertaken in three Australian ski resorts — Perisher-Blue, Thredbo and Charlotte Pass. All three resorts are located in Kosciuszko National Park in New South Wales. The survey questioned:

- whether skiers in Australian ski resorts think they would continue to ski if future winters had little natural snow and, if so, asked where they would ski.
- what tourists would consider as important when choosing a resort if there were more future winters with little natural snow; and
- whether skiers perceive that climate change could threaten skiing in Australia.

The first part of the questionnaire asked respondents if they would continue to ski if future winters had little natural snow and, if so, where they would ski. In the second part of the questionnaire, those respondents who would still ski in Australia in future winters with little natural snow were asked what factors they would consider as important in deciding where to ski. The third part of the questionnaire asked respondents whether global warming due to enhanced greenhouse effect could threaten skiing in Australia.

Questionnaires were handed out to respondents and collected by the author in restaurants and coffee shops in the above ski resorts. Restaurants and coffee shops were chosen so as to receive responses from beginner, intermediate and advanced downhill skiers and snowboarders. The survey was conducted during two periods in the peak visitor months of July and August, when most ski lifts were expected to be open to public use. The first period covered 18 July to 21 July, 1996, inclusive. The second period was from 18 August to 24 August, 1996, inclusive. Of the 1,000 questionnaires distributed, 938 were returned (response rate of 93.8%). Of these 938 questionnaires, 12 were considered as not valid because of a lack of information, leaving 926 questionnaires to be analysed.

It is important to note that in order to obtain a representative sample of the population of skiers in NSW ski resorts, the number of questionnaires distributed at each resort was proportional to the number of skier days spent in each resort. In brief, Perisher-Blue accounts for approximately 65% of all skier days spent in New South Wales resorts. The corresponding figures for Thredbo and Charlotte Pass are 30% and 5% respectively (Osborn 1991). Of the 926 questionnaires analysed, 595 (64%) came from Perisher-Blue, 282 (30%) from Thredbo and 49 (6%) from Charlotte Pass. These numbers closely approximate the proportion of total skier days in NSW spent in each resort. Further weighting of the sample was therefore not required.

TABLE 1
Respondents Characteristics.

Type of skier	Number of skiers	Percent of total
Downhill	801	87 %
Snowboarders	94	10 %
Cross country	31	3 %
Total	926	100 %
NUMBER OF DAYS PER YEAR SPENT SKIING		
1 to 2	177	19 %
3 to 5	344	37 %
6 to 10	245	26 %
11 to 15	74	8 %
16 to 20	36	4 %
21 to 30	17	2 %
more than 30	33	4 %
Total	926	100 %
SKIING SKILL		
Novice	51	6 %
Beginner	117	13 %
Low intermediate	328	35 %
High intermediate	276	29 %
Advanced	109	12 %
Expert	45	5 %
Total	926	100 %

The characteristics of respondents are shown in Table 1. The great majority of respondents (801 or 87%) described themselves as being mainly downhill skiers, followed by 94 (10%) who described themselves as being mainly snowboarders, and 31 (3%) who described themselves as being mainly cross country skiers. With respect to skiing frequency, 19% of respondents spend, on average, 1 to 2 days per year skiing, 37% spend 3 to 5 days, and 26% spend 6 to 10 days per year skiing. The great majority of skiers (82%) therefore spend an average of 10 days or less per year skiing. Finally, almost two thirds (64%) of all respondents described themselves as intermediate skill skiers. Novice and beginner skiers comprised 19% of all respondents, and 17% of all respondents were advanced or expert skiers. These findings are supported by earlier market research in Australian ski fields (Travers Morgan, 1990, Buckby *et al.* 1993). In short, the sampling frame adopted for the purpose of this study closely approximates the population of skiers in Australian resorts.

Findings and discussion

If and where to ski

Respondents were asked, assuming that the next five winters would have very little natural snow, how this would influence their decision about if and where to ski and snowboard. Figure 1 shows that a quarter of all respondents (25%) would still ski in Australia, and would do so as often as they used to, whilst a third of all respondents (31%) would keep on skiing in Australia but would ski less often. The majority of respondents (56%) would therefore keep on skiing in Australia if there were more future winters with little natural snow. However, more than a third of all respondents (38%) would not ski anymore in Australia but would instead ski overseas (mainly in New Zealand and Canada), while 6% would give up skiing. Consequently, Australian ski fields would lose 44% of all respondents if there were more snow-deficient winters ahead. Note that despite the fact that some respondents may ski less in Australia *and* would ski more overseas, respondents were asked to only choose one answer in response to this question. This was done because it was felt that it would be difficult for respondents to describe how much they would ski in Australia and how much they would ski overseas. In order to receive a clearer picture, respondents were therefore asked to decide *between* skiing in Australia and skiing overseas.

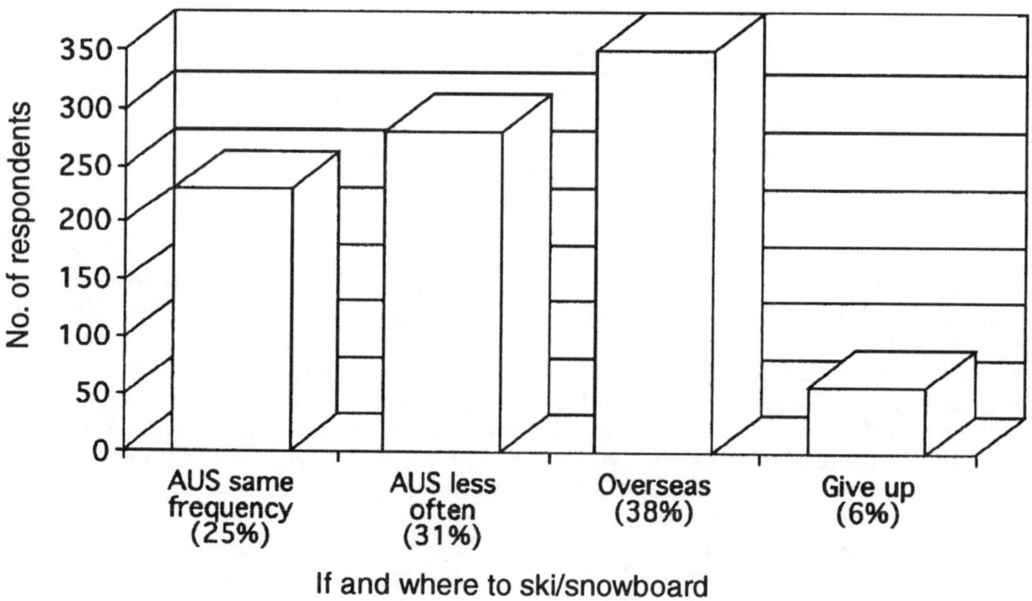

Fig. 1. Where respondents would ski/snowboard if the next five winters had little natural snowcover.

It was found that significantly more high skill skiers than low skill skiers would ski overseas, and significantly more low skill skiers would remain skiing in Australia (includes both categories "Australia, same frequency" and "Australia, less often") (Chi square, $p < 0.05$). For example, only 18% of the novice respondents would ski overseas but half of all advanced skiers (50%) would do so. The pattern of those who would give up skiing showed that good skiers are more determined to ski than beginners. Only 3% of advanced skiers would give up their sport, whilst 16% of novice skiers would do so if there were future winters with little natural snow.

Significantly, more cross country skiers than downhill skiers or snowboarders would keep on skiing in Australia rather than going overseas (Chi square, $p < 0.05$). It was found that almost half of all cross country skiers would ski in Australia, but less often (48%), and that 77% of all cross country skiers would continue skiing in Australia (both with the same frequency and less often). The corresponding figure for downhill skiers and snowboarders was only 55% and 48% respectively. This may be due to the fact that cross country skiers are more flexible in their snow pack requirements than downhill skiers and snowboarders, or this may be influenced by the small sample size for cross country skiers. Cross country skiers are able to adapt to snow deficient winters in lower areas by travelling to any snow covered area with reasonable access. Only 13% of all cross country skiers would ski overseas instead of in Australia, compared to 38% of all downhill skiers and 50% of all snowboarders. The snowboarders also contained the lowest percentage (2%) of those who would give up their sport, while cross country skiers contained the highest percentage (10%) of those who would give up skiing.

Therefore, cross country skiers and downhill skiers are most likely still to ski in Australia if there were more future winters with little snow (77% of respondents and 56% of respondents respectively), whereas a majority of snowboarders (52% of respondents) are likely to snowboard overseas or give up snowboarding. This may be the most unfortunate combination for the resorts, as cross country skiers spend in general much less money in the resorts (e.g. no lift tickets are required) than downhill skiers and snowboarders. Moreover, snowboarders, who are most likely to snowboard overseas, are — in contrast to skiers — a rapidly growing market segment, which is expected to increase significantly the number of visitor days spent in Australian ski resorts over the next decade.

Perception of possible response strategies

In the second part of the questionnaire, possible response strategies of the resorts to climate change were presented (possibilities and limitations of these response strategies under climate change are discussed below). Those respondents, who would still ski in Australia if there were more future winters with little natural snow, were asked which response strategies they would consider as important when deciding where to ski in future winters with little natural snow ("If you were to ski/snowboard in Australia in winters with very little natural snow, which of the following would be important/not important for you when deciding in which resort to ski at?") (see Table 2).

TABLE 2
Perception of possible response strategies.

Strategy	Important	Not important
lots of snow-making	89 %	11 %
cheaper lift tickets	88 %	12 %
expansion lifts into higher terrain	53 %	47 %
night life/lots of fun events	49 %	51 %
availability of non-snow activities	31 %	69 %

The two most important factors respondents looked for when deciding at which resort to ski in winters with little natural snow are plentiful snow-making (89% of all respondents considered it as important) and cheaper lift tickets (88% of all respondents considered this as important). The importance of cheaper lift tickets in snow-deficient winters was further illustrated by the fact that 65% of respondents would increase the frequency of their visits to the ski fields in winters with little natural snow if the lift tickets were half-price. Extensive expansion of snowmaking and a flexible ticket price policy (according to the number of lifts operating) are therefore the two most promising strategies for resorts to attract visitors in snow-deficient winters.

Approximately half of all respondents (53%) considered expansion of ski lifts into higher terrain as important when deciding where to ski in future winters with little natural snow, while slightly less than half of all respondents (49%) felt that an attractive night life/lots of fun, and other events are important when deciding where to ski during future snow-deficient winters. Indeed, 31% of all respondents considered the availability of non-snow activities as important when deciding where to ski in future winters with little natural snow.

It was found that significantly more high skill respondents considered the importance of expansion of lifts into higher terrain as important, whilst significantly more low skill respondents considered an attractive night life lots of fun events and the availability of non-snow activities as important (Chi square, $p > 0.05$). This clearly shows that response strategies of resorts which directly improve skiing conditions (such as expansion of lifts into higher terrain) are relatively more successful in attracting skiers with higher skill levels. Non snow-related strategies, which aim at improving the entertainment value of the resort (availability of indoor sports, attractive night life/ fun events) but do not influence the skiing conditions, prove to be relatively more successful in attracting visitors with lower skiing skills. For example, 70% of novice skiers considered an attractive night life/lots of fun events as important when deciding where to ski.

It was shown in the previous section that the likelihood that respondents would still ski in Australia if there were more future winters with little natural snow increases with decreasing skill level. Whereas most advanced respondents would ski overseas, the majority of novices, beginners and intermediate skiers would remain in Australia. The findings reported here therefore suggest that response strategies which focus on the market segment of the low and intermediate skill level respondents would be most promising in attracting visitors to Australian ski resorts if predicted climate change was to occur. In other words, response strategies of resorts have to match the demand of those who remain visiting Australian ski resorts (mainly skiers with lower skill levels) and not the demand of those who will travel overseas regardless of strategic responses.

Perception of climate change

Respondents were asked whether they believe that global warming due to enhanced greenhouse effect could threaten skiing in Australia. It was found that the majority of respondents (78%) believed that skiing in Australia could be threatened. This indicates that most skiers perceive global warming as a problem which is expected to have severe impacts on skiing in Australia. Cross country skiers were significantly more likely to believe that skiing in Australia could be threatened by global warming than snowboarders or downhill skiers (Chi square, $p < 0.05$).

Approximately 9% of those who stated that skiing in Australia could be threatened by global warming believed that skiing in Australia will be severely affected before the year 2000, 62% of respondents (cumulative percentage) believed that this will occur before 2030, and 80% believed that skiing in Australia will be severely affected before the year 2060. The remaining 20% of respondents believed that skiing in Australia will be severely affected between the years 2060 and 2100. The results of this survey indicate that respondents who believed that skiing in Australia could be threatened by global warming, do not see global warming as an abstract notion but rather as a process which will have significant impacts over the next 40 to 100 years. It is worth mentioning that this perception of skiers concerning the time frame of possible impacts of climate change on snow in Australia, matches the projections of research on this topic surprisingly well (see e.g. Whetton *et al.* 1996).

In sum, the skier survey found that Australian ski resorts would lose 44% of their skiers if there were more future winters with little natural snow. High levels of snow-making and cheaper lift tickets are most important to skiers when choosing a resort in future snow-deficient winters. However, a lower skiing skill of respondents was associated with a higher importance of non-snow related activities. Seventy-eight percent of respondents believed that climate change due to an enhanced greenhouse effect could threaten skiing in Australia. Most of these respondents, believed that this will be the case before 2030 (62%).

POSSIBLE ADAPTATION STRATEGIES OF THE INDUSTRY TO CLIMATE CHANGE

Figure 2 shows the range of possible adaptation strategies of ski resort management to climate change. They can be subdivided into (1) those which maintain ski tourism (including artificial snow-making, flexible ticket price policy, ski slope design/super

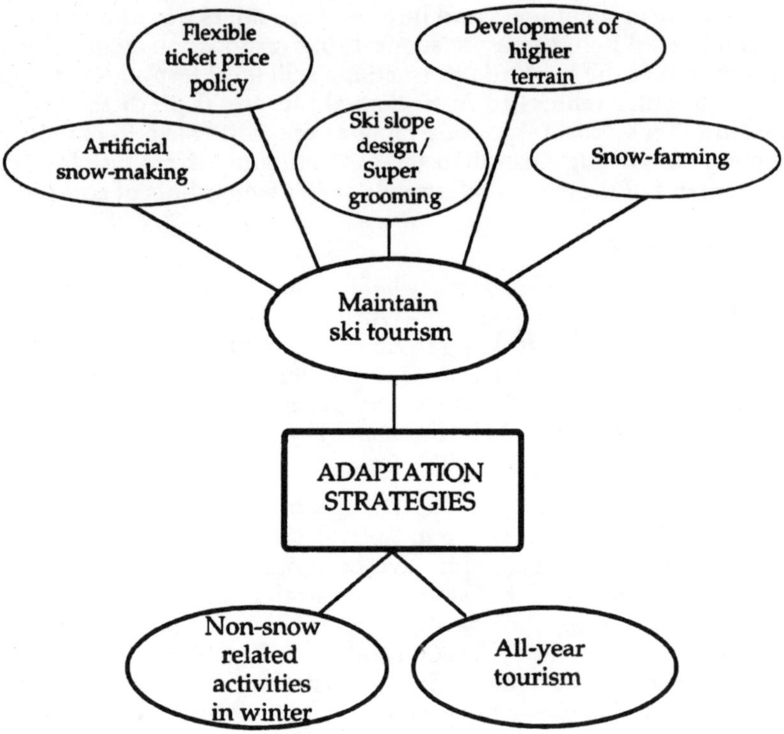

Fig. 2. Potential and actual response strategies of the alpine tourism industry to climate change.

grooming, development of higher terrain, snow farming), (2) non-snow related activities in winter, and (3) all-year tourism.

It is important to stress that there are many uncertainties when discussing and recommending adaptation strategies to climate change. One of the main sources of uncertainty is the almost total lack of information about future non-climatic factors (e.g. improvement in snow-making technology, changing leisure trends etc.). However, it should be noted that most of the discussed strategies are "no regret" strategies and are, because of the high variability of snowfalls, worthy of consideration even if climate change does not occur. Overall, those adaptation strategies summarised under "maintain ski tourism" (top of Fig. 2) are relatively well developed in Australia. Clearly, not enough has been done in the development of "non-snow related activities" and an enhanced "all-year tourism".

Maintaining ski tourism

Flexible ticket price policy

The skier survey discussed above showed that a flexible ticket price policy related to the number of lifts running is (together with "lots of snow-making") the most promising strategy for resorts in order to attract visitors in poor snow seasons. Approximately 88% of respondents considered cheaper lift tickets as important when choosing a resort in future snow-deficient winters.

Resorts which do not adapt their ticket prices to the number of lifts running may argue that it is most expensive to operate ski fields during periods with little natural snow (high costs of snow-making, snow-farming) and that they do not want to decrease their revenue during these periods with lower ticket prices. However, it is possible that the higher number of visitors (due to lower ticket prices) would actually compensate for a potentially decreasing income due to lower ticket prices. The price elasticity of skiing requires investigation, as information is lacking. Further arguments against lower ticket prices and a higher number of skiers during snow-deficient winters include (1) congestion at the lower number of ski lifts, and (2) overcrowded slopes which may lead to more accidents. Resorts may argue that they would rather have a lower number of skiers paying higher prices than vice versa. Also, higher altitude resorts may see no reason to reduce ticket prices if they are the only resort offering skiing in Australia (Unique Selling Positions, USP).

The frequency of poor snow seasons is expected to increase with predicted climate changes, and competition among resorts is therefore also expected to intensify. The development and implementation of a ticket price strategy for poor seasons requires more skier market research, including research into the price elasticity of skiing. Because each resort has different prerequisites and a different skier market segment, these strategies will differ between resorts and may even involve, for the reasons mentioned above, the decision not to reduce ticket prices in poor snow seasons.

Artificial snow-making

The area of artificial snow-making in the Australian Alps has increased dramatically over the last 10 years. In 1996, 16% of the lift-serviced skiable area in Australia (271 ha) was covered with artificial snow. It is interesting to note that this percentage is significantly higher than in many European countries such as Switzerland which, has 3.6% or 790 ha (Theus 1995). In 1997, Lake Mountain was the last Australian ski field to install snow-making. The area covered with snow-making is expected to grow further. Snow-making projects in Victorian ski fields, for example, cover up to 30% of the alpine resort area (Hewitt 1997). Most Australian ski resorts start producing snow two to three weeks prior to the beginning of the ski season (June long weekend). The importance of snow-making during snow-deficient winters was illustrated in 1988 when Mt. Buller lost 70% of its visitor numbers compared to the average of previous years due to the non-existence of snow-making. Falls Creek on the other hand increased its visitor numbers because of extensive snow-making. Another good example of the importance of artificial snow-making is Thredbo. Prior to snow-making, top to bottom skiing in Thredbo was only possible for approximately 17 days per season. This figure increased with snow-making up to approximately 100 days per season.

The results of the skier survey shows that "lots of snow-making" is the most important factor skiers are looking for when choosing a resort in future snow-deficient winters (89% of respondents considered it as important). Extensive expansion of snow-making therefore remains the most promising strategy in order to attract skiers in poor snow seasons. However, it is important to stress that snow-making facilities such as fixed high pressure snow guns, water pipelines and water dams affect the appearance of the resorts in summer, and are thus expected to have a negative impact on the growth of all-year tourism. Also, a number of economic, climatic and ecological factors, all of which are interrelated, may limit the future expansion of snow-making.

Favourable climate conditions for snow-making are low temperatures and low humidity. Temperatures below –5°C and humidity of less than 65% are ideal (Ruffini 1993). Snow-making additives (e.g. Snowmax), which are widely used in the Australian Alps (NSW National Parks and Wildlife Service 1994, Darby 1994), and advances in snow-making technology, however, allow the production of snow under more marginal conditions. Most Australian resorts are able to produce snow in temperatures up to –1°C. However, even the most advanced technology to date can not guarantee snow-making under any condition. The downhill ski world championships of the International Ski Federation (FIS) in the Sierra Nevada (Spain) in 1994, for example, had to be cancelled for the first time in its history because it was too warm to produce snow.

Artificial snow-making is expensive. High costs and water shortage (see below) are the main reasons for resorts not extending their area of snow-making. The high investment costs are mainly determined by the size of the coverage, the snow-making capacity and the water supply situation. Economic models have been developed in order to find the ideal coverage and capacity for different ski areas. Following Farwell (1987), for example, the "snow-making decision" must be based upon a quantitative analysis of the impact of snow-making on:

- the length of the operating season;
- the increases of the average utilisation of the ski field and the resort facilities (e.g. restaurants);
- the increases in ticket prices; and
- the decreases in revenue losses during snow-deficient winters.

Only if the positive effects on these revenue sources outweigh the additional costs of snow-making is its installation justified. The maintanance costs of snow-making are mainly determined by water and electricity costs. Despite the high maintanance costs, most Australian resorts produce snow whenever the weather conditions permit (even during heavy snowfalls). The reason for this "producing snow on stock" lies in the unpredictable nature of weather conditions and the high variability of snow falls (see Ruddell *et al.* 1990).

Artificial snow-making requires enormous amounts of water. Water shortage may constrain the extension of snow-making and often results in an increase of investment costs. At Mt. Hotham and Mt. Buller, for example, both resorts are located on top of the mountain, and water has to be pumped up from rivers in nearby valleys and is stored in artificial dams in the ski field.

The impacts of the bacteria *Psedomonas syringae*, commonly known as Snowmax, on the alpine vegetetation has been the subject of exensive winter monitoring programs in Victoria in 1989 (Hallam 1991) and in NSW in 1991 (NSW National Parks and Wildlife Service 1994). However, little has been written on general ecological problems associated with snow-making in Australia. European studies suggest that snow-making has negative ecological impacts in terms of changing hydrology of rivers, flora diversity and noise pollution (see e.g. CIPRA 1989, Mosimann 1994).

There is no doubt that snow-making has improved the skiing conditions and therefore stabilised the revenue of the Australian resorts over the last 10 years. However, it is unlikely that artificial snow-making can, for the reasons discussed above, replace natural snow in the future. To operate the entire resort, the ski industry will still rely largely on the presence of a sufficient natural snow pack. Nevertheless, even under predicted future climates, there will be times when snow-making will be possible. A large snow-making capacity is therefore expected to remain an important market advantage if predicted climate changes were to occur. The shorter the time period when snow-making will be possible, the more important will a high snow-making capacity be. This will be a clear disadvantage for smaller resorts with less financial means and will enhance the concentration of the ski industry on the larger, financially stronger resorts. It is worth repeating that there is a trade-off between more snow-making facilities and an "undisturbed", visually attractive alpine environment which is the base for more all-year tourism. It will be argued below that an increase in all-year visitation is one of the most effective response strategies of the alpine tourism industry to climate change.

Ski slope design and super grooming

Ski slope design and super grooming both aim at maintaining a long lasting snow cover. Ski slope design includes consideration of the micro climate effects associated with aspect (wind direction, sun), sub-surface drainage, as well as the space requirements of skiers. In particular the installation of sub-surface drainage systems has been useful in increasing the longevity of the snow pack by diverting sub-surface meltwater (snow does not last as long on wet ground as on dry) (Keage 1990). Super grooming involves the removal of surface irregulations such as rocks and bushes to provide an even ground surface. Groomed slopes require substantially less snow to be skiable than slopes with natural surface and vegetation. While the latter need up to 50 cm or more (Bezzola 1975), groomed slopes require under ideal conditions (frozen ground) as little as 5–10 cm to be skiable.

The two main drawbacks of super grooming are (1) ecological problems and (2) negative impacts on all-year tourism. Firstly, super grooming can result in erosion and may also affect vegetation diversity. It also has been shown to have negative effects on alpine fauna such as the endangered Mountain Pygmy-possum *Burramys parvus* (Mansergh *et al.* 1989). Secondly, in a similar fashion to snow-making facilities, super groomed slopes have a negative impact on the visual apperance of the resorts in summer and therefore impact the growth of all-year tourism in a negative way. Under predicted climate change, increasing pressure for further super grooming is expected. This again will be a clear disadvantage for smaller resorts with less financial means. In a similar fashion to more snow-making facilities, there is a trade off between more super groomed ski slopes and an attractive alpine environment which is a key element for more all-year tourism.

Snow farming

Snow farming techniques include (1) snow fencing, (2) transportation of snow from surplus accumulation areas to high-wear areas, and (3) snow grooming. The deposition and accumulation of snow can be significantly increased by snow fences. The transport of snow from areas with surplus accumulation (e.g. fenced areas) to high wear areas or areas of patchy snow cover is mainly done with snow grooming machines. Where slopes cross roads (e.g. at Mt. Hotham) snow is also transported by trucks. Snow grooming compacts and scarifies the snow surface in order to increase its impact strength. Keage (1990) showed that, depending on snow surface temperature, snow grooming may increase snow surface hardness by 300% to 400%. The compaction and scarifaction of snow associated with grooming increases snow cover durability to skier wear and reduces snow loss.

The improvement of snow-farming techniques is of high priority in all resorts. Assuming climate change, it can be expected that the importance of snow farming will further increase. It is important to mention that, unlike other adaptation strategies discussed above, snow farming has minor impacts on the visual attractiveness of the alpine environment (providing snow fences are removed after the ski season) and has therefore no negative impacts on all-year tourism.

Development of higher terrain

Higher terrain generally offers steeper slopes, greater snow-depth and drier snow and is therefore an attractive option for ski field development. In the European Alps, for example, a trend towards the extension of ski fields into higher terrain has been observed over the last 5 to 10 years. This is seen as being a direct consequence of the three poor snow seasons between 1987/88 and 1989/90 (e.g. see König 1994, Abegg

1996). In Australia, however, the possibilities of extending and developing ski fields in higher sections of the Alps are limited:

- Many resorts are located on top of mountain ridges and the summit areas are already developed.
- Most Australian resorts are located within National Parks. Any extension projects of ski fields beyond the existing resort boundaries are potential threats for extremely sensitive ecological areas and would undoubtedly have negative impacts on the alpine flora and fauna. Such projects would therefore face strong resistance from conservation groups (e.g. see Grenier 1992, Watermann 1996).
- Steep terrain and rough surfaces (rocks, bushes) make the building of lifts and slopes difficult and result in high construction costs. The rougher surface in higher areas also requires greater snow-depth to be skiable.
- Severe weather conditions (e.g. high wind) in higher areas may result in frequent closure of lifts.

The skier survey showed that more than half of all respondents (53%) considered the extension of lifts into higher terrain as important when deciding where to ski in future poor snow seasons. Nevertheless, despite increasing visitor pressure on higher ski fields due to climate change, it is unlikley, for the reasons mentioned above, that ski fields will be developed beyond today's resort boundries. However, assuming climate change, it can be expected that skiing activities will concentrate on the higher areas of the current lease area of skifields. Several resorts with a larger altitudinal range of lifts have already built middle stations for T-bars and chairlifts so that lifts can operate only in the higher sections if snow conditions at the bottom station are unfavourable (e.g. Mt. Buller, Thredbo, Perisher-Blue).

Another observed trend, not only in Australia but world wide, is the replacement of T-bars with chairlifts. The German expression "Bodenunabhängige Tranportanlagen" (="ground independant transport facilities") describes the advantage of chairlifts over T-bars very well. While T-bars require snow on the ground, chairlifts (and cable cars) can transport skiers into higher areas even if there is no snow on the ground. This advantage is important if there is no natural snow at all and if skiing is only possible on slopes prepared with artificial snow (already often the case for example at Friday Flat in Thredbo). This is also an advantage if there is no snow at the bottom station and skiers need to be tranported to higher areas where there is enough snow to ski. Further reasons for replacement of T-bars by chairlifts are higher security standards, higher transport capacity and greater comfort (in particular for snowboarders), and that they are easier to ride. However, in high wind areas, T-bars will continue to be constructed (Osborn 1991).

In summary, a shorter snow-cover duration in lower areas due to climate change is expected to increase the pressure on development into higher terrain. The skier survey showed that a small majority of skiers consider development of higher terrain as important when deciding where to ski in future snow-deficient winters. Nevertheless, from an ecological point of view, the extension of commercial ski fields beyond existing resort boundaries has to be rejected.

Supply of non-snow related activities in winter

The supply of non-snow related activities in winter aims at reducing the snow-reliance of the alpine tourism industry. Abegg (1996) stressed in his study in the Swiss Alps the distinction between "alternative activities in snow-deficient winters for skiers" and "new activities for non-skiers". While the former aims at offering skiers alternative activities if skiing is not possible, the latter aims at attracting non-skiers for a winter holiday in the mountains. It it is argued below that in Australia there is a demand for more alternative activities in winter. To attract more non-skiers in winter to the Australian alpine resorts, however, seems most difficult.

Offering alternative activities in poor snow seasons for skiers aims at preventing skiers from returning home by making their holidays enjoyable even if skiing is not possible. The supply of alternative activities in poor snow seasons is therefore only useful for resorts with a high proportion of tourists staying for several nights rather than day trippers. Skiing may not be possible because of lack of snow or due to adverse weather (e.g. high winds). It is important to note that resort managers generally believe that a considerable percentage of their winter guests only ski occasionally. In Thredbo for example it is believed that up to 60% of the guests in Thredbo village only ski occasionally. Alternative activities besides skiing would therefore also be of great benefit for those guests who only ski occasionally, as well as for all skiers during poor snow conditions.

The skier survey discussed showed that 49% of respondents considered an attractive night life/lots of fun events as important when deciding where to ski in future snow-deficient winters, while 31% felt that the availability of non-snow related activities would be important when deciding in which resort to ski in future poor snow seasons. Significantly more low-skill skiers than high-skill skiers considered the availability of non-snow related activities as important, indicating that the supply of non-snow related activities in poor snow seasons would be significantly more successful in attracting low skill skiers than skiers with higher skiing skills.

With the exception of Thredbo and Mt. Buffalo very little has been done in terms of offering non-snow related activities in snow deficient winters. Typically, these two resorts also have the highest all-year visitation, demonstrating that all-year tourist facilities and supplies are of benefit during poor snow seasons.

Although the skier survey indicated that there is a market for more non-snow related activities, the potential of these supplies should not be overestimated. The main reason for most skiers booking a holiday in the resorts is skiing. Nevertheless, non-snow related activities for skiers can help to keep regular guests in the resorts during poor snow seasons, and may also increase all-year tourism. In addition, Swiss research showed that goodwill and a positive image of the resort was gained among tourists in those resorts which offered alternative activities in poor snow seasons (Burren *et al.* 1995, Abegg 1996).

All year tourism

The most effective strategy in order to reduce snow-reliance of the alpine tourism industry is an enhanced all-year tourism. This includes tourism in spring, summer and autumn and all climate-independant tourism such as conference tourism, educational tourism and health tourism. Although all-year patronage in the Australian alpine resorts has increased over the last few years, it is, with the exception of Mt. Buffalo and Thredbo, still marginal. However, the successful examples of the two resorts Thredbo and Mt. Buffalo, the general observed trend away from beach holidays towards activity and education holidays, and the growth of adventure sports such as mountain biking and paragliding (e.g. Lynch and Veal 1996) all indicate a promising future for all-year tourism in the Australian Alps.

Clearly, those resorts with an attractive appearance, offering a lively village centre with facilities and services such as shops, restaurants, coffee shops, news agencies etc. available, have good prerequisites to increase all-year patronage. It is not surprising that Thredbo with its "village atmosphere" has the highest percentage of overnight stays and day trippers in summer of all NSW resorts.

While the mountain scenery and the mainly undisturbed alpine environment is certainly the basis for an increased all-year visitation, it is not a sufficient condition for increasing non-winter visitation. Resorts need to attract potential all-year visitors with "something special". All alpine resorts can offer access to an undisturbed alpine environment and most resorts offer activities such as mountain biking, guided walks, etc. It is important to stress that with an expected increase in all-year tourism and hence competition among resorts, it will be necessary to have one or several Unique Selling Positions (USP) in order to survive. This USP can be a natural or artificial site, or a specific activity exclusively offered in the resort. Thredbo, for example, offers the closest walking distance to the highest peak in Australia (Mt. Kosciuszko) and is the only resort with the so called "Alpine Slide". Mt. Buffalo is one of the few places in Australia where hangliders train and competitions are organised regularly (e.g. women's world championships in 1996). Another USP of Mt. Buffalo is the Chalet Hotel, built in 1910, a unique building in the Alps. It is one of Australia's oldest and largest historic guesthouses. Also, both resorts offer cultural events which are unique in the Alps. Thredbo holds annual Blues, Jazz, and Film festivals and Mt. Buffalo organises an annual "Opera Chorus in the Alps". All events report remarkable success.

CONCLUSIONS

This chapter discussed the results of a skier survey undertaken in three ski resorts in New South Wales and presented possible response strategies of the Australian ski industry to climate change. The skier survey showed that climate change would have significant implications for the Australian snow resorts in terms of visitor numbers. It was shown that Australian ski resorts would lose 44% of their skiers if there were more future winters with little natural snow. These respondents would either ski overseas rather than in Australia (38%), or give up their sport (6%); 56% of respondents would remain skiing in Australia (25% as often as they used to and 31% less often). The amount of artificial snow-making and price of lift tickets are most important to skiers when choosing a resort in future snow-deficient winters. However, a lower level of skiing skills of respondents was associated with a higher importance of non-snow related activities. Of all respondents 78% believed that climate change due to an enhanced greenhouse effect could threaten skiing in Australia. Most of these respondents, believed that this will be the case before 2030 (62%).

The range of possible adaptation strategies of the ski industry to climate change were discussed. It was shown that the more technical strategies such as snow-making, super grooming, and snow farming, which aim at maintaining ski tourism, are well developed in Australia. This is most likely due to the already high variability of the snow-pack under present climate conditions. The increasing importance of these expensive technical response strategies (i.e. snow-making) under climate change will be a clear disadvantage for smaller resorts with less financial means. This will enhance the concentration of the ski industry, on those resorts with the best natural snow conditions and with the greatest financial resources. In the long run, however, those alpine resorts which reduce their snow-reliance through the provision of non-snow related activities and an enhanced all season visitation will be in the best position.

Increasing all-year visitation in the alpine resorts is certainly a difficult task. It was pointed out in Thredbo and Mt. Buffalo, that it requires "a lot of hard work". Nevertheless, there is a market for more all-year visitation in the Australian Alps. The two succesful examples, Thredbo and Mt. Buffalo, show that with (1) a visually attractive resort, (2) the availability of suitable facilities and services (3) at least one USP, and (4) a well researched all-year marketing strategy, all-year visitation to alpine resorts can be increased. It is worth repeating that this is, despite the smaller financial turnover in summer, the most effective strategy in order to cope with a diminishing snow-pack due to climate change.

REFERENCES

Abegg, B., 1996. Klimaänderung und Tourismus. Klimafolgenforschung am Beispiel des Wintertourismus in den Schweizer Alpen. Vdf Verlag, Zürich, Switzerland.

Barringer, J. R. F., 1990. Modelling Present and Past Snowline Altitude and Snowfalls on the Remarkables. *Weather and Climate* **11:** 43–47.

Bezzola, A., 1975. Probleme der Eignung und Aufnahmekapazität touristischer Bergregionen der Schweiz. Haupt, Berne, Stuttgart.

Breiling, M., 1993. Die zukünftige Umwelt- und Wirtschaftssituation peripherer alpiner Gebiete. Ph.D. thesis, Universität für Bodenkultur, Vienna, Austria.

Buckby, M., Burgan, B., Molloy, J. and Mc Donald, S., 1993. The Economic Significance of Alpine Resorts. The Centre for South Australian Economic Studies, Adelaide, Australia.

Bultot, F., Gellens, D., Schädler, B. and Sprefico, M., 1994. Effects of Climate Change on Snow Accumulation and Melting in the Broye Catchment (Switzerland). *Climatic Change* **28**: 339–363.

Burren, A., Kaufmann, K. and Tamàs, O., 1995. Schneemangel - Erfolgsanalyse von konkreten Alternativen. Assignment at the Höhere Fachschule für Tourismus Samedan. Samedan, Switzerland.

CIPRA (Commission international pour la protection des regions alpines), 1989. Beschneiungsanlagen im Widerstreit der Interessen. Schaan, Lichtenstein.

Darby, J., 1994. Skiing in Victoria. Melbourne, Australia.

Farwell, T., 1987. The Economics of Snowmaking. Paper presented at the Perisher Valley Seminar, 11 June 1987, Perisher, Australia.

Grenier, P., 1992. Skiing in the Australian Alps: development and conflict. *The Australian Alps. Revue de Géographie Alpine* **2–3**: 227–275.

Hallam, M., 1991. The environmental Impact of 'Snowmax'. Paper presented at the 1991 Ski Areas Trade Show and Conference 12–14 June 1991, Thredbo, Australia.

Haylock, M., Whetton, P. H. and Deborough, C. 1994. Climate Change and Snow Cover Duration in the Victorian Alps. Report to the Environment Protection Authority, Publication Nr. 403, Melbourne, Australia.

Hewitt, S., 1997. An impact analyisis of an enhanced greenhouse climate change on the Australian alpine snow pack. Ph.D. thesis, University of Melbourne, Australia.

Keage, P., 1990. Skiing into the Greenhouse. *Trees and Natural Resources* **2**: 15–18.

Koch, E. and Rudel, E., 1990. Mögliche Auswirkungen eines verstärkten Treibhauseffekts auf die Schneeverhältnisse in Österreich. *Sonderdruck aus Wetter und Leben* **3/4**: 137–153.

König, U., 1994. Entwicklung und Zukunft des Gletscherskitourismus in der Schweiz, Zürich.

König, U., 1997 (in press). Tourism in a warmer world: Implications of climate change due to an enhanced greenhouse effect for the Australian ski industry. CRES Working Paper Series, Australian National University, Canberra, Australia.

KPMG Management Consulting, 1994. New South Wales Alpine Industry: Economic Impact Study. Report to the Australian Ski Areas Association, Adelaide, Australia.

Lynch, R. and Veal, A. J., 1996. Australian Leisure. Longman press: Melbourne.

Mosimann, T., 1994. Skigebiete raumverträglich planen. Amt für Ramplanung Graubünden, Chur, Sitzerland.

NSW (New South Wales) National Parks and Wildlife Service, 1994. Plan of Management Amendments 1994, Sydney, Australia.

Osborn, D. B., 1991. Industry perspective. Paper presented at the 1991 Ski Areas Trade Show and Conference, 12–14 June 1991, Thredbo, Australia.

Ruddell, A. R., Budd, W. F., Smith, I. N., Keage, P. L. and Jones, R., 1990. The South East Australian Alpine Climate Study. Report to the Alpine Resort Commission, Melbourne, Australia.

Ruffini, F., 1993. Geregelter Einsatz von Beschneiungsanlagen. *Mitteilungen zum Natur- und Umweltschutz in Südtirol, Naturschutzblatt* **4**: 3–9.

Theus, R., 1995. Klima und Seilbahnen im Wandel: Struktur und Strategien der Seilbahnunternehmungen. Paper presented at the Status Seminar Nationales Forschunsprojekt 31, 15 November 1995, Zürich, Switzerland.

Travers Morgan, 1990. Skier Market Survey. Working papers 1–7. Report to the New South Wales National Parks and Wildlife Service.

Wall, G., 1993. Impacts of Climate Change on Recreation and Tourism in North America. Report to the Office of Technology Assessment, Congress of the United States, Washington.

Waterman, R., 1996. The uphill battle for Mt. Stirling — time to act. *Park Watch* **185**: 16–21.

Whetton, P. H., Haylock, M. R. and Galloway, R. W., 1996. Climate Change and Snow-Cover Duration in the Australian Alps. *Climatic Change* **32**: 447–479.

CHAPTER 13

The implications of ozone depletion for the Australian Alps: a review

SARA BROOMHALL

INTRODUCTION

HABITAT disturbance is widely acknowledged as the most significant threat to biodiversity to date (e.g. Saunders *et al.* 1991), and currently human disturbances are altering the environment on a global scale. Climate warming is one such example. Similarly, the observed depletion of the ozone layer (Farman *et al.* 1985, Manney *et al.* 1994, Russell *et al.* 1996) has led to fears that the corresponding increase in solar ultraviolet-B (UV-B), 280–320 nm radiation (e.g. Blumthaler and Ambach 1990, Zerefos *et al.* 1995) may have serious ecological consequences.

The rapid and widespread nature of population declines evident in a diverse array of amphibian species at high altitudes, in apparently undisturbed environments (e.g. Bradford 1991, Wake 1991, Crump *et al.* 1992, Fellers and Drost 1993, Osborne *et al.* 1996), has indicated a ubiquitous causal agent common to all areas of the globe (Wyman 1990), one perhaps particularly manifest at higher altitudes. High altitudes receive relatively large amounts of UV-B radiation in comparison to lower elevations (Blumthaler *et al.* 1997). Accordingly, a link between increases in ultraviolet-B (UV-B) radiation and diminishing amphibian populations was proposed in North America (Blaustein *et al.* 1994, 1995), with implications for the decline of other amphibian species. The credibility of this hypothesis is strengthened by the results of laboratory studies on the effects of UV radiation on amphibians (e.g. Gurdon 1960, Worrest and Kimeldorf 1976), and by reports of the biologically damaging effects of UV-B on numerous marine organisms (e.g. Bidigare 1989, Beardall *et al.* 1997).

In south-eastern Australia, serious declines have occurred in nine species of frogs at high-altitudes (Osborne 1989, 1996, Gillespie and Hollis 1996, Mahony 1996, Osborne *et al.* 1996, Hunter *et al.* 1997). Six of these have been listed nationally as threatened species (Tyler 1997). It is possible, therefore, that the link between UV-B and amphibian mortality may also apply to declines in Australian anuran species at high altitudes. Recent research has indicated that this may be the case for the Alpine Tree Frog (*Litoria verreauxii alpina*) (Broomhall 1997). If this is so, then concern must be held for other organisms which may be sensitive to increases in UV-B levels.

While a great deal of research effort has been expended on studying the effects of UV-B radiation on marine organisms (e.g. Smith and Baker 1980, Schofield *et al.* 1995, Häder *et al.* 1996), on terrestrial crop plants (e.g. Yu and Björn 1997), and on molecular mechanisms for DNA damage (e.g. Peak and Peak 1982, Hurks *et al.* 1997), little is known in the context of natural land-based ecosystems, particularly on terrestrial animals. Given concern about increased levels of ultraviolet radiation, particularly at high elevations in the southern hemisphere, there is a surprising paucity of research on the implications of increasing UV-B radiation on alpine regions, and particularly whether increases in UV-B radiation are likely to impact upon the Australian alpine environment. In this chapter I review the state of knowledge of UV-B radiation at high elevations and summarise the relevant research, with emphasis on amphibians.

OZONE DEPLETION

Ozone (O_3) is created by the dissociation of molecular oxygen by wavelengths less than 242 nm (Stolarski *et al.* 1992), and forms an important barrier to substantial amounts of solar radiation (Boucher and Prezelin 1996). Concentrations of ozone are at their highest in the lower stratosphere at about 15–25 km above the Earth's surface (Solomon 1990, Stolarski *et al.* 1992).

There have been steady, almost linear decreases in ozone levels at middle and high latitudes in both hemispheres (Kerr 1991, Stolarski *et al.* 1992, Herman and Larko 1994) since the Antarctic "hole" was first noticed in 1984 (Farman *et al.* 1985). Using satellite data, ozone declines have been traced back as early as 1970 (Farman *et al.* 1985, Jones and Shanklin 1995). Independent measurement systems, including both satellite (Total Ozone Mapping Spectrophotometer, or TOMS) and ground based systems (Stratospheric Aerosol and Gas Experiment, Dobson stations and ozonesondes), all calibrated or corrected for aerosol, solar cycle and quasi-biennial oscillation error, have confirmed these declines (e.g. Proffitt *et al.* 1989, Kerr *et al.* 1993, Miller *et al.* 1996). Ozone decreases have been reported to be about 10% per decade at northern middle latitudes, with larger losses at high latitudes (Stolarski *et al.* 1992). Depletion of ozone has been especially severe in the southern hemisphere (Liu *et al.* 1991, Stolarski *et al.* 1992, Madronich *et al.* 1995), particularly above Australia and New Zealand (Seckmeyer and McKenzie 1992), while tropical latitudes appear exempt from statistically significant decreases (Madronich 1992, Stolarski *et al.* 1992, Herman and Larko 1994). In the Antarctic vortex, O_3 depletion is nearly complete (Proffitt *et al.* 1989).

Several theories have been proposed to explain the ozone hole. The chlorofluorocarbon theory (Stordal *et al.* 1982, McElroy *et al.* 1986, Solomon *et al.* 1986), which linked anthropogenic chlorofluorocarbons (CFCs) with reported ozone losses, appears to be extremely accurate, with the latest measurements and models able to account for virtually all observed ozone destruction (e.g. Waters *et al.* 1993, Manney *et al.* 1994, Russell *et al.* 1996).

The mechanisms by which CFCs destroy ozone are also the reason that ozone loss is highest in Antarctica, and therefore of concern in southern Australia. The extremely low temperatures of the Antarctic stratosphere lead to a greater ozone depletion than in the northern hemisphere (Schoeberl and Hartmann 1991). The rate of ozone decline over Antarctica is increasing; not only is the Antarctic springtime ozone hole continuing to deepen, but it is lasting longer, well into January and February (Jones and Shanklin 1995). Moreover, there is a possibility of future increases, as much as 100-fold, in the size of the Antarctic ozone hole (Schoeberl and Hartmann 1991).

The depletion of ozone evident in the Antarctic has direct effects on ozone levels at mid-latitudes. When solar heating returns to the polar regions, the temperature gradient begins to weaken; far from being a gradual dispersion, the vortex wind system is thought to transit in an abrupt fashion by a final warming cycle that shatters the vortex into smaller fragments of chemically perturbed air (Atkinson *et al.* 1989, Schoeberl and Hartmann 1991), which can then drift to mid-latitudes (Tuck *et al.* 1992, Jones and MacKenzie 1995). This transport of ozone poor air is the reason that ozone depletion is thought to be particularly severe in Australia and New Zealand (Roy *et al.* 1990).

Stratospheric ozone losses are expected to continue (Madronich *et al.* 1995, Montzka *et al.* 1996). Despite the Vienna Agreement for the Protection of the Ozone Layer in 1985 and the amended Montreal Protocol On Substances that Deplete the Ozone Layer in 1987 and 1992, concentrations of atmospheric contaminants are still increasing (Madronich and de Gruijl 1993, Russell *et al.* 1996). Furthermore, developing countries have been granted permission to continue to produce and use unlimited amounts of CFCs until at least 2010 (MacKenzie 1995). Anthropogenic ozone destruction is undisputed, and the projected increase in ozone loss raises serious concerns about the amount of solar ultraviolet radiation penetrating the atmosphere, with corresponding biological consequences. As ozone declines are largest at high latitudes in the southern hemisphere, countries such as Chile, Argentina, New Zealand, and Australia may be most affected.

OZONE AND ULTRAVIOLET-B RADIATION

The UV-B (280–320 nm) region of the spectrum is particularly absorbed by ozone (Calkins and Thordardottir 1980, Smith 1989, Blumthaler and Ambach 1990), and consequently is of specific significance when considering ozone depletion. One of the most important impacts of decreased ozone levels is an increase in the amounts of UV-B radiation reaching the Earth's surface (Liu *et al.* 1991, Madronich 1992, Stamnes *et al.* 1992). UV-C radiation, incorporating wavelengths shorter than 280 nm, is almost entirely absorbed by ozone and molecular oxygen (Boucher and Prezelin 1996), and UV-A (320–400 nm) is not appreciably attenuated by atmospheric ozone at all (Lubin *et al.* 1992). UV-A is important in photorepair mechanisms and is not proportionally enhanced by decreased ozone levels (Smith 1989, Smith *et al.* 1992). Effective surface UV-B irradiance is also influenced by cloud cover, solar zenith angle, atmospheric turbidity and albedo of the terrain (McKenzie *et al.* 1991, Kerr and McElroy 1993, Zerefos *et al.* 1995). Cloud cover reduces UV-B levels (Lubin and Jensen 1995) by as much as 50% (Ilyas 1987). A non-linear relationship between ozone and ultraviolet-B radiation has been reported by a number of researchers (Stordal *et al.* 1982, Lubin *et al.* 1989, McKenzie *et al.* 1991, Madronich *et al.* 1995).

Numerous studies have illustrated that increases in UV-B levels, independent of cloud cover, albedo and atmospheric aerosol content, correlate with decreases in ozone concentrations in both hemispheres (e.g. McKenzie *et al.* 1991, Lubin and Jensen 1995, Zerefos *et al.* 1995). For example, in 1987, radiation levels in Antarctica at 305 nm and near 300 nm were found to be enhanced by a factor of 5 to 6 and 18, respectively, as compared with a background reference level (Lubin *et al.* 1989). More recently, UV levels in Antarctica at the 305 nm wavelength exceeded values expected under normal, undepleted ozone levels by a factor of between 3 and 6 in October 1990 (Stamnes *et al.* 1992). In Chile during 1992, UV-B radiation was at least ten times higher during October than expected under undepleted ozone conditions in the same time period (Kirchhoff *et al.* 1997). Furthermore, the highest increases were at wavelengths around 297 nm — those that are most biologically damaging (Kirchhoff *et al.* 1997). It has also been shown that long-term mild ozone depletions later in the year at Ushuaia, Argentina, during low solar zenith angles, can lead to higher UV-B irradiation than more severe, yet shorter, depletion events earlier in the spring season (Diaz *et al.* 1994). In this way, even less serious ozone depletion events can have potentially substantial biological consequences.

RELATIONSHIP BETWEEN UV-B AND ALTITUDE

It has been established that irradiance increases almost linearly with solar elevation (Bjorn and Vogelmann 1996), and hence the UV-B dose is much larger when the sun is highest in the sky (Caldwell *et al.* 1980, Diffey 1991). UV-B levels also increase with altitude (Blumthaler *et al.* 1992), because radiation at high altitudes travels a shorter path length through the atmosphere and therefore experiences less scattering and absorption (Blumthaler *et al.* 1997). In Germany, Blumthaler *et al.* (1994) measured solar UV simultaneously in a valley and on a mountaintop vertically separated by 1000m (730m a.s.l. and 1730m a.s.l., respectively), using four spectroradiometers. Irradiances were found to be higher at all wavelengths on the mountaintop, changing gradually from 9% higher at 370 nm, to 11% at 320 nm and 24% at 300 nm, independent of zenith angle, and cloud and snow cover (Blumthaler *et al.* 1994). Further research in alpine regions of Austria and Switzerland revealed that the altitude effect depended slightly on solar elevation, in that UV-B irradiance increased by 15.1% ± 1.8% per 1000m at 60° solar elevation, and 18.6% ± 2.9% per 1000m at 20° solar elevation (Blumthaler *et al.* 1997). Previous research has measured a 32% to 42% increase in effective UV-B from 1500 m to 3350 m, (Caldwell *et al.* 1980). Alpine regions are therefore particularly subject to UV-B radiation.

Differences in UV-B irradiance between altitudes are considerably larger when the higher altitude is snow-covered and the lower site free of snow (Blumthaler *et al.* 1997). Snow reflects strongly in the UV-B spectrum (Blumthaler *et al.* 1992), and the UV-B albedo of old and new snow has been measured at 74.4% and 94.4%, respectively (Blumthaler and Ambach 1988). By comparison, the UV-B albedo of primitive rock was 3.7%, while alpine pasture reflected 4.9% of incident UV-B radiation (Blumthaler and Ambach 1988). Furthermore, a large proportion of UV-B is received as a diffuse component, as opposed to direct irradiance (Caldwell *et al.* 1980). Snow cover can double the daily incident UV-B dose through an increase in UV-B diffuse irradiance (Caldwell *et al.* 1980). Organisms inhabiting snow covered areas would probably experience large doses of reflected UV-B radiation in addition to incident UV-B. Given the comparatively low altitude of the Australian Alps, any increase is potentially important.

BIOLOGICAL EFFECTS OF UV-B RADIATION

Molecular Effects, Skin Damage and Cataract Formation

UV-B contains much of the wavelengths most damaging to living organisms (Calkins and Thordardottir 1980, Smith *et al.* 1992, Jordan 1996), as this is the region where proteins and nucleic acids, particularly DNA, show strong absorption (Buma *et al.* 1995, Jordan 1996). The mechanisms for DNA damage appear to be manifold. For example, UV-B radiation inhibits DNA synthesis and reduces the rate of cells entering mitosis (Olsen and Kirkhus 1989); it can cause substantial DNA damage (Hansen *et al.* 1997; also see review by Ananthaswamy and Pierceall 1990), and various kinds of DNA lesions (Okaichi *et al.* 1989), often through the formation of thymine (Hurks *et al.* 1997) or pyrimidine dimers (Kusewitt *et al.* 1996). Recent research has shown that UV-B will interact directly with DNA in the presence of molecular oxygen to induce modifications (Rosen *et al.* 1996). Ultraviolet-B has also been reported to cause DNA damage by inhibiting metallothionein transcription activity, high levels of which are thought to potentially mediate further UV damage in the human epidermis 24 hours post UV exposure (Hansen *et al.* 1997). DNA lethality, mutagenesis, and dimer induction in *Escherichia coli* cells have been shown to decline rapidly at wavelengths longer than 280 nm, with the observation that a 1 nm shift in the solar spectrum to shorter wavelengths resulting from ozone reduction would increase deleterious effects by 20 to 35% (Peak *et al.* 1984). One study has indicated that DNA damage may occur under normal field conditions of UV-B (Buma *et al.* 1995). In regions of escalating UV-B, such as at high latitudes, the possibility of serious DNA damage is increasingly likely.

Solar UV is also thought to contribute strongly to photochemical cataractogenesis, or cataract formation, in ocular tissues (Kusewitt *et al.* 1996). While animals living in snow covered regions have presumably evolved some mechanisms to minimize phototoxic corneal disorders, recent, rapid and large fluctuations in ozone and UV-B amounts may be beyond their adaptive tolerance.

The epidermal effects of UV-B are undisputed and have been the subject of considerable research (see review by Ananthaswamy and Pierceall 1990). Ultraviolet-B radiation has been observed to induce malignant melanomas in marsupials (Ley *et al.* 1989) and cause squamous and basal cell carcinoma (van der Leun 1996). Furthermore, it has been recently reported that 22 of 31 opossums of both sexes developed multiple skin tumors such as fibrosarcomas, hemangiosarcomas, seromas and cutaneous lymphosarcomas after chronic exposure to UV radiation (Kusewitt *et al.* 1996). The biological effectiveness of UV-B for causing skin damage decreases logarithmically with increasing wavelength; it is less than 0.1% as effective at 330 nm as it is at 297 nm (Cole *et al.* 1986, Scotto *et al.* 1988).

Ozone depletion in the period 1979 to 1992 was projected to cause increases in basal cell carcinoma and squamous cell carcinoma incidence of 12.7% ± 4.8% and 23.9% ± 9.2%, respectively, at 45 °S (Madronich and de Gruijl 1993). Such increases were calculated to be smaller at lower southern latitudes and at similar latitudes in the northern hemisphere (Madronich and de Gruijl 1993). Furthermore, models have shown that with no restrictions on ozone-depleting substances, excess human skin cancer rates related exclusively to ozone depletion will quadruple by the year 2100 (Slaper *et al.* 1996). Under the Montreal Protocol, excess skin cancer cases would double in the same time period (Slaper *et al.* 1996). Overall, it is expected that UV-B increases will significantly impact upon incidence of skin cancer in humans (Urbach 1989) and, by extension, on other organisms.

Immune System Effects

It is widely acknowledged that UV radiation induces immunosuppression (Krutmann *et al.* 1996), both local and systemic (Roberts and Beasley 1997), and through several mechanisms and pathways (Hurks *et al.* 1997, and references therein). This has been shown in plants (Orth *et al.* 1990) and mammals (Garssen *et al.* 1996). Another study on human cells has demonstrated that UV-B radiation down-regulates endothelial cell functions, which are involved in immune and inflammatory responses (Yamawaki *et al.* 1996).

While there is the possibility of immediate adaptive and protective behaviour in humans, many terrestrial and aquatic species do not have such options, and changes in UV-B could therefore have serious repercussions. Terrestrial plants, and organisms inhabiting freshwater environments, such as invertebrates and fish, are also potentially at risk from UV-B enhancements. In particular, both larval and adult amphibians are likely to be especially vulnerable. The remainder of this review addresses what is known of the likely impacts on these groups.

AQUATIC ECOSYSTEMS

UV-B Attenuation in Water

Ultraviolet radiation can reach considerable depths in natural waters (Smith 1989). In clear freshwater lakes harmful intensities of UV-B radiation can penetrate to several metres (Schindler *et al.* 1996, Sommaruga and Psenner 1997). Concern about the effects of UV-B enhancements on planktonic communities has been expressed as phytoplankton are a primary sink for atmospheric carbon dioxide (Häder *et al.* 1995, Worrest and Häder 1997), and marine ecosystems have substantial climatic impact on global carbon cycles (Santas *et al.* 1996). Biologically damaging effects of UV-B on numerous marine organisms have been recorded in sea water (see review by Cullen and Neale 1994) where penetration can also be to depths of many tens of metres (Diffey 1991, Smith *et al.* 1992, Häder *et al.* 1995). An extensive list of references on the UV-photobiology of aquatic organisms has been compiled in an article by Karentz *et al.* (1994), who concluded that UV-B is an important factor in aquatic ecosystems, with both direct and indirect effects, none of which should be dismissed.

Penetration by UV-B is modified by a number of factors, such as surface waves (Booth and Morrow 1997), and the dissolved organic matter in the water (Smith 1989, Diffey 1991, Booth and Morrow 1997). Dissolved organic carbon (DOC) has been found to be of great importance in UV-B attenuation (Scully and Lean 1994, Williamson *et al.* 1996). Research has revealed that UV-B penetration increases exponentially as DOC declines (Schindler *et al.* 1996, Yan *et al.* 1996). Further research on this relationship has been conducted in Ontario, with indications that both acidification and climate warming can lead to decreases in DOC in freshwater streams, leading in turn to a much greater penetration of UV-B into the water (Schindler *et al.* 1996). Such increases in the depth of underwater UV-B levels enhances the stress already on biota (Yan *et al.* 1996). Alpine lakes and streams are considered to be of greatest concern, as they are likely to have very low DOC concentrations (Schindler *et al.* 1996, Leavitt *et al.* 1997) and high amounts of ambient UV-B (Schindler *et al.* 1996). Furthermore, future changes in UV radiation will probably have higher impact on low DOC water bodies (Williamson *et al.* 1996), simply because they block so little ambient UV-B radiation. This will be further exacerbated by any climate modifications which decrease DOC concentrations. The possibility exists, therefore, that anthropogenic increases in UV-B radiation could be coupled with drought or acidification events, leading to extremely high underwater UV-B penetration of lakes and streams. In the shallow, extremely clear pools found in alpine regions (e.g. Burgess *et al.* 1988, Sommaruga and Psenner 1997) high levels of UV-B are already likely to penetrate to all depths used as egg development sites by amphibians, and to the areas used by many aquatic species such as fish, invertebrates, plankton, and algae.

This supposition gains credence from underwater UV measurements in Lake Gossenköllesee in the Austrian Alps at 2417 m a.s.l., which indicated that high water transparency allowed substantial UV penetration to the bottom, at a depth of 10 m (Sommaruga and Psenner 1997). The lowest DOC concentration measured in this lake was 0.3 mg L^{-1}, comparable to values in a number of pools within the Snowy Mountains region of south-eastern Australia (Broomhall 1997). Such values appear to be exceedingly low, as other studies have rarely reported values less than around 2.5 mg L^{-1} (e.g. Scully and Lean 1994, Schindler *et al.* 1996, Williamson *et al.* 1996). The 10% attenuation depth at 305 nm was 9.6 m in Gossenköllesee, whilst 60% of incident UV-B radiation was present at a depth of 5 m (Sommaruga and Psenner 1997), certainly much deeper than most of the water bodies in the Snowy Mountains, apart from glacial lakes.

It has been reported that ambient UV-B can suppress periphyton development at alpine sites such as Snowflake Lake in Canada (Vinebrook and Leavitt 1996), which has a DOC concentration of 3.3 mg L^{-1} (Scully and Lean 1994). Moreover, dissolved organic carbon appears to have a substantial influence on the dynamics of the planktonic community of Redberry Lake, a saline lake in Canada, through its effect on UV-B radiation (Ferreyra *et al.* 1997). Mean DOC levels in this oligotrophic lake have been measured as 35.3 mg L^{-1} at 2 m deep, and 37.4 mg L^{-1} at a depth of 16 m (Tumber *et al.* 1993). Given that UV-B-related effects have been found at such relatively high DOC concentrations, the low DOC levels measured in waterbodies within Kosciuszko National Park, Australia (Broomhall 1997) would be expected to provide even less protection from UV-B, with commensurate implications for the communities inhabiting them. Furthermore, in still, shallow waters, steep temperature gradients can form, and greatly hinder the movement of plankton to greater depths, thereby exacerbating the time spent in regions of high UV-B (Kirk 1994, Milot-Roy and Vincent 1994). However, aquatic vegetation may partially shield anuran embryos and some aquatic animals from incident UV-B radiation in lowland and subalpine water bodies. The clear, oligotrophic ponds found at alpine altitudes, such as within the Snowy Mountains, are often depauperate in aquatic vegetation (pers. obs.), and thus impart no such protection.

Biological Effects on Freshwater Organisms

Ambient levels of UV-B in freshwater lakes above 500 m a.s.l. have been reported to kill the cladocerans *Dahnia galeata* and *Daphnia pulex obtusa* at shallow depths within one day (Siebeck and Bohm 1994). UV-B-induced effects can also result in sub-lethal consequences. For example, UV-B has been observed to impair graviperception, swimming velocity, and/or motility in some strains of the dinoflagellate *Prorocentrum* (Sebastian *et al.* 1994). This was thought to have potentially detrimental effects on growth and survival (Sebastian *et al.* 1994). Lahontan Cutthroat Trout, *Oncorhynchus clarki henshawi,* have been shown to experience darkening of the dorsal skin, or sunburn, followed by fungal infection and mortality within six days of exposure to realistic levels of UV-B radiation (Little and Fabacher 1994). In this study, *Oncorhynchus mykiss,* or Rainbow Trout, also exhibited sunburn and fungal infection, but survived, while few

Apache Trout (*Oncorhynchus apache*) were observed to be sunburnt (Little and Fabacher 1994). This illustrates not only the stress that UV-B can place on organisms, but that such stresses and effects can differ between species.

Different species often have different sensitivities to UV-B radiation, and those species living in high ambient UV-B environments may be expected to have evolved some sort of mechanism to minimize UV-B-induced damage (Gleason and Wellington 1995). Nevertheless, at least in some cases, concentrations of protective compounds cannot be sufficiently augmented in the short-term to compensate for rapid UV light increases (Gleason and Wellington 1993). As such, even species acclimatized to high ambient UV-B may be affected by increases associated with ozone depletion events.

UV-A and UV-B

Under undepleted ozone conditions, UV-B was shown to be responsible for a 4.9% loss of the daily primary production in marine phytoplankton assemblages; this increased by 3.8% under decreased ozone levels (Holm-Hansen *et al.* 1993). This figure was found to be higher in the investigations of Smith *et al.* (1992), who found a 6–12% reduction in water column productivity, corresponding to a yearly loss of production for the Southern Ocean conservatively estimated at 2%. On the other hand, UV-A has been found to be even more deleterious to phytoplankton than UV-B (Smith *et al.* 1992, Holm-Hansen *et al.* 1993), and it has been argued that enhanced UV-B would therefore have a relatively small impact when compared with the inhibition experienced under normal ozone conditions from both UV-A and UV-B together (Holm-Hansen *et al.* 1993). Nevertheless, Boucher and Prezelin (1996) observed that laboratory cultures of marine diatoms can be at least two orders of magnitude more sensitive to UV-A than field communities, and speculated that the effects of UV-A may therefore be overestimated in many studies. Moreover, despite the fact that UV-B accounts for only a small fraction of solar radiation, photoinhibition in the Mediterranean green alga *Halimeda tuna* was much more pronounced for the shorter UV-B wavelengths, while UV-A was found to contribute little to the observed photoinhibition (Hader *et al.* 1996). This pattern was similar to that reported in phytoplankton (Milot-Roy and Vincent 1994).

In amphibians, Hays *et al.* (1996) found severe effects of both UV-A and UV-B light on *Hyla regilla* and *Rana cascadae* tadpoles and metamorphs in the laboratory; these included developmental abnormalities and high mortalities. On the other hand, laboratory studies have found that UV-A irradiance at levels twice that of outdoors had no effect on the eggs or larvae of *Bufo americanus, Hyla versicolor, Rana clamitans* or *R. sylvatica* (Grant and Licht 1995). Further research would be required to elucidate the full effects of UV-A radiation, especially in conjunction with elevated UV-B levels. The importance of this is highlighted by the concern that organisms could have evolved a UV-B avoidance strategy based on detecting and responding to levels of UV-A irradiance (Karentz *et al.* 1994), as in the case of some protozoa (Barcelo and Calkins 1979). Ozone depletion would therefore increase UV-B, but not intensify the avoidance stimulus (Barcelo and Calkins 1979, Karentz *et al.* 1994).

Long-term cycles

Scant information is available concerning potentially long-term changes that may have been caused by recent increases in UV-B radiation. It has been reported that fossil pigments (with absorbance maxima in the UV-A region, at 381 nm and 441 nm) produced by benthic algae indicate that UV levels in the past have, on occasion, been higher in northern mid-latitude lakes than that observed so far under ozone depletion (Leavitt *et al.* 1997). However, stratospheric ozone destruction increases ambient radiation at wavelengths somewhat shorter than those maximally absorbed by these compounds, as mentioned previously, and therefore one might question the comparison. Such events of enhanced penetration of UV were thought to be linked to reductions in DOC during drought events (Leavitt *et al.* 1997). In this way, there is the potential for future large fluctuations in UV-B due to climate driven effects, in addition to that derived from ozone depletion. Stratigraphic sequence research in Antarctica found that changes in the composition of the diatom component of the phytoplankton community over the past 20 years were indistinguishable from natural long-term variability, although there may have been a decline in the production of some sea-ice diatoms (McMinn *et al.* 1994). While this indicated no significant effects of the ozone hole, the authors remarked that coastal areas such as their study site are under a deep ice cover when ozone depletions are at their highest, perhaps mitigating UV-B damage, an observation also made by Bidigare (1989). Other researchers have discovered that UV-B penetrates ice to a varying degree, depending on ice thickness and type (Stamnes *et al.* 1992), together with extent and depth of snow on the ice. For example, the sea ice cover surrounding Antarctica is thought to be less turbid in early spring than later in the season, leading to enhanced UV-B levels under the ice (a 20-fold increase compared to other seasons) (Trodahl and Buckley 1989). In this way, organisms which are seemingly protected under an ice or snow cover may also potentially be affected by ozone depletion.

TERRESTRIAL PLANTS

The potential for UV-B mediated interactions and the possibility of ecosystem change are also evident in terrestrial situations. Bacteria, mycorrhizal fungi and other below-ground organisms are reported to be affected by UV-B through changes in the carbon flow of the plant soil system (Klironomos and Allen 1995). Similarly, insect herbivory has been found to be modified by UV-B exposure of the host plant (Grant-Petersson and Renwick 1996). More specifically, larvae of *Pieris rapae* consumed less foliage from plants that had been exposed to UV-B than they did of controls, and also weighed less (Grant-Petersson and Renwick 1996). The other species of herbivorous insect, *Trichoplusia ni* , did not exhibit preferential consumption (Grant-Petersson and Renwick 1996). From this, it might be expected that enhanced UV-B may lead to changes in herbivore feeding patterns. Research in Sweden on leaves of *Vaccinium uliginosum* and *V. myrtillus* collected from the subarctic, and subjected to UV-B both in the field and in the laboratory, revealed that litter quality was altered under UV-B conditions (Gehrke *et al.* 1995). In addition, microbial respiration decreased, and there was substantially less colonization of leaf litter in the UV-B treatments (Gehrke *et al.* 1995).

Two of three fungal decomposers were sensitive to UV-B, while the other species remained unaffected (Gehrke *et al.* 1995). This suggested that UV-B led to changes in decomposer community structure, both directly and indirectly (Gehrke *et al.* 1995, Gehrke *et al.* 1996). Enhanced UV-B is therefore likely to slow down litter turnover and nutrient cycling (Gehrke *et al.* 1996). Any increase in UV-B radiation due to ozone depletion will almost undoubtedly impact at many trophic levels.

While it is thought that marine organisms could possibly acclimate to increasing ultraviolet radiation, any counteracting strategies would in turn have an associated energy cost which may result in reduced growth (Calkins and Thordardottir 1980, Cullen and Neale 1994). This has also been proposed for terrestrial plants (Day *et al.* 1992). Non-lethal exposures to UV-B may entail costs in terms of protection and repair, perhaps leading to a compromise in reproductive success (Karentz *et al.* 1994). Production of UV-B protective compounds seemed to incur a cost to plant growth in *Calamagrostis lapponica,* a species of subarctic grass (Gwynn-Jones and Johanson 1996). Furthermore, while plant growth is directly affected by UV-B (e.g. Mark *et al.* 1996), such as in the case of the pea *Pisum sativum* L. (Mepsted *et al.* 1996); indirect changes to plant biomass allocation, form and developmental timing by UV-B (e.g. Teramura 1983, Teramura and Sullivan 1994) may be more important than immediately damaging effects (Caldwell *et al.* 1995, Rozema *et al.* 1997). For example, UV-B levels associated with ozone depletion have been found to reduce viable pollen production substantially in *Brassica rapa,* despite increased flower numbers (Demchik and Day 1996), with commensurate implications for reproductive success. This finding encapsulates the misleading results which may be obtained by examining only one response variable, in this case, numbers of flowers. The incorporation of factors such as pollen viability provided more insight. Similarly, long-term perspectives are required to understand the implications of elevated UV-B.

Accumulated UV-B doses over three generations had a greater effect on the performance of the plant *Dimorphotheca sinuata* than immediate UV-B irradiation (Musil 1996). Such accumulated UV-B effects were manifested as reductions in dry mass, earlier reproductive effort, and decreased chlorophyll a concentrations, seed germination, stem and inflorescence production, and pollen-tube growth (Musil 1996). Moreover, stimulative effects, for example, may be short-term responses whereby... "stressed plants exhaust themselves in growth or reproduction in order to survive genetically" (Gehrke *et al.* 1996; p 197). The study by Gehrke *et al.* (1996) also illustrated the importance of long-term testing. Four species of subarctic dwarf shrubs were grown under ambient and enhanced (Å 15% ozone depletion) UV-B levels for several years. Relative growth after one year was unaffected by enhanced UV-B in two deciduous species and *Vaccinium vitis-idaea* (although it was reduced in *Empetrum hermephroditum*), but after two years, all plants under enhanced UV-B had a lower relative growth than those in the ambient treatments (Gehrke *et al.* 1996). It is possible that such a response may also be found in Australian alpine species. As low temperatures tend to result in low biomass production and slow nutrient cycling, responses to perturbations may take time to manifest in high latitude ecosystems (Gehrke *et al.* 1996).

Aquilegia caerulea, an alpine columbine, responded differently to enhanced UV-B than did *A. canadensis*, a lower-elevation species (Larson *et al.* 1990). Although both became significantly shorter and produced a greater number of leaves when compared to controls grown without UV-B after seven weeks, this growth response was greater in *A. canadensis*, while the number of leaves was greater in the alpine species, *A. caerulea* (Larson *et al.* 1990). UV-B treatment led to increased flavonoid concentrations and decreased water-use efficiency in both species (Larson *et al.* 1990). Latitude may also play a role in plant sensitivities. Earlier research has indicated that some Arctic/alpine plant species collected from alpine regions of the equator exhibited no UV-B induced damage in field trials, while inhibition of photosynthesis was detected in certain species from higher latitudes (Barnes *et al.* 1987). Similarly, accumulation of UV absorbing pigments was only noted in plants from high latitudes, although even this did not entirely mitigate UV-B-induced damage (Barnes *et al.* 1987). Populations at higher latitudes have generally experienced a lower ambient UV-B than those in equatorial regions, as mentioned previously, and therefore presumably have not adapted to high levels of UV-B. Such an Arctic/alpine latitudinal gradient in UV-B adaptation may in part explain these results, however, patterns of both photosynthetic inhibition and of leaf thickness in relation to UV-B sensitivity indicated that plants had different inherent sensitivities to UV-B (Barnes *et al.* 1987). Similarly, Rau and Hofman (1996) found that various plant genera growing in the European Alps exhibited marked differences in their sensitivities to UV-B. Furthermore, plant foliage has been found to be differentially effective at screening out UV-B radiation; the leaf epidermis of herbaceous dicots allowed 18–41% transmittance, while conifer needles attenuated practically all incident UV-B (Day *et al.* 1992).

Such differential sensitivity to UV-B may lead to shifts in species composition (Häder *et al.* 1995, Worrest and Häder 1997), which may in turn have greater and more complex impacts on the ecosystem than simple changes in community numbers (e.g. Siebeck *et al.* 1994, Kiffney *et al.* 1997, Rozema *et al.* 1997). In any case, increases in UV-B radiation due to depletion of the ozone layer is likely to lead to shifts in the dominance of genotypes, and selection for UV-B resistant variants (Schick *et al.* 1996). The molecular effects of UV-B radiation on terrestrial plants has been comprehensively reviewed by Jordan (1996) with the overall conclusion that plant communities may suffer in response to future increases in UV-B. This has obvious implications for all regions of the globe.

AMPHIBIANS AND UV-B

Recent studies at high altitudes in North America have suggested a link between declining frog populations and increases in UV-B radiation (Blaustein *et al.* 1994). Species-specific differences in the ability of anuran eggs to repair radiation damage to DNA via the photolyase enzyme were demonstrated (Blaustein *et al.* 1994). The researchers also observed reduced hatching success of embryos exposed to solar radiation when compared to controls shielded from UV-B radiation at natural oviposition sites (Blaustein *et al.* 1994). A criticism of the research was that Blaustein *et al.* (1994) failed to investigate if UV-B levels had actually risen in the last ten years (Roush 1995). This criticism could be considered somewhat irrelevant: there is a large body of literature on ozone depletion and concomitant UV-B increases, as has been discussed in the preceding sections, and an effect of UV-B was clearly shown. Analogous field experiments conducted at lower elevations (183 m) in the Oregon Coast Range, on the salamander *Ambystoma gracile*, detected similar results (Blaustein *et al.* 1995). However, hatching success in the red-legged frog *Rana aurora* was unaffected by UV-B, possibly due to high levels of photolyase activity (Blaustein *et al.* 1996).

Evidence for a link between UV-B radiation and developmental abnormalities has been substantiated by the earlier laboratory work of Higgins and Sheard (1926), Gurdon (1960), Worrest and Kimeldorf (1976), and more recently by Grant and Licht (1995). Worrest and Kimeldorf (1976) reported that boreal toad tadpoles, *Bufo boreas boreas*, exposed to UV-B at a late stage in their development (Gosner stages 26–35), had dramatically higher abnormalities and mortalities compared to shielded individuals. Anomalies often took the form of "lordosis", an abnormally increased concavity of the spine (Worrest and Kimeldorf 1976).

Gurdon (1960) discovered that the unfertilized egg nucleus of *Xenopus laevis* was completely destroyed by small doses of UV radiation. Further, Gurdon (1960) found that eggs of *X. laevis* developed into haploid individuals if irradiated soon after fertilization, and these individuals developed abnormally. *Rana pipiens* eggs and embryos exposed to UV suffered between 70–100% mortality; the remaining tadpoles were all aberrant in some way, with the most common characteristics being curvature of the tail as previously noted, and large globular swellings of the ventral body wall (Higgins and Sheard 1926).

More comprehensive laboratory experiments by Grant and Licht (1995) ascertained that the hatching success of embryos of *Rana sylvatica* was unaffected by exposure to high-intensity UV-B for 15 mins, while 30 mins of UV-B exposure resulted in 100% mortality. Moreover, lower temperatures produced higher proportions of both morphological and developmental abnormalities in *R. sylvatica* for a given amount of UV-B radiation (Grant and Licht 1995). Included in the research were studies on the larvae of *Bufo americanus, Hyla versicolor, R. sylvatica* and *R. clamitans,* which demonstrated that the former three species suffered 100% mortality under two or three minute exposures to UV-B, while 30% of larval *R. clamitans* survived (Grant and Licht 1995). These survivors exhibited abnormalities similar to those described by Worrest and Kimeldorf (1976). Metamorphosed juveniles of *Bufo americanus* survived high-intensity UV-B longer than those of *R. sylvatica* and *R. clamitans*, although even these animals died some weeks later (Grant and Licht 1995). There was also some indication that embryos may be more tolerant of UV-B than larvae, possibly due to protection by the jelly surrounding ova (Grant and Licht 1995).

Although Grant and Licht (1995) concluded that current natural levels of UV were unlikely to be detrimental, the authors remarked that areas of relatively high UV-B levels, such as high altitudes, might present a greater risk to anurans and consequently required investigation (Grant and Licht 1995, Licht and Grant 1997). This concern was also raised by Blaustein *et al.* (1995) who further asserted that at relatively high elevations, development times in amphibians may be extended, and they would therefore undergo prolonged UV-B exposure. Moreover, high altitude amphibians are often heliothermic, and actively seek out sunlight to increase developmental temperatures (Stebbins and Cohen 1995). However, once again, it is expected that higher elevation populations are possibly more resistant to UV-B, or there may be selection pressure for the evolution of efficient repair mechanisms or behavioural adaptations to minimize UV damage (Blaustein *et al.* 1995). Even so, rapid anthropogenic environmental changes may occur too quickly for evolutionary adaptation (Blaustein *et al.* 1995).

A number of anuran species in south-eastern Australia have declined dramatically over the past 15 years from high altitude areas, lending strong circumstantial support to a possible link with increases in UV-B radiation. As the UV monitoring network in Australia only commenced in 1987 (S. Toomey, pers. comm., Australian Radiation Laboratories), the exact trends in UV-B levels cannot be ascertained for the full extent of observed Australian frog declines. While research has indicated that UV-B is unlikely to be responsible for the decline of *Litoria aurea* in the low elevation Southern Tablelands of Australia (van de Mortel and Buttemer 1996; note only eggs and embryos were studied), the decline of *Litoria verreauxii alpina* in the Snowy Mountains region has provided an opportunity to examine the possible role of UV-B in population declines in a high-altitude species. *Litoria v. alpina* is a subspecies of the widespread Whistling Tree Frog (*Litoria verreauxii*) (Green and Osborne 1994), and has been listed as nationally vulnerable (Tyler 1997).

The Alpine Tree Frog (*L. v. alpina*) is one of only two species of frog which occur at alpine altitudes in the Snowy Mountains: the other is the Common Eastern Froglet (*Crinia signifera*) (Green and Osborne 1994). Although both *C. signifera* and *L. v. alpina* were previously common at all alpine altitudes, populations of *L. v. alpina* have collapsed in the last 15 years (Hunter *et al.* 1997). *Litoria v. alpina* has disappeared entirely from the alpine zone (above about 1850 m), and few recent records exist above 1500 m (W. Osborne and D. Hunter, pers. comm.). By contrast, *C. signifera* remains abundant at all altitudes including over 2000 m (W. Osborne and D. Hunter pers. comm.). These regions are subject to little local human impact, although the region was extensively grazed during the summer up until the late 1960s. Both frog species lay their eggs in still, shallow, open pools (Green and Osborne 1994). Given the exposed nature of the water bodies at alpine altitudes, it is possible that UV-B radiation could be contributing to the disappearance of *L. v. alpina*.

Research has examined the sensitivity of these two species of frog in the Snowy Mountains to current ambient levels of ultraviolet radiation (Broomhall 1997). Tadpoles raised in the open showed differential levels of mortality depending on whether they were protected from ultraviolet radiation (Broomhall 1997). Blocking UV-B radiation significantly enhanced survival for the declining species *L. v. alpina* and the non-declining species *C. signifera*, with the effect being particularly pronounced in *L. v. alpina*. This finding has implications for many other species occurring at high altitudes in Australia.

If ozone depletion is implicated in the severe decline of *L. v. alpina*, it suggests that the species may have been on the periphery of its adaptive tolerance for UV-B, such that even small increases in radiation were sufficient to cause elevated mortality. Similarly, it has been suggested that in clear ponds and lakes in subarctic and arctic regions, populations of *Daphnia pulex* are exposed to natural UV-doses close to their tolerance limits, despite melanisation in Arctic populations (Hessen 1994). An intensified UV-stress is therefore expected to be detrimental to exposed populations (Hessen 1994). UV-B increases would have occurred so quickly on an evolutionary scale that species have presumably had little time to adapt. Moreover, at certain times of the year, organisms may be experiencing UV-B levels up to 15 times higher than in previous years (Stamnes *et al.* 1992), even though the overall yearly increase may be much smaller. In this way, a few days of exceedingly high UV in an otherwise "normal" season may be sufficient to stress or eradicate a high proportion of a given population.

Interactions between UV-B and other factors: temperature

It is also possible that lower temperatures, such as those in the Australian Alps, may stress organisms, increasing their vulnerability to damage by UV-B radiation. Existing research lends credence to this hypothesis. Embryonic mortality of *Rana temporaria* in the field was reported to be mostly caused by low water temperatures (Beattie and Tyler-Jones 1992), or low minimum pond temperatures (Beattie *et al.* 1991). Furthermore, low water temperatures caused high proportions of abnormalities in the tailed frog *Ascaphus truei*; development proceeded normally at 7.6°C, whereas 50% were abnormal when kept at 5°C (Brown 1975). Cooke (1981) suggested that exposure to environmental stress in the form of pollutants or extremes in temperature could increase the incidence of deformities in frog embryos. Other studies have shown that conditions stressful to developing anuran embryos (such as low temperatures or pH) increase susceptibility to *Saprolegnia* infection (Banks and Beebee 1988, Beattie *et al.* 1991), and to *Pseudomonas aeruginosa* infection (Brodkin *et al.* 1992). Moreover, a similar link between low temperatures and immunosuppression was argued by Carey (1993).

A comparable synergism was discovered by Long *et al.* (1995), who reported that survival of *Rana pipiens* embryos in the laboratory was unaffected by either low pH or UV-B in isolation, but was significantly decreased by the two acting in concert. Additionally, UV-B radiation and *Saprolegnia* infection together caused a much greater reduction in the hatching success of *Rana* and *Bufo* embryos than either of these factors alone (Kiesecker and Blaustein 1995). Elsewhere, it has been suggested that lower temperatures may impede photorepair in anuran embryos (Grant and Licht 1995). Milot-Roy and Vincent (1994) observed that low temperatures such as those in subarctic lakes may slow recovery processes while photochemical damage may continue at temperature independent rates, and hence amplify damage.

Grant and Licht (1995) found that eggs of *Rana sylvatica* exposed in the laboratory to a given amount of UV-B at 12°C subsequently hatched with an extremely high proportion of behavioural and morphological abnormalities, when compared with eggs exposed to exactly the same amount of UV-B at 20°C. Specifically, after five minutes exposure to UV-B at 12°C, less than 7% were abnormal, after ten minutes, 60% were abnormal, and after fifteen minutes 96% were abnormal (Grant and Licht 1995). By contrast, less than 7% were abnormal for all exposure times at 20°C (Grant and Licht 1995). The authors hypothesized that the eggs maintained at lower temperatures would receive a greater cumulative dose of UV-B by the time they hatched, due to the extended developmental period (Grant and Licht 1995). Another consideration is that some stages in amphibian development are possibly more vulnerable to UV-B damage than other stages. Consequently, the lower temperatures experienced at higher altitudes may prolong the amount of time spent at such developmental stages, with a commensurate effect on mortality. Given the existing research on synergistic interactions for both UV-B and lower temperatures, it is possible that lower temperatures may combine with the effects of UV-B radiation to cause an elevated stress.

CONCLUSION

The expected increases in ultraviolet-B radiation due to anthropogenic ozone depletion appear likely to have particular impact on the Australian Alps, as it is situated at high elevations, middle to high latitudes, and in the southern hemisphere. A number of studies have affirmed that ozone-related changes in UV-B will probably be most pronounced at such locations. While complete clarification of the effects of UV-B on many different organisms remains elusive, existing research indicates that in many cases, UV-B can have deleterious consequences. For example, UV-B radiation has been directly linked to skin cancer, corneal tumors, and immunosuppression. This is of specific significance in alpine regions where levels of UV-B radiation are expected to be high, particularly given the high albedo of snow in the UV-B wavelengths. Many plant species have also been shown to be negatively impacted by UV-B, although some of these impacts may take a number of years to appear. Furthermore, complex interactions between trophic levels and differential UV-B sensitivities may lead to substantial changes in species composition. As there appears to be a paucity of long-term data on the effects of increases in UV-B radiation, it seems that the most acceptable solution is to initiate rigorous sampling and monitoring studies, and simultaneously assess and test the effects of UV-B in conjunction with other stresses, such as low temperatures. Finally, research has determined that existing levels of UV-B in the south-eastern alpine region of Australia are likely to be a significant causative factor in the decline of populations of a high altitude frog species. This finding, in conjunction with previous research, leads to concern over the potential vulnerability of other species in the Australian Alps to pervasive increases in UV-B radiation.

REFERENCES

Ananthaswamy, H. N. and Pierceall, W. E., 1990. Molecular mechanisms of ultraviolet radiation carcinogenesis. *Photochem. Photobiol.* **52:** 1119–1136.

Atkinson, R. J., Matthews, W. A., Newman, P. A. and Plumb, R. A., 1989. Evidence of mid-latitude impact of Antarctic ozone depletion. *Nature* **340:** 290–294.

Banks, B. and Beebee, T. J. C., 1988. Reproductive success of Natterjack Toads *Bufo calamita* in two contrasting habitats. *J. Anim. Ecol.* **57:** 475–492.

Barcelo, J. A. and Calkins, J., 1979. Positioning of aquatic microorganisms in response to visible light and simulated solar UV-B irradiation. *Photochem. Photobiol.* **29:** 75–83.

Barnes, P. W., Flint, S. D. and Caldwell, M. M., 1987. Photosynthesis damage and protective pigments in plants from a latitudinal arctic/alpine gradient exposed to supplemental UV-B radiation in the field. *Arctic Alp. Res.* **19:** 21–27.

Beardall, J., Berman, T., Markager, S., Martinez, R. and Montecino, V., 1997. The effects of ultraviolet radiation on respiration and photosynthesis in two species of microalgae. *Can. J. Fish. Aquat. Sci.* **54:** 687–696.

Beattie, R. C., Aston, R. J. and Milner, A. G. P., 1991. A field study of fertilization and embryonic development in the common frog *Rana temporaria* with particular reference to acidity and temperature. *J. Appl. Ecol.* **28:** 346–357.

Beattie, R. C. and Tyler-Jones, R., 1992. The effects of low pH and aluminium on breeding success in the frog *Rana temporaria*. *J. Herp.* **26:** 353–360.

Bidigare, R. R., 1989. Potential effects of UV-B radiation on marine organisms of the Southern Ocean: distributions of phytoplankton and krill during Austral Spring. *Photochem. Photobiol.* **50:** 469–477.

Bjorn, L. A. and Vogelmann, T. C., 1996. Quantifying light and ultraviolet radiation in plant biology. *Photochem. Photobiol.* **64:** 403–406.

Blaustein, A. R., Edmond, B., Kiesecker, J. M., Beatty, J. J. and Hokit, D. G., 1995. Ambient ultraviolet radiation causes mortality in salamander eggs. *Ecol. Appl.* **5:** 740–743.

Blaustein, A. R., Hoffman, P. D., Hokit, D. G., Kiesecker, J. M., Walls, S. C. and Hays, J. B., 1994. UV repair and resistance to solar UV-B in amphibian eggs: A link to population declines? *Proc. Natl. Acad. Sci. USA* **91:** 1791–1795.

Blaustein, A. R., Hoffman, P. D., Kiesecker, J. M. and Hays, J. B., 1996. DNA repair activity and resistance to solar UV-B radiation in eggs of the Red-legged frog: *Cons. Biol.* **10:** 1398–1402.

Blumthaler, M. and Ambach, W., 1988. Solar UVB-albedo of various surfaces. *Photochem. Photobiol.* **48:** 85–88.

Blumthaler, M. and Ambach, W., 1990. Indication of increasing solar ultraviolet-B radiation flux in alpine regions. *Science* **248:** 206–208.

Blumthaler, M., Ambach, W. and Ellinger, R., 1997. Increase in solar UV radiation with altitude. *J. Photochem. Photobiol. B: Biol.* **39:** 130–134.

Blumthaler, M., Ambach, W. and Rehwald, W., 1992. Solar UV-A and UV-B radiation fluxes at two alpine stations at different altitudes. *Theor. Appl. Clim.* **46:** 39–44.

Blumthaler, M., Webb, A. R., Seckmeyer, G., Bais, A. F., Huber, M. and Mayer, B., 1994. Simultaneous spectroradiometry: a study of solar UV irradiance at two altitudes. *Geophys. Res. Lett.* **21:** 2805–2808.

Booth, C. R. and Morrow, J. H., 1997. The penetration of UV into natural waters. *Photochem. Photobiol.* **65:** 254–257.

Boucher, N. P. and Prezelin, B. B., 1996. Spectral modeling of UV inhibition of *in situ* Antarctic primary production using a field-derived biological weighting function. *Photochem. Photobiol.* **64:** 407–418.

Bradford, D. F., 1991. Mass mortality and extinction in a high-elevation population of *Rana muscosa*. *J. Herp.* **25:** 174–177.

Brodkin, M. A., Simon, M. P., DeSantis, A. M. and Boyer, K. J., 1992. Response of *Rana pipiens* to graded doses of the bacterium *Pseudomonas aeruginosa*. *J. Herp.* **26:** 490–495.

Broomhall, S. D.,1997. "Comparative effects of ultraviolet-B (UV-B) radiation on two sympatric species of Australian anurans, *Litoria verreauxii alpina* and *Crinia signifera*". Unpublished Honours thesis, University of Canberra.

Brown, H. A., 1975. Temperature and development of the tailed frog, *Ascaphus truei*. *Comp. Biochem. Phys.* **50A:** 397–405.

Buma, A. G. J., Van Hannen, E. J., Roza, L., Veldhuis, M. J. W. and Gieskes, W. W. C., 1995. Monitoring ultraviolet-B-induced DNA damage in individual diatom cells by immunofluorescent thymine dimer detection. *J. Phycol.* **31:** 314–321.

Burgess, J., Gillieson, D. and Spate, A., 1988. On the thermal stratification of freshwater lakes in the Snowy Mountains, Australia, and Larsemann Hills, Antarctica. *Search* **19:** 147–149.

Caldwell, M. M., Robberecht, R. and Billings, W. D. 1980. A steep latitudinal gradient of solar ultraviolet-B radiation in the Arctic-alpine life zone. *Ecology* **61:** 600–611.

Caldwell, M. M., Teramura, A. H., Tevini, M., Bornman, J. F., Bjorn, L. O. and Kulandaivelu, G., 1995. Effects of increased solar ultraviolet radiation on terrestrial plants. *Ambio* **24:** 166–173.

Calkins, J. and Thordardottir, T., 1980. The ecological significance of solar UV radiation on aquatic organisms. *Nature* **283:** 563–566.

Carey, C., 1993. Hypothesis concerning the causes of the disappearance of Boreal toads from the mountains of Colorado. *Cons. Biol.* **7:** 355–362.

Cole, C. A., Forbes, P. D. and Davies, R. E., 1986. An action spectrum for photocarcinogenesis. *Photochem. Photobiol.* **43:** 275–284.

Cooke, A. S., 1981. Tadpoles as indicators of harmful levels of pollution in the field. *Environ. Poll. Ser.* **A 25:** 123–133.

Crump, M. L., Hensley, F. R. and Clark, K. L., 1992. Apparent decline of the Golden Toad: underground or extinct? *Copeia* **1992:** 413–420.

Cullen, J. J. and Neale, P. J., 1994. Ultraviolet radiation, ozone depletion, and marine photosynthesis. *Photosyn. Res.* **39:** 303–320.

Day, T. A., Vogelmenn, T. C. and DeLucia, E. H., 1992. Are some plant life forms more effective than others at screening out ultraviolet-B radiation? *Oecologia* **92:** 513–519.

Demchik, S. M. and Day, T. A., 1996. Effect of enhanced UV-B radiation on pollen quantity, quality, and seed yield in *Brassica rapa (Brassicaceae)*. *Amer. J. Bot.* **8:** 573–579.

Diaz, S. B., Booth, C. R., Lucas, T. B. and Smolskaia, I., 1994. Effects of ozone depletion on irradiances and biological doses over Ushuaia. *Ergeb. Limnol.* **43:** 115–122.

Diffey, B. L., 1991. Solar ultraviolet radiation effects on biological systems. *Phys. Med. Biol.* **36:** 299–328.

Farman, J. C., Gardiner, B. G. and Shanklin, J. D., 1985. Large losses of total ozone in Antarctic reveal a seasonal ClO_x/NO_x interaction. *Nature* **315:** 207–210.

Fellers, G. M. and Drost, C. A., 1993. Disappearance of the Cascades frog *Rana cascadae* at the southern end of its range, California, USA. *Biol. Cons.* **65:** 177–181.

Ferreyra, G. A., Demers, S., del Giorgio, P. and Chanut, J-P., 1997. Physiological responses of natural plankton communities to ultraviolet-B radiation in Redberry Lake (Saskatchewan, Canada). *Can. J. Fish. Aquat. Sci.* **54:** 705–714.

Garssen, J., Goettsch, W., de Gruijl, F., Slob, W. and Van Loveren, H., 1996. Risk assessment of UVB effects on resistance to infectious diseases. *Photochem. Photobiol.* **64:** 269–274.

Gehrke, C., Johanson, U., Gwynn-Jones, D., Björn, L. O., Callaghan, T. V. and Lee, J. A., 1996. Effects of enhanced ultraviolet-B radiation on terrestrial subarctic ecosystems and implications for interactions with increased atmospheric CO_2. *Ecol. Bull.* **45:** 192–203.

Gehrke, C., Johanson, U., Callaghan, T. V., Chadwick, D. and Robinson, C., 1995. The impact of enhanced ultraviolet-B radiation on litter quality and decomposition processes in Vaccinium leaves from the Subarctic. *Oikos* **72:** 213–222.

Gillespie, G. R. and Hollis, G. J., 1996. Distribution and habitat of the Spotted Tree Frog, *Litoria spenceri* Dubois Anura: Hylidae, and an assessment of potential causes of population declines. *Wildl. Res.* **23:** 49–75.

Gleason, D. F. and Wellington, G. M., 1993. Ultraviolet radiation and coral bleaching. *Nature* **365:** 836–838.

Gleason, D. F. and Wellington, G. M., 1995. Variation in UVB sensitivity of planula larvae of the coral *Agaricia agaricites* along a depth gradient. *Mar. Biol.* **123:** 693–703.

Grant, K. P. and Licht, L. E., 1995. Effects of ultraviolet radiation on life-history stages of anurans from Ontario, Canada. *Can. J. Zool.* **73:** 2292–2301.

Grant-Petersson, J. and Renwick, J. A. A., 1996. Effects of ultraviolet-B exposure of *Arabidopsis thaliana* on herbivory by two crucifer-feeding insects (Lepidoptera). *Environ. Entomol.* **25:** 135–142.

Green, K. and Osborne, W.S., 1994. 'Wildlife of the Australian Snow Country.' Reed, Chatswood, NSW.

Gurdon, J. B., 1960. The effects of ultraviolet irradiation on uncleaved eggs of *Xenopus laevis*. *Quart. J. Micros. Sci.* **101:**: 299–311.

Hader, D-P., Porst, M., Herrmann, H., Schafer, J. and Santas, R., 1996. Photoinhibition in the Mediterranean Green Alga *Halimeda tuna* Ellis et Sol measured *in situ*. *Photochem. Photobiol.* **64:** 428–434.

Hansen, C., Abiett, E., Green, A., Sturm, R. A., Dunn, I. S., Fairlie, D. P., West, M. L. and Parsons, P. G., 1997. Biphasic response of the metallothionein promoter to ultraviolet radiation in human melanoma cells. *Photochem. Photobiol.* **65:** 550–555.

Hays, J. B., Blaustein, A. R., Kiesecker, J. M., Hoffman, P. D., Pandelova, I., Coyle, D. and Richardson, T., 1996. Developmental responses of amphibians to solar and artificial UVB sources: A comparative study. *Photochemi. Photobiol.* **64:** 449–456.

Herman, J. R. and Larko, D., 1994. Low ozone amounts during 1992–1993 from Nimbus 7 and Meteor 3 total ozone mapping spectrometers. *J. Geophys. Res.* **99(D2):** 3483–3496.

Hessen, D. O., 1994. *Daphnia* responses to UV-light. *Ergeb. Limnol.* **43:** 185–195.

Higgins, G. M. and Sheard, C., 1926. Effects of ultraviolet radiation on the early larval development of *Rana pipiens*. *J. Exp. Zool.* **46:** 333–343.

Hollis, G. J., 1995. Reassessment of the distribution, abundance and habitat of the Baw Baw Frog *Philoria frosti* Spencer: preliminary findings. *Vict. Nat.* **112:** 190–201.

Holm-Hansen, O., Helbling, E. W. and Lubin, D., 1993. Ultraviolet radiation in Antarctica: inhibition of primary production. *Photochem. Photobiol.* **58:** 567–570.

Hunter, D., Osborne, W. S. and Smith, M. J., 1997. Distribution and abundance of the alpine tree frog (*Litoria verreauxii alpina*) in the Australian Alps National parks: Report on the first seasons survey (1996–97). Australian Alps National Parks Liaison Committee, Canberra.

Hurks, H. M. H., Out-Luiting, C., Vermeer, B. J., Claas, F. H/ J. and Mommaas, A. M., 1997. *In-situ* action spectra suggest that DNA damage is involved in ultraviolet radiation-induced immunosuppression in humans. *Photochem. Photobiol.* **66:** 76–81.

Ilyas, M., 1987. Effect of cloudiness on solar ultraviolet radiation reaching the surface. *Atmos. Environ.* **21:** 1483–1484.

Jones, R. L. and MacKenzie, A. R., 1995. Observational studies of the role of polar regions in mid-latitude ozone loss. *Geophys. Res. Lett.* **22:** 3485–3488.

Jones, A. E. and Shanklin, J. D., 1995. Continued decline of total ozone over Halley, Antarctica, since 1985. *Nature* **376:** 409–411.

Jordan, B. R., 1996. The effects of ultraviolet-B radiation on plants: a molecular perspective. *Adv. Bot. Res.* **22:** 97–162.

Juterbock, J. E., 1990. Comments about the amphibian decline symposium. *Herp. Review* **21:** 75–76.

Karentz, D., Bothwell, M. L., Coffin, R. B., Hanson, A., Herndl, G. J., Kilham, S. S., Lesser, M. P., Lindell, M., Moeller, R. E., Morris, D. P., Neale, P. J., Sanders, R. W., Weiler, C. S. and Wetzel, R. G., 1994. Impact of UV-B radiation on pelagic freshwater ecosystems: Report of working group on bacteria and phytoplankton. *Ergeb. Limnol.* **43:** 31–69.

Kiesecker, J. M. and Blaustein, A. R., 1995. Synergism between UV-B radiation and a pathogen magnifies amphibian embryo mortality in nature. *Proc. Natl. Acad. Sci. USA* **92:** 11049–11052.

Kiffney, P. M., Little, E. E. and Clements, W. H., 1997. Influence of ultraviolet-B radiation on the drift response of stream invertebrates. *Freshwater Biol.* **37:** 485–492.

Kirchhoff, V. W. J. H., Zamorano, F. and Casicca, C., 1997. UV-B enhancements at Punta Arenas, Chile. *J. Photochem. Photobiol. B:Biol.* **38:** 174–177.

Kirk, J. T. O., 1994. Optics of radiation in natural waters. *Ergeb. Limnol.* **43:** 1–16.

Klironomos, J. N. and Allen, M. F., 1995. UV-B-mediated changes on below-ground communities associated with the roots of *Acer saccharum*. *Funct. Ecol.* **9:** 923–930.

Krutmann, J., Ahrens, C., Roza, L. and Ariett, C. F., 1996. The role of DNA damage and repair in ultraviolet B radiation-induced immunomodulation: Relevance for human photocarcinogenesis. *Photochem. Photobiol.* **63:** 394–396.

Kusewitt, D. F., Lewis, J. L., Griffith, W. C. Jr. and Ley, R. D., 1996. Effect of chronic exposure to ultraviolet radiation and photoreactivation on life span and tumor development in the marsupial *Monodelphis domestica*. *Radiation Res.* **146:** 187–197.

Larson, R. A., Garrison, W. J. and Carlson, R. W., 1990. Differential responses of alpine and non-alpine *Aquilegia* species to increased ultraviolet-B radiation. *Plant Cell Environ.* **13:** 983–987.

Leavitt, P. R., Vinebrooke, R. D., Donald, D. B., Smol, J. P. and Schindler, D. W., 1997. Past ultraviolet radiation environments in lakes derived from fossil pigments. *Nature* **388:** 457–459.

van der Leun, J. C., 1996. UV radiation from sunlight: summary, conclusions and recommendations. *J. Photochem. Photobiol. B:Biol.* **35:** 237–244.

Ley, R. D., Applegate, L. A., Padilla, R. S. and Stuart, T. D., 1989. Ultraviolet radiation-induced malignant melanoma. *Photochem. Photobiol.* **50:** 1–5.

Licht, L. E. and Grant, K. P., 1997. The effects of ultraviolet radiation on the biology of amphibians. *Amer. Zool.* **37:** 137–145.

Little, E. E. and Fabacher, D. L., 1994. Comparative sensitivity of rainbow trout and two threatened salmonids, Apache trout and Lahontan cutthroat trout, to ultraviolet-B radiation. *Ergeb. Limnol.* **43:** 217–226.

Liu, S. C., McKeen, S. A. and Madronich, S., 1991. Effect of anthropogenic aerosols on biologically active ultraviolet radiation. *Geophys. Res. Lett.* **18:** 2265–2268.

Long, L. E., Saylor, L. S. and Soule, M. E., 1995. A pH/UV-B synergism in amphibians. *Cons. Biol.* **9:** 1301–1303.

Lubin, D., Frederick, J. E. and Krueger, A. J., 1989. The ultraviolet radiation environment of Antarctica: McMurdo Station during September–October 1987. *J. Geophys. Res.* **94(D6):** 8491–8496.

Lubin, D. and Jensen, E. H., 1995. Effects of clouds and stratospheric ozone depletion on ultraviolet radiation trends. *Nature* **377:** 710–713.

Lubin, D., Mitchell, B. G., Frederick, J. E., Alberts, A. D., Booth, C. R., Lucas, T. and Neuschuler, D., 1992. A contribution toward understanding the biospherical significance of Antarctic ozone depletion. *J. Geophys. Res.* **97(D8):** 7817–7828.

MacKenzie, D., 1995. Ozone's future is up in the air. *New Scient.* **2008:** 14–15.

Madronich, S., 1992. Implications of recent total atmospheric ozone measurements for biologically active ultraviolet radiation reaching the Earth's surface. *Geophys. Res. Lett.* **19:** 37–40.

Madronich, S. and de Gruijl, F. R., 1993. Skin cancer and UV radiation. *Nature* **366:** 23.

Madronich, S., McKenzie, R. L., Caldwell, M. M. and Bjorn, L. O., 1995. Changes in ultraviolet radiation reaching the Earth's surface. *Ambio* **24:** 143–152.

Mahony, M., 1996. The decline of the green and golden bell frog *Litoria aurea* viewed in the context of declines and disappearances of other Australian frogs. *Aust. Zool.* **30:** 237–247.

Manney, G. L., Froidevaux, L., Waters, J. W., Zurek, R. W., Read, W. G., Elson, L. S., Kumer, J. B., Mergenthaler, J. L., Roche, A. E., O'Neill, A., Harwood, R. S., MacKenzie, I. and Swinbank, R., 1994. Chemical depletion of ozone in the Arctic lower stratosphere during winter 1992–93. *Nature* **370:** 429–434.

Mark, U., Saile-Mark, M. and Tevini, M., 1996. Effects of solar UVB radiation on growth, flowering and yield of Central and Southern European maize cultivars *Zea mays* L. *Photochem. Photobiol.* **64:** 457–463.

McElroy, M. B., Salawitch, R. J., Wofsy, S. C. and Logan, J. A., 1986. Reductions of Antarctic ozone due to synergistic interactions of chlorine and bromine. *Nature* **321:** 759–762.

McKenzie, R. L., Matthews, W. A. and Johnston, P. V., 1991. The relationship between erythemal UV and ozone, derived from spectral irradiance measurements. *Geophys. Res. Lett.* **18:** 2269–2272.

McMinn, A., Heijnis, H. and Hodgson, D., 1994. Minimal effects of UVB radiation on Antarctic diatoms over the past 20 years. *Nature* **370:** 547–549.

Mepsted, R., Paul, N. D., Stephen, J., Corlett, J. E., Nogués, S., Baker, N. R., Jones, H. G. and Ayres, P. G., 1996. Effects of enhanced UV-B radiation on pea (*Pisum sativum* L.) grown under field conditions in the UK. *Global Change Biol.* **2:** 325–334.

Miller, A. J., Hollandsworth, S. M., Flynn, L. E., Tiao, G. C., Reinsel, G. C., Bishop, L., McPeters, R. D., Planet, W. G., DeLuisi, J. J., Materr, C. L., Wuebbles, D., Kerr, J. and Nagatani, R. M., 1996. Comparisons of observed ozone trends and solar effects in the stratosphere through examination of ground-based Umkehr and combined solar backscattered ultraviolet SBUV and SBUV 2 satellite data. *J. Geophys. Res.* **101(C4):** 9017–9021.

Milot-Roy, V. and Vincent, W. F., 1994. UV radiation effects on photosynthesis: The importance of near-surface thermoclines in a subarctic lake. *Ergeb. Limnol.* **43:** 171–184.

Montzka, S. A., Butler, J. H., Myers, R. C., Thompson, T. M., Swanson, T. H., Clarke, A. D., Lock, L. T. and Elkins, J. W., 1996. Decline in the tropospheric abundance of halogen from halocarbons: implications for stratospheric ozone depletion. *Science* **272:** 1318–1322.

van de Mortel, T. F. and Buttemer, W. A., 1996. Are *Litoria aurea* eggs more sensitive to ultraviolet-B radiation than eggs of sympatric *L. peronii* or *L. dentata*? *Austr. Zool.* **30:** 150–157.

Musil, C. F., 1996. Accumulated effect of elevated ultraviolet-B radiation over multiple generations of the arid-environment annual *Dimorphoheca sinuata* DC. (Asteraceae). *Plant Cell Environ.* **19:** 1017–1027.

Newman, P. A., 1986. The final warming and polar vortex disappearance during the southern hemisphere Spring. *Geophys. Res. Lett.* **13:** 1228–1231.

Okaichi, K., Kajitani, N., Nakajima, K., Nozu, K. and Ohnishi, T., 1989. DNA damage and its repair in *Dictyostelium discoideum* irradiated by health lamp light UV-B. *Photochem. Photobiol.* **50:** 69–73.

Olsen, W. M. and Kirkhus, B., 1989. The epidermal cell kinetic response to ultraviolet B irradiation combines regenerative proliferation and carcinogen associated cell cycle delay. *Photochem. Photobiol.* **50:** 391–397.

Orth, A. B., Teramura, A. H. and Sisler, H. D., 1990. Effects of ultraviolet-B radiation on fungal disease development in *Cucumis sativus*. *Amer. J. Bot.* **77:** 1188–1192.

Osborne, W. S., 1989. Distribution, relative abundance and conservation status of corroboree frogs *Pseudophryne corroboree* Moore Anura: Myobatrachidae. *Aust. Wildl. Res.* **16:** 537–547.

Osborne, W.S., 1996. Recovery plan for the Southern Corroboree Frog *Pseudophryne corroboree*. NSW National Parks and Wildlife Service, Sydney.

Osborne, W. S., Littlejohn, M. J. and Thomson, S. A., 1996. Former distribution and apparent disappearance of the *Litoria aurea* complex from the Southern Tablelands of New South Wales and the Australian Capital Territory. *Aust. Zool.* **30:** 190–198.

Peak, M. J. and Peak, J. G., 1982. Lethal effects on biological systems caused by solar ultraviolet light: Molecular considerations. Pp. 325–336 *in* "The Role of Solar Ultraviolet Radiation in Marine Ecosystems." ed by J. Calkins. Plenum Press, New York.

Peak, M. J., Peak, J. G., Moehring, M. P. and Webb, R. B., 1984. Ultraviolet action spectra for DNA dimer induction, lethality, and mutagenesis in *Escherichia coli* with emphasis on the UVB region. *Photochem. Photobiol.* **40:** 613–620.

Proffitt, M. H., Steinkamp, M. J., Powell, J. A., McLaughlin, R. J., Mills, O. A., Schmeltekopf, A. L., Thompson, T. L., Tuck, A. F., Tyler, T., Winkler, R. H. and Chan, K. R., 1989. *In situ* ozone measurements within the 1987 Antarctic ozone hole from a high-altitude ER-2 aircraft. *J. Geophys. Res.* **94(D14):** 16547–16555.

Rau, W. and Hofmann, H., 1996. Sensitivity to UV-B of plants growing in different altitudes in the Alps. *J. Plant Physiol.* **148:** 21–25.

Richards, S. J., McDonald, K. R. and Alford, R. A., 1993. Declines in populations of Australia's endemic tropical rainforest frogs. *Pac. Cons. Biol.* **1:** 66–77.

Roberts, L. K. and Beasley, D. G., 1997. Sunscreens prevent local and systemic immunosuppression of contact hypersensitivity in mice exposed to solar-simulated ultraviolet radiation. *J. Photochem. Photobiol. B: Biol.* **39:** 121–129.

Roche, A. E., Kumer, J. B. and Mergenthaler, J. L., 1993. CLAES observations of $CLONO_2$, and HNO_3 in the Antarctic stratosphere, between June 15 and September 17, 1992. *Geophys. Res. Lett.* **20:** 1223–1226.

Rosen, J. E., Prahalad, A. K. and Williams, G. M. 1996. 8-Oxodeoxyguanosine formation in the DNA of cultured cells after exposure to H_2O_2 alone or with UVB or UVA irradiation. *Photochem. Photobiol.* **64:** 117–122.

Roush, W., 1995. When rigour meets reality. *Science* **269:** 313–315.

Roy, C. R., Gies, H. P. and Graeme, E., 1990. Ozone depletion. *Nature* **347:** 235–236.

Rozema, J., van de Staaij, J., Bjorn, L. O. and Caldwell, M., 1997. UV-B as an environmental factor in plant life: stress and regulation. *TREE* **12:** 22–28.

Russell, J. M. III, Luo, M., Cicerone, R. J. and Deaver, L. E., 1996. Satellite confirmation of the dominance of chlorofluorocarbons in the global stratospheric chlorine budget. *Nature* **379:** 526–529.

Santas, R., Hader, D-P. and Lianou, C., 1996. Effects of solar UV radiation on diatom assemblages of the Mediterranean. *Photochem. Photobiol.* **64:** 435–439.

Saunders, D. A., Hobbs, R. J. and Margules, C. R., 1991. Biological consequences of ecosystem fragmentation: A review. *Cons. Biol.* **5:** 18–32.

Schindler, D. W., Curtis, P. J., Parker, B. R. and Stainton, M. P., 1996. Consequences of climate warming and lake acidification for UV-B penetration in North American boreal lakes. *Nature* **379:** 705–708.

Schoeberl, M. R. and Hartmann, D. L., 1991. The dynamics of the stratospheric polar vortex and its relation to springtime ozone depletions. *Science* **251:** 46–52.

Schofield, O., Kroon, B. M. A. and Prezelin, B. B., 1995. Impact of ultraviolet-B radiation on photosystem II activity and its relationship to the inhibition of carbon fixation rates for Antarctic ice algae communities. *J. Phycol.* **31:** 703–715.

Scotto, J., Cotton, G., Urbach, F., Berger, D. and Fears, T., 1988. Biologically effective ultraviolet radiation: Surface measurements in the United States, 1974 to 1985. *Science* **239:** 762–764.

Scully, N. M. and Lean, D. R. S., 1994. The attenuation of ultraviolet radiation in temperate lakes. *Ergeb. Limnol.* **43:** 135–144.

Sebastian, C., Scheuerlein, R. and Häder, D.-P., 1994. Graviperception and motility of three *Prorocentrum* strains impaired by solar and artificial ultraviolet radiation. *Mar. Biol.* **120:** 1–7.

Seckmeyer, G. and McKenzie, R. L., 1992. Increased ultraviolet radiation in New Zealand 45ûS relative to Germany 48°N. *Nature* **359:** 135–137.

Siebeck, O. and Böhm, U., 1994. Challenges for an appraisal of UV-B effects upon planktonic crustaceans under natural radiation conditions with a non-migrating (*Daphnia pulex obtusa*) and a migrating cladoceran (*Daphnia galeata*). *Ergeb. Limnol.* **43:** 197–206.

Siebeck, O., Vail, T. L., Williamson, C. E., Vetter, R. Hessen, D., Zagarese, H., Little, E., Balseiro, E., Modenutti, B., Seva, J. and Shumate, A., 1994. Impact of UV-B radiation on zooplankton and fish in pelagic freshwater ecosystems. *Ergeb. Limnol.* **43:** 101–114.

Slaper, H., Velders, G. J. M., Daniel, J. S., de Gruijl, F. R. and van der Leun, J. C., 1996. Estimates of ozone depletion and skin cancer incidence to examine the Vienna Convention achievements. *Nature* **384:** 256–258.

Smith, R. C., 1989. Ozone, middle ultraviolet radiation and the aquatic environment. *Photochem. Photobiol.* **50:** 459–468.

Smith, R. C. and Baker, K. S., 1980. Stratospheric ozone, middle ultraviolet radiation, and carbon-14 measurements of marine productivity. *Science* **208:** 592–593.

Smith, R. C., Prezelin, B. B., Baker, K. S., Bidigare, R. R., Boucher, N. P., Coley, T., Karentz, D., MacIntyre, S., Matlick, H. A., Menzies, D., Ondrusek, M., Wan, Z. and Waters, K. J., 1992. Ozone depletion: Ultraviolet radiation and phytoplankton biology in Antarctic waters. *Science* **255:** 952–959.

Solomon, S., 1990. Progress towards a quantitative understanding of Antarctic ozone depletion. *Nature* **347:** 347–354.

Solomon, S., Garcia, R. R., Rowland, F. S. and Wuebbles, D. J., 1986. On the depletion of Antarctic ozone. *Nature* **321:** 755–758.

Sommaruga, R. and Psenner, R., 1997. Ultraviolet radiation in a high mountain lake of the Austrian Alps: air and underwater measurements. *Photochem. Photobiol.* **65:** 957–963.

Stamnes, K., Jin, Z. and Slusser, J., 1992. Several-fold enhancement of biologically effective ultraviolet radiation levels at McMurdo Station Antarctica during the 1990 ozone "hole". *Geophys. Res. Lett.* **19:** 1013–1016.

Stebbins, R. C. and Cohen, N. W., 1995. 'A Natural History of Amphibians.' Princeton University Press, Princeton, New Jersey.

Stolarski, R., Bojkov, R., Bishop, L., Zerefos, C., Staehelin, J. and Zawodny, J., 1992. Measured trends in stratospheric ozone. *Science* **256:** 342–349.

Stordal, F., Hov, O. and Isaksen, I. S. A., 1982. The effect of perturbation of the total ozone column due to CFC on the spectral distribution of UV fluxes and the damaging UV doses at the ocean surface: A model study. Pp. 93–107*in* "The Role of Solar Ultraviolet Radiation in Marine Ecosystems." ed by J. Calkins. Plenum Press, New York.

Teramura, A. H., 1983. Effects of ultraviolet-B radiation on the growth and yield of crop plants. *Physiologia Plantarum* **58:** 415–427.

Teramura, A. H. and Sullivan, J. H., 1994. Effects of UV-B radiation on photosynthesis and growth of terrestrial plants. *Photosyn. Res.* **39:** 463–473.

Trodahl, H. J. and Buckley, R. G., 1989. Ultraviolet levels under sea ice during the Antarctic spring. *Science* **245:** 194–195.

Tuck, A. F., Davies, T., Hovde, S. J., Noguer-Alba, M., Fahey, D. W., Kawa, S. R., Kelly, K. K., Murphy, D. M., Proffitt, M. H., Margitan, J. J., Loewenstein, M., Podolske, J. R., Strahan, S. E. and Chan, K. R., 1992. Polar stratospheric cloud processed air and potential vorticity in the northern hemisphere lower stratosphere at mid-latitudes during Winter. *J. Geophys. Res.* **97(D8):** 7883–7904.

Tumber, V. P., Robarts, R. D., Arts, M. T., Evans, M. S. and Caldwell, D. E., 1993. The influence of environmental factors on seasonal changes in bacterial cell volume in two prairie saline lakes. *Microb. Ecol.* **26:** 9–20.

Tyler, M. J., 1997. 'The Action Plan for Australian Frogs.' Wildlife Australia, Endangered Species Program: Canberra.

Urbach, F., 1989. Potential effects of altered solar ultraviolet radiation on human skin cancer. *Photochem. Photobiol.* **50:** 507–513.

Wake, D. B., 1991. Declining amphibian populations. *Science* **253:** 860.

Waters, J. W., Froidevaux, L., Read, W. G., Manney, G. L., Elson, L. S., Flower, D. A., Jarnot, R. F. and Harwood, R. S., 1993. Stratospheric ClO and ozone from the Microwave Limb Sounder on the Upper Atmosphere Research Satellite. *Nature* **362:** 597–602.

Williams, M. W. and Melack, J. M., 1991. Solute chemistry of snowmelt and runoff in an alpine basin, Sierra Nevada. *Water Resources Res.* **27:** 1575–1588.

Williamson, C. E., Stemberger, R. S., Morris, D. P., Frost, T. M. and Paulsen, S. G., 1996. Ultraviolet radiation in North American lakes: Attenuation estimates from DOC measurements and implications for plankton communities. *Limnol. Oceanogr.* **41:** 1024–1034.

Worrest, R. C. and Häder, D.-P., 1997. Overview of the effects of increased solar UV on aquatic microrganisms. *Photochem. Photobiol.* **65:** 257–259.

Worrest, R. C. and Kimeldorf, D. J., 1976. Distortions in amphibian development induced by ultraviolet-B enhancement (290-315 nm) of a simulated solar spectrum. *Photochem. Photobiol.* **24:** 377–382.

Wyman, R. L., 1990. What's happening to the amphibians? *Cons. Biol.* **4:** 350–352.

Yamawaki, M., Futamura, S. and Horio, T., 1996. UVB radiation suppresses the TNF-alpha-induced expression of E-selectin and ICAM-1 on cultured human umbilical vein endothelial cells. *J. Derm. Sci.* **13:** 11–17.

Yan, N. D., Keller, W., Scully, N. M., Lean, D. R. S. and Dillon, P. J., 1996. Increased UV-B penetration in a lake owing to drought-induced acidification. *Nature* **381:** 141–143.

Yu, S-G. and Björn, L. O., 1997. Effects of UV-B radiation on light-dependent and light-independent protein phosphorylation in thylakoid proteins. *J. Photochem. Photobiol. B:Biol.* **37:** 212–218.

Zerefos, C. S., Bais, A. F., Meleti, C. and Ziomas, I. C., 1995. A note on the recent increase of solar UV-B radiation over northern middle latitudes. *Geophys. Res. Lett.* **22:** 1245–1247.

Index